Analysis of Sedimentary Successions

A Field Manual

Analysis of Sedimentary Successions

A Field Manual

Ajit Bhattacharyya
and
Chandan Chakraborty

A.A.BALKEMA / ROTTERDAM / BROOKFIELD / 2000

A.A. Balkema, P.O. Box 1675, 3000 BR Rotterdam, Netherlands
Fax: +31.10.4135947; E-mail: balkema@balkema.nl
Internet site: http://www.balkema.nl

Distributed in USA and Canada by
A.A. Balkema Publishers, 2252 Ridge Road, Brookfield, Vermont 05036 USA
Fax: 802.276.3837; E-mail: Info@ashgate.com

ISBN 90 5809 209 7 (HC)
ISBN 90 5809 227 5 (PB)

Foreword

Harking back to Robert R. Shrock (1948) "one of the most important problems facing a field geologist is determining order of succession". Hence this book, titled "Analysis of Sedimentary Successions - A Field Manual", is concerned with sedimentary geology as practiced in the field, or in other words as Shrock expressed it " the sequence of bedded ...... rocks". Shrock's emphasis of importance, already predicted in 1948, materialized in the 1980's and 1990's with the dynamic concepts of sequence stratigraphy and basin analysis. The current approach to basin analysis is based upon the concepts of plate tectonics.

Likewise, even as early as 1922, L. Dudley Stamp noted that the geologist in the field has been expected to remember " Successions" in "varied often obscure and imperfectly known localities in his own or other countries", "too often he has been left to form some connected idea of the whole almost undecided". Hence even today, 77 years later, publication of this field manual, especially at the turn of the new century, is needed and timely to provide the geologist with the tools for field study.

This book is a summary of field features of sedimentary deposits progressing from the analysis of the important sedimentary lithologies to facies (paleogeographic settings), sequence stratigraphy, and basin analysis (tectonic signatures in the basin fill). It emphasizes the core of sedimentary geology and its coverage is deliberately weighted in favor of sedimentary lithologies, such as shale, sandstone, conglomerate, and carbonate rocks that compose 99% of the sedimentary record.

Sedimentary geology, consisting of sedimentology and stratigraphy, is a broad, comprehensive discipline within the field of geology. No single volume can deal satisfactorily with all the many details of sedimentary geology. This book recognizes this limitation and specifically includes in its title the words "a field manual". In my current experience I see more geologists sitting at a computer than walking in the field examining rock exposures. In part the purpose of this book is to bring the geologist back into the field. It selects for the student and practicing geologist the areas of knowledge which the field offers—enormous opportunities beckon out there.

REFERENCES

Shrock, Robert R., 1948, Sequence in layered rocks. New York, McGraw-Hill Book Company, 507p.

Stamp, L. Dudley, 1923, An introduction of stratigraphy, London, Thomas Muskyzlo, 381p.

Gerald M Friedman
Distinguished Professor of Geology
Brooklyn College and Graduate School
City University of New York, Brooklyn, N.Y.
C/o Northeastern Science Foundation, Inc.,
Rensselaer Center of Applied Geology
15 Third Street, P.O. Box 746, Troy, N.Y. 12181, U.S.A.

Preface

The motivation of preparing a monograph of this kind stemmed from our experience of fieldwork in sedimentary terrains with undergraduate and postgraduate students of geology. The primary hurdle the students faced in the field was what to observe in a sedimentary succession. Geology is a scientific discipline that observes the earth by decoding the *signatures* in rocks, sediments and fossils. What does the term 'sign' mean to geologists? Is it a material signifier - a mind-independent geometrical entity? Were it so, anybody could be a geologist. Geology has its own *art* of observation. Meaningful observation of geological features in rocks begins with a tentative genetic interpretation of the object, or a working hypothesis, grounded in the geological sense of the observer (Baker, 1999). This transforms the geometrical object into a sign - a symbol of significance. Once this process begins derivative signs and hypotheses with more significant meaning progressively emerge from the initial sign. Thereafter a planned search for newer signs may follow, enabling more and more meaningful observation and understanding of the earth. This monograph aims at introducing sedimentary geology with this perspective to facilitate students of geology while at fieldwork in sedimentary basins.

The concept of scale and time is another important aspect of geological studies. Geological observations can be made in megascopic, mesoscopic and microscopic scales. All scales of observations should ideally converge towards a uniform understanding of the earth. The present book deals mainly with the mega- and mesoscopic features of sedimentary successions. Field-based observations of *soft-rocks* with only hints of their geological significance are emphasized so as to encourage the students to build up their own web of thoughts.

Greater emphasis has been given to two topics discussed in this manual, namely paleosols and trace fossils. This is because none of the textbooks on sedimentology discuss paleosols in any detail and that on trace fossils very scantily. This forced the authors to discuss the theoretical aspects of these two topics in some detail before embarking on the field aspects thus taking the students comfortable.

Code of stratigraphic nomenclatures are appended at the end of this manual; this topic is totally field-based and it is hoped that this will help the students and professionals alike in establishing and correctly defining lithostratigraphic, biostratigraphic and time-stratigraphic units.

Sufficient references have been provided on various topics discussed in the manual with the objective that the students may browse through them to develop a sound foundation on the subject.

Reference: *Baker, V.R., 1999, Geosemiosis. Geol. Soc. Am., Bull., 111: 633-645.*

Ajit Bhattacharyya
Department of Geological Sciences
Jadavpur University
Calcutta, India

Chandan Chakraborty
Geological Studies Unit
Indian Statistical Institute
Calcutta, India

Acknowledgements

This book has been possible thanks to a generous grant from the Department of Science and Technology, Government of India, New Delhi, to one of the authors (AB), under the "Utilization of Scientific Expertise of Retired Scientist (USER) scheme" (Project # HR/UR/26/95). Numerous people have helped in the preparation of this book. We would especially like to thank Prof. Gerald M. Friedman, U.S.A., Prof. V. Paul Wright, U.K., Prof. R. Goldring, U.K., Prof. T. Aigner, Germany, Prof. D.H. Yaalon, Israel, Prof. R. Steel, U.S.A., Prof. A. Choudhuri and Dr. Tapan Chakraborty, Calcutta India, Prof. Asoke Sahni, Chandigarh, India, for taking the trouble by going through various chapters of the manuscripts and offering suggestions for improvement. Their comments have brought greater coherence, accuracy and balance of various chapters of the book. The authors are also indebted to Prof. S. Dzulynnski, Poland, Dr. J.N.J. Visser, South Africa, Prof. Paul M. Myrow, U.S.A., Prof. R.G.C. Bathurst, U.K., Prof. J.F. Read, U.S.A., Prof. F.R. Lucchi, Italy, Dr. Robert Wray, Australia, Prof. V. Paul Wright, U.K., Dr. D. Weston, New Zealand, Dr. J. Forsyth, New Zealand, Prof. S.K. Tandon, University of Delhi, India, Dr. Tapan Chakraborty and Dr. Partha Ghosh, Geological Studies Unit, Indian Statistical Institute, Calcutta, for offering photographs for the various chapters of the book. Last but not least, Jadavpur University provided us an office in the Department of Geological Sciences and all other logistic support to enable us to complete this work. The second author acknowledges the infrastructural facilities provided by Indian Statistical Institute, Calcutta. Diagrams and photographs were prepared by Mr. T. Bhattacharya, Department of Geological Sciences, Jadavpur University, Calcutta and Mr. K. Das, Photo Hobby Center, Dhakuria, Calcutta, respectively. We are grateful to all of them.

Contents

1. PREAMBLE

A sedimentary succession is the result of sediment accumulation on the upper surface of the earth's crust during a geologically significant time interval under the influence of tectonism, eustasy and climate. Sediments preferentially accumulate in a geographic domain where there is supply of sediments via different earth-surface depositional processes. The accumulated sediment pile would have a potential for preservation to result in a sedimentary succession if the area undergoes tectonic subsidence during sedimentation. The site of sediment accumulation and preservation is known as a sedimentary basin, which is represented in the geological record as a sedimentary succession.

The thickness of a sedimentary succession is dominantly controlled by the net tectonic subsidence that a basin suffered during its lifetime. Basin tectonics, coupled with climate, again influences the rate of sediment supply, types of sediment (i.e., lithology) and the depositional systems and hence the paleogeography of a sedimentary basin via its control on topography and geomorphology. A sedimentary succession, therefore, has the potential to unravel the dynamics of evolution of a part of the earth's crust.

Analysis of a sedimentary succession begins with characterizing the lithologies and interpretation of their depositional process. This would lead to delineation of lithosomes, representatives of different depositional systems (i.e. sedimentary environments), which prevailed in the basin during its lifetime. In long terms, a sedimentary succession does not develop continuously, but in pulses with intervening phases of non-deposition and erosion. These phases of non-deposition and erosion are represented in the sedimentary succession as discontinuity surfaces. Discontinuity surfaces, in many cases, represent more geologic time than that taken by the sedimentary pile to accumulate. Therefore, it is necessary to subdivide the sedimentary succession into units bounded by discontinuity surfaces. Each such unit would then represent a continuum of sedimentation during the evolution of the sedimentary basin.

In a sedimentary succession the products of different laterally disposed, coeval depositional systems are stacked vertically. Understanding the mechanism of vertical superposition of different lithological packages representing different depositional environments, either of a continental or a marine setting is of fundamental importance in the analysis of a sedimentary succession. The next step is to identify groups of packages in the succession, each representing a set of depositional systems that prevailed in the basin contemporaneously over its space. Within such a group of package, the different marine depositional systems may be stacked either showing a shallowing-upward trend or a deepening-upward trend. Each such group of packages would reveal the paleogeography of the basin at different periods of its evolution and their stacking pattern reflects the temporal change in the paleogeographic setting of the basin. The immediate aim of analyzing a sedimentary succession is to interpret the information, so far obtained in terms of the variables, such as rate of subsidence, rate of eustatic sea level change, rate of sediment supply and climate. The ultimate aim is to understand the dynamics of the sedimentary basin, i.e., the subsidence mechanism and the tectonic setting for which the following data on the basin fill are required: 1) thickness distribution and geometry of the basin fill, 2) understanding of the depositional mechanisms based on precise characterization of the lithologies, 3) structural architecture of the strata, 4) nature and occurrences of unconformities in the succession, 5) the sediment dispersal pattern and 6) evidence of syn-sedimentary tectonism and magmatism. The procedure of analyzing a sedimentary succession, as outlined above, is described in detail in the following chapters.

2. SEDIMENTARY SUCCESSION: ITS COMPONENTS

2.1 Conglomerate

Coarse terrigenous sediments (dominant clast size >2mm) are found in recent sedimentary deposits as well in stratigraphic records. Unlithified products are commonly known as gravel deposits, whereas lithified equivalents are either known as CONGLOMERATE or SEDIMENTARY BRECCIA, depending upon the degree of roundness of the coarse clast fractions. They are known as conglomerate when the coarse clast fractions are dominantly rounded and as sedimentary breccia when these are dominantly angular. The composition of the coarser clasts may be either siliciclastic or carbonate. The classification of conglomerate, as proposed by Folk (1965), appears to be more meaningful sedimentologically amongst many such classifications (see Pettijohn 1975). Following Folk's classification, when clast fraction (>2 mm) constitutes 80% or more, the deposit is classified as conglomerate. When the coarser fraction varies between 30 and 80%, the deposit is grouped as sandy conglomerate/muddy conglomerate, depending upon the proportion of sand and mud components.

Gravel deposits and their lithified equivalent product, conglomerate/breccia, form volumetrically an insignificant part of the stratigraphic record. Nevertheless, much geological information (e.g. tectonic activity of the basin, provenance, paleo-transport direction, depositional environment, mechanism of clast emplacement, rheology of gravel) can be gleaned from conglomerates when these deposits are properly looked into. The field features, that are to be carefully documented are: (a) clast count and clast lithology, (b) size, shape and roundness of the coarser fractions, (c) sorting and size distributions of the clasts, (d) clast *vs* matrix ratio, (e) clast imbrication and fabric of a gravel/ conglomerate deposit, (f) spatial size distribution of coarser fractions, and, (g) internal structure of a gravel/conglomerate bed.

Field Features of Conglomerates

Clast Count and Clast Lithology

Clast count and clast lithology are the techniques of counting individual clasts and their lithology within a statistically determined area till a desired sample size is obtained. Before undertaking clast count, it is important to reconnoitre the areas to be studied and sampled.

This is necessary to pre-determine the most effective sampling plan and at the same time, while reconnoitering, collect representative clasts for thin-section studies in the laboratory for correct identification of clasts at the time of actual sampling. If the reconnaissance is not undertaken before the actual study and sites are selected randomly from published geological maps, sites so selected may be logistically too difficult to access, or the area may be under thick cover of soil and vegetation. Since each site is expected to have a unique set of circumstances, the only hard and fast rule is to have an element of random sampling at each level of the sampling process (Howard 1993). For counting the clasts, the ribbon or area method, as suggested by Howard (1993), is recommended. The technique is to count each individual clast within a desired area until the desired sample size is obtained. The size of the area should be at least 2.5 times the long axis diameter of the largest clast. A marker pen should be used, while counting, so that the same clast is not counted twice.

After clast count and clast lithology are determined, it is necessary to divide the clasts into their respective lithologies, e.g., extrusive igneous, intrusive igneous, sedimentary and metamorphic rocks and plotting of the data on a triangular diagram (with end members as extrusive igneous, intrusive igneous, and sedimentary and metamorphic clasts clubbed together). Such triangular plots will instantly indicate the influence of lithologies that contributed significantly in the making of a gravel/conglomerate bed(s). It is, however, to be always kept in mind that gravel-forming capability is not the same for all lithologies. Limestone will rarely produce gravel-size fraction under a humid tropical setting, and is never reflected as one of the source rocks despite its presence at the source (Pettijohn 1975, p. 161). For example, the Rohtas Formation (dominantly limestone) of the Semri Group, India, and the overlying Kaimur Group (dominantly siliciclastic), Middle Proterozoic Vindhyan Supergroup, India, is separated by regional unconformity. The Kaimur Group, a product of aeolian and fluvial environments, is totally devoid of coarse clast fraction of the underlying Rohtas Formation. However, conglomerates, bearing limestone clasts, eroded from carbonate rocks can form under restricted climatic conditions with less moisture (Friedman et al., 1992, his Fig. 2-11). Similar is the case with shale clasts, which being mechanically fragile, are rarely preserved in rock record.

Determination of lithology of clasts helps to determine whether a conglomerate/gravel bed is petromict or oligomictic in nature (see, Pettijohn 1975).

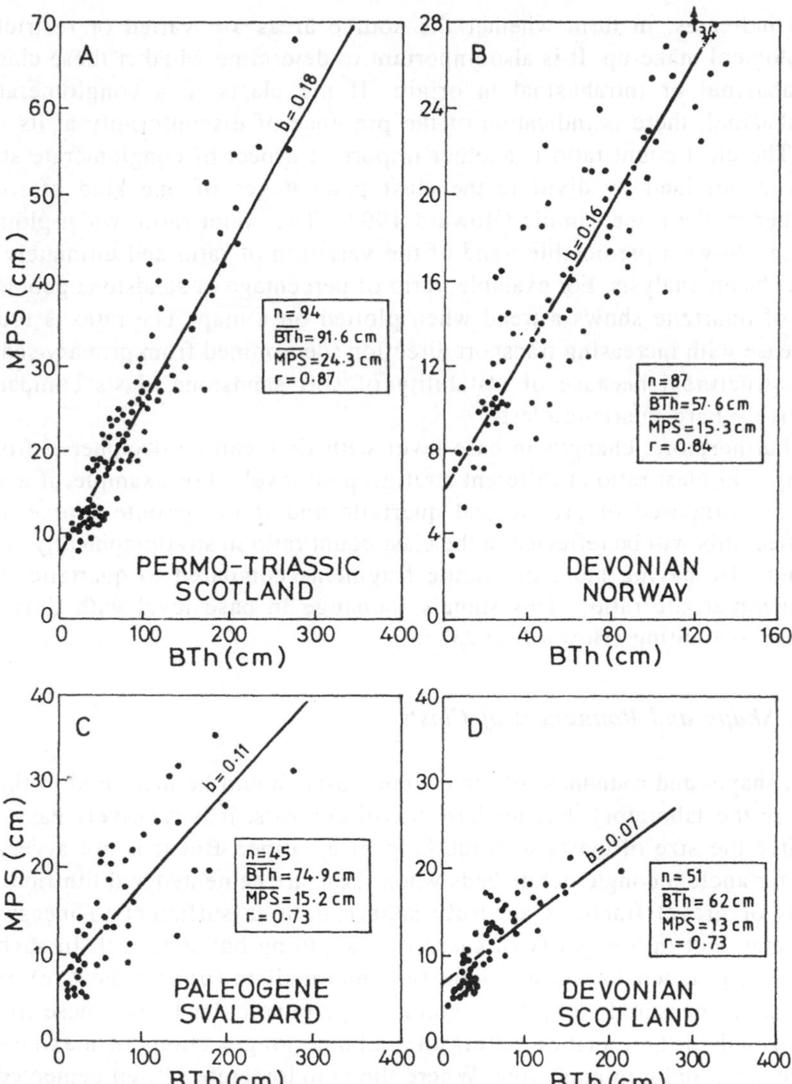

Fig. 2.1.1: Examples of MPS/BTh data from subaerial deposits of cohesive gravity flows. Data from: A, New Red Sandstone fanglomerate of Stornway and Inch Kenneth, West Scotland; B, Devonian fanglomerates of the northern part of Hornelen Basin, S.W. Norway; C, Paleogene fanglomerates of Prince Karl foreland, Spitsbergen; D, Old Red Sandstone fanglomerates of Clyde area, Scotland. The slant lines are least square regressive lines fitted to the data. Explanation of symbols: b, regressive coefficient (line gradient); n, numbers of data; Bth = mean bed thickness; MPS = mean maximum particle size; r, Pearsonian Correlation coefficient. (Reproduced from W. Nemec and R.J. Steel: Alluvial and coastal conglomerates: Their significant features and some comments on gravelly mass-flow deposits, Fig. 20, *in*, E.H. Koster and R.J. Steel, eds., Sedimentology of Gravels and Conglomerates. Canadian Soc. of Petroleum Geologists, Memoir 10, © 1984, Canadian Soc. of Petroleum Geologists, Alberta, Canada with kind permission)

This indicates, in turn, whether the source areas are varied or restricted in lithological make-up. It is also important to determine whether these clasts are extrabasinal or intrabasinal in origin. If the clasts in a conglomerate are intrabasinal, there is indication of the presence of disconformity at its base.

The clast-count ratio is another important aspect of conglomerate studies. This is obtained by dividing the clast percentages of one kind of rock by another in the same sample (Howard 1993). The count ratio, when plotted on a map, shows a predictable trend of the variation of ratio and ultimately helps in the basin analysis. For example, ratio of percentage of sandstone pebbles and that of quartzite shows a trend when plotted on a map. The ratio is found to decrease with increasing transport direction (determined from primary sedimentary structures) because of instability of soft sandstone clasts compared to relatively hard quartzite clasts.

Furthermore, changes in base level with time can be deciphered from the changes in clast ratio at different stratigraphic levels. For example, if a source area is composed of granite and quartzite and if the granite source area is uplifted, this will be reflected in the clast count ratio in stratigraphically younger sections by having more of granite fragments compared to quartzite (higher granite/quartzite ratio). This signals a change in base level with time in response to faulting (Howard, 1993).

Size, Shape and Roundness of Clasts

Size, shape, and roundness of the coarse clasts should be measured in the field than in the laboratory. For modern gravel deposits, it is relatively easy to determine the size of clasts with the help of a caliper. But it is not a very easy task for ancient conglomerate beds when these are cemented and lithified. If the clasts of gravel fractions are truly spherical, it is sufficient to measure the diameter. But in a majority of cases it is anything but spherical. In such situations, it is better to measure long (a), intermediate (b) and short (c) axes of the clasts, measured along three mutually perpendicular lines. These lines are orthogonal to three planes cutting the pebble along sections of maximum, intermediate, and minimum area. Where the conglomerate is well cemented, it is not easy to measure these. In such situations, it is convenient to determine the maximum particle size (MPS) by measuring the size (apparent long axis, as exposed) of the ten largest clasts and taking their mean. It is also necessary to determine the bed thickness (BTh) of gravel/conglomerate beds. Plotting of these two data (MPS vs. BTh) helps to discriminate a gravel / conglomerate bed of gravity flow origin from that of fluvial origin. While gravity flow deposits show a positive correlation between MPS vs. BTh (Figs.2.1.1, 2.1.2 and 2.1.3), this is not the case for stream deposits. This absence of correlation between MPS vs BTh is suggested to be due to erosion and to 'grain-by-grain' mode of

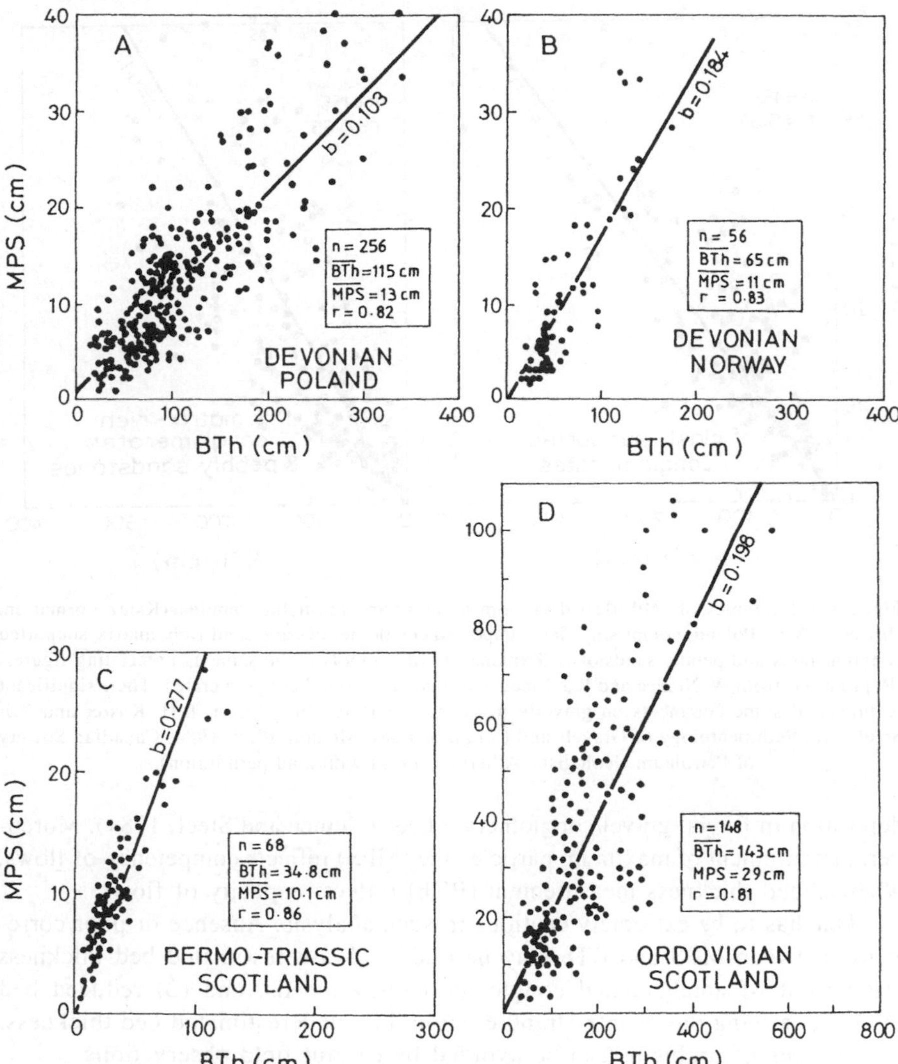

Fig. 2.1.2: Examples of MPS/BTh data from subaerial and subaqueous deposits of cohesionless debris flows. Data from: **A.** Chawliszow Formation (submarine fan-delta complex, Famennian-? lowest Tournaisian) of Swiebodzice depression, S.W. Poland); **B.** Domba Conglomerate Member (data from subaerial 'proximal' part of lacustrine fan delta from Devonian Hornelen Basin, S.W. Norway; **C.** New Red Sandstone fanglomerates of Rhum and Scalpay Islands, W. Scotland; **D.** Ordovician conglomerates (Caradox submarine canyon-fan complex) of South Scotland. Explanation of symbols same as the preceding figure. (Reproduced from W. Nemec and R.J. Steel: Alluvial and coastal conglomerates: Their significant features and some comments on gravelly mass-flow deposits, Fig. 21, in, E.H. Koster and R.J. Steel, eds., Sedimentology of Gravels and Conglomerates, Memoir 10, © 1984, Canadian Society of Petroleum Geologists, Alberta, Canada with kind permission).

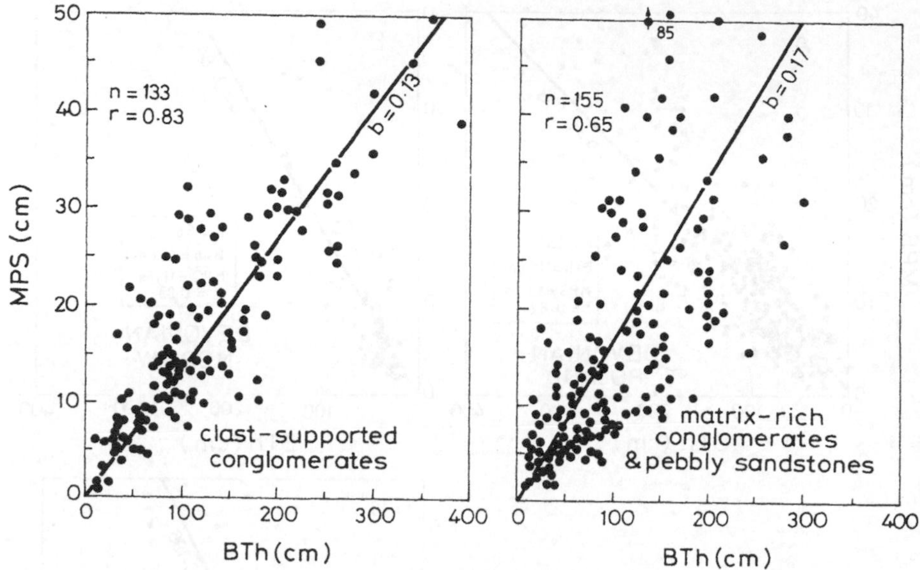

Fig. 2.1.3: Examples of MPS/BTh data from a submarine fan-delta complex (Ksiaz Formation, Devonian, S.W. Poland) comprising clast-supported conglomerates and sand-rich, matrix-supported conglomerates and pebbly sandstones. Explanation of symbols is the same as preceeding figures. (Reproduced from W. Nemec and R.J. Steel: Alluvial and coastal conglomerates: Their significant features and some comments on gravelly mass-flow deposits, Fig. 22, in, E.H. Koster and R.J. Steel, eds., Sedimentology of Gravels and Conglomerates, Memoir 10, © 1984, Canadian Society of Petroleum Geologists, Alberta, Canada with kind permission).

deposition in fluvial gravel/conglomerate beds (Nemec and Steel, 1984). Moreover, measurement of maximum particle size (MPS) reflects competency of flows, whereas bed thickness measurement (BTh) reflects capacity of flows.

One has to be extremely cautious in such analysis. Absence or poor correlation between MPS and BTh may be due to (1) overestimated bed thickness arising out of amalgamated conglomerate/gravel beds, and (2) reduced bed thickness arising out of interstratal erosion, i.e., underestimated bed thickness. These types of problems can be avoided by careful field observations.

Rheological behavior of such gravelly/conglomeratic deposits of gravity flow origin can be estimated (Nemec and Steel 1984). If the linear regression line of the plot (MPS vs BTh) passes through the origin (Figs. 2.1.2 and 2.1.3), it represents a cohesionless debris flow of either sub-aerial or sub-aqueous nature. If it does not, but intersects at some point on the ordinate (Fig. 2.1.1), emplacement as sub-aerial cohesive debris flow is suggested.

Roundness of the clasts is estimated with the help of silhouette chart for pebbles as proposed by Krumbein (1941) and should be determined in the field.

Geological implications are: (i) well-worn rounded clasts indicate long history of transport before their final burial, (ii) clasts with angular edges and corners indicate local derivation and short history of transport before their final burial. Conglomerate and breccia are differentiated on the basis of roundness of the clasts.

Effects of mode of transport and/or environment on the shape of clasts in gravel are uncertain since certain other factors, e.g., nature of bedding, jointing and rock cleavage are also involved. However, a close look at these clasts may yield some information: i) wind-faceted einkanter and dreikanter pebbles, ii) faceted pebbles with etchings on some of them due to ice movement, iii) percussion marks and spalls resulting from impact in fluvially carried clasts are some of the features that provide a tell-tale evidence in deciphering depositional environment.

Sorting, Size Distribution and Clast-Matrix Ratio

These are the properties that are best studied in the field. Determine (i) if the coarser clasts are in touch with each other forming a gravitationally stable structure, or (ii) the coarser clasts are separated from each other by matrix (Fig. 2.1.4 a). In case of the former it is known as clast-supported conglomerate [orthoconglomerate of Pettijohn (1975)] and in the second case, it is known as matrix-supported conglomerate [paraconglomerate of Pettijohn (1975)].

In the first case, presence of sand as matrix in the interstices may not necessarily mean a high-velocity fluctuation of the depositing current; it may mean either (i) simultaneous deposition of clasts and sand matrix, or (ii) later infiltration of sand into the interstices of the coarser mode. The problem can be resolved by the use of the curve (Fig. 2.1.5), relating the size of the clasts transported by rolling to the size of sand (matrix) suspended by the same flow.

When the clasts are not in touch with each other but are separated by a higher percentage of matrix, it may mean either (i) sub-aerial glacial deposits, or (ii) sub-aerial/sub-aqueous gravity flow deposits. To discriminate between the two, one has to search for field evidences of glacial deposits, e.g., varves, dropstones (Fig. 2.1.6 a, b) striated pebbles and/or pavements (Figs. 2.1.7 a, b), faceted pebbles, tillites.

Amongst these, dropstone and tillites form the best evidences of a glacial environment. Various glacial geomorphological features are hard to find in rock records because of low preservation potential. On the other hand, presence of a Bouma sequence, laminations of grain-flow origin, inverse and inverse to normal grading, imbrication of the clasts (>2 mm) and positive correlation of the plot MPS vs. BTh are the positive features of gravity-flow deposits in case of matrix-supported and clast-supported gravel/conglomerate.

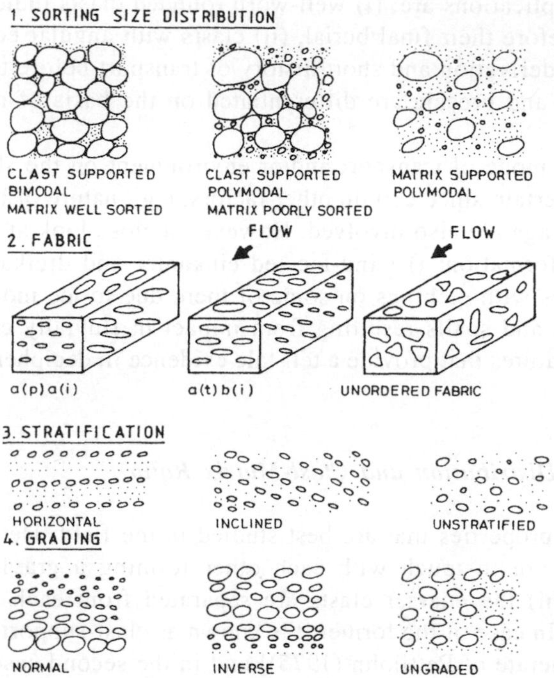

Fig. 2.1.4 a: Descriptive features for conglomerates, Under fabric, the coding a(p)a(i) means that the a-axis (long diameter) of the gravels is parallel to the flow and imbricated; the coding a(t)b(i) means that the a-axis is transverse to the flow direction with b-axis imbricated. (Reproduced from J.C. Harms, J.B. Southard and D.R. Spearing: Depositional Environments as Interpreted from Primary Sedimentary Structures and Stratification Sequences, SEPM Short Course No. 2, Fig. 7-1, © 1975, Society for Sedimentary Geology (SEPM), Tulsa, Oklahoma, U.S.A. with kind permission)

Clast Imbrication and Fabric

Analysis of the imbrication of the clasts is another important feature of conglomerate/gravel studies that are to be undertaken in the field and helps in understanding their mode of transport and emplacement. In fluvial clast-supported conglomerate, the long axes of the clasts statistically orient parallel to the depositional strike (long axis of the clast perpendicular to the flow direction) and the plane containing "a" and "b" axes will dip upcurrent. This can be written as "a" axis transverse and "b" axis imbricate or in short, as a(t) b(i) (Fig. 2.1.4 a). This fabric of the anisotropic clasts suggests that individual clasts are free to interact with the moving fluid without any interference from neighbouring clasts. Imbrication is also found in pebbly beaches produced by the oscillatory motion of waves. The asymmetry of the orbital motion, with the predominance of landward push, causes the pebbles to dip seaward.

Fig. 2.1.4 b: Photograph showing clast imbrication a(t) b(i) in gravels. Gravel pit in Pleistocene littoral deposits, Marchr region, Italy (courtesy Dr. Ricci Luchi, Italy).

Another type of clast imbrication is observed in conglomerates, where the long "a" axis is oriented parallel to flow and imbricated upstream and can be expressed as a(p) a(i) fabric (Fig.2.1.4 a). This kind of fabric suggests that coarser clasts are not free to interact with the entrained flow and are a product of gravity flow.

Sub-aerial and sub-aqueous mudflows (i.e., gravity flows) do not, however, show any clast fabric; the clasts remain separated from each other by matrix and are frozen when the flow ceases to move and solidifies. Since such high-density flows suppress internal turbulence and since the clasts are not free to interact with the flow, imbrication of clasts does not develop. One may, however, find clast imbrication near the bottom of such flows resulting from the shearing effect of such high-density flows.

How to go about in determining clast orientation in a gravel/conglomerate bed in the field? Following are the techniques suggested by Collinson and Thompson (1982):

i. In case of strongly indurated conglomerate, determine the dip and strike of the apparent a-b plane. When, however, the clasts project out, the strike and dip of a-b plane can, with reasonable certainty, be determined by placing a notebook on the plane.

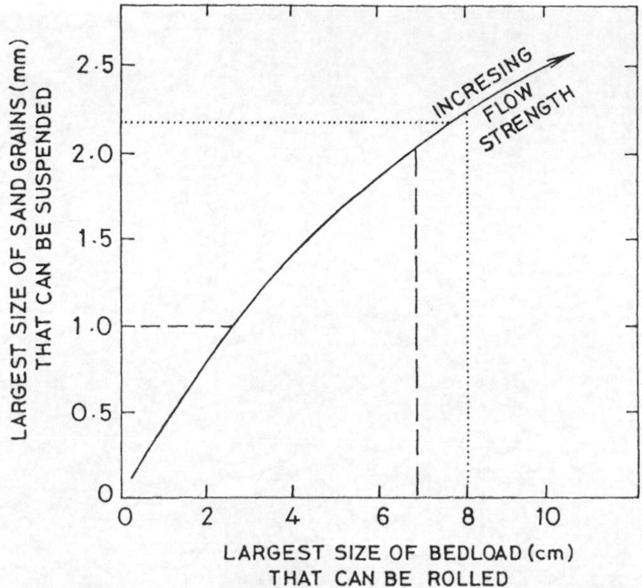

Fig. 2.1.5: Curve relating size of clasts transported by rolling to size of sand suspended by the same flow. For example, the dotted lines show simultaneous rolling of 8 cm cobbles and suspension of 2.2 mm granules. The dashed lines indicate a misfit between bedload and rolling of 7 cm cobbles, followed by a decrease in flow strength and subsequent infiltration of pore spaces of 1 mm sand (Reproduced from J.C. Harms, J.B. Southard and D.R. Spearing: Depositional Environments as Interpreted from Primary Sedimentary Structures and Stratification Sequences, SEPM Short Course no. 2, Fig. 7-3, © 1975, Society for Sedimentary Geology (SEPM), Tulsa, Oklahoma, U.S.A. with kind permission).

ii. In less indurated / unindurated rocks, measure the dip and strike of the clasts and then carefully remove them to measure the orientation of the axes "a" and "b". One can then plot the data on a stereonet for display or show them on proper plan and sectional views.
iii. If three-dimensional reconstruction is not possible in the field, obtain the statistical orientation of preferred elongation in plan view and measure preferred imbrication in vertical cross-section; Davies and Walker (1974) have applied this technique successfully.

If the gravel layers are 'welded', where beds have similar matrix, it means that the two depositional episodes followed each other closely in time and energy for the second episode partially reworked the materials deposited during the first episode, thus having a common matrix and welded contact.

Gravel/conglomerate beds, having a single clast thickness, may represent a product of a current lag deposit. There is another class of conglomerate that is neither clast-supported nor are the clasts in them oriented and imbricated. This class of conglomerate has been termed by Pettijohn (1975) as laminated pebbly

Fig. 2.1.6 a: Large dropstone in fluvio-glacial cross-bedded unit of Talchir Formation, Permo-Carboniferous Gondwana Supergroup, Ramgarh Coalfield, Bihar, India.

b: A large dropstone in diamictite, Dwyka tillite, Kransgat River, South Africa. Note that the overlying and underlying laminations are lapping around the dropstone (Courtesy J.N.J. Visser, University of Orange Free State, Bloemfontein, South Africa)

mudstone. This is produced by dropping of pebbles (dropstones) in to still bottom water where laminated silt and mud are accumulating. It is important to look carefully into the nature of laminae in and around such dropstones. The laminae around the clasts are distorted and bend down beneath those dropstones as well as lap on or arch over them (Fig. 2.1.6 a, b). Distortion of laminae

Fig. 2.1.7a: Soft-sediment pavement in Dwyka tillite. Note four levels (marked as 1, 2, 3 and 4 in ascending order) of striated soft-sediment pavements overlain by ripple sand (Courtesy J.N.J. Visser, University of Orange Free State, Bloemfontein, South Africa).

b: A broken piece from glacially striated pavement, Penganga Limestone, Irai, Chandrapur District, Maharastra, India (Courtesy: Prof. A. Choudhuri, Indian Statistical Institute, Calcutta)

around dropstones indicates dropping of clasts when the sedimentation process was still continuing. As Pettijohn (1975, p. 171) has indicated "in general, laminated argillites with rafted blocks are closely associated with tillites and are, perhaps, the best evidence of the presence of glacial ice and of a glacial origin for associated strata".

Spatial Size Distribution of Clasts

It has been observed that clast size in conglomerate/gravel decreases downcurrent. Such a decline in size of the clasts can be used to determine the direction of sediment transport. Determine the maximum size of the clasts and plot them on a map, which when completed over a large area, shows a definite trend of size reduction of the clasts and consequently the paleocurrent direction (Meckel 1967; Pelletier 1958; Plumley 1948; Potter 1955; Schlee 1957; Yeakel 1959).

Internal Structure of Gravel Deposits

It is often difficult to detect bedding in a thick gravel/conglomerate deposit. Following features in a gravel or conglomerate deposit may help one to detect bedding: a) compositional changes of clasts, b) colour, c) variation in gravel size, d) sorting, and e) fabric in different layers in an apparently massive-looking conglomerate/gravel bed (Fig. 2.1.4 a). The above features define crude layering in contrast to sharp contact in fine-grained terrigenous rocks. Nature of contact between the conglomerate beds may reveal more information. Collinson and Thompson (1982) have suggested the following:

i) If the gravel layers are sharply bounded, it is suggested that the different processes are responsible for each layer,
ii) gravel layers grading into each other, indicate that the depositional process is either fluctuating or pulsatory,
iii) gravel layers inter-bedding with sandy layers, do not necessarily mean high current fluctuation. This association can well be the result of slight fluctuations of flow strength.

Characteristics of Gravel and Conglomerate Deposits of Different Environments

Coarse-grained gravelly/conglomeratic deposits are found in different terrestrial and marine environments and it is necessary that one should be able to discriminate them in the field and laboratory. Common association of features in gravel /conglomerate, formed in different sedimentary environments, are described below, as compiled from data in Bourgeois and Leithold, 1984; Clifton, 1973, 1981; Ethridge and Wescott, 1984; Hein, 1982, 1984; Houseknecht and Ethridge, 1978; Nemec et al., 1980, 1984.

Glacial Gravel/Conglomerate

Poorly sorted, poorly rounded, non-stratified extensive sheet of gravel/conglomerate having multi-lithologic composition of clasts are common, occasionally with exotic clasts; elongated clasts are oriented parallel to ice flow direction; clasts and pavements may be striated and are usually associated with varves with complex internal rhythms; presence of dropstone within stratified and laminated varves is common; tabular cobbles with pentagonal outline along with faceted and striated cobbles and pebbles are common; striations on pebbles and cobbles are mutually sub-parallel to one another and parallel to the long axis of the pebbles and cobbles and are characteristic features of glacial gravel/conglomerate deposits.

Fluvial Gravel/Conglomerate

Braided stream gravel/conglomerate forms the largest group within alluvial conglomerate; pebbles and cobbles are poorly sorted and poorly segregated; individual gravel/conglomerate bed is lenticular with common basal erosional surface; gravel/conglomerate is commonly inter-bedded with gravelly sandstone; gravel/conglomerate bed shows small-scale fining-upward sequence; thickness of gravel/conglomerate bed ranges between 1-2 m; graded bed is not common; clasts show high sphericity but low roundness; horizontal stratification with layers of different clast sizes occur that are interspersed with beds having large-scale planar cross-beds; fabric is commonly clast supported and usually display a(t) b(i) fabric whereas smaller clasts may display a(p) a(i) fabric; random fabric is also reported towards the top of the gravel bar possibly because of reworking; voids between clasts may be matrix-filled; style of stratification changes with grain size or sorting, reflecting variable flow and discontinuous accretion; thickness of strata/cross-strata of gravel/conglomerate beds provide rough estimate of water depth; upward-fining facies motif is characteristic of braided alluvial deposits; association with pedosols or non-marine fauna or ichnofacies are charcateristic of alluvial gravel/conglomerate deposits.

Near-Shore Gravel / Conglomerate

Beach gravel/conglomerate is well-sorted and well-segregated and is clast-supported; individual gravel beds are laterally more continuous with rare basal erosional contacts; inter-bedding with sandstone is rare; sphericity of clasts is low but show high roundness; imbrication is well-developed seaward showing a fabric a(t) b(i); fining-up or coarsening-up sequences are rare; associated sandstone are planar bedded and rarely graded; swash cross-stratification and streaks of heavy minerals in sandstone are characteristic of near-shore sandstone; faunal diversity is low but population density may be quite high; upper shoreface gravel/conglomerate may be crudely graded whereas lower shoreface gravel/conglomerate is characterized by low-angle stratified pebbly sandstone and lenticular gravel/conglomerate, reflecting scour deposits; tabular beds and trough cross-stratified pebbly sandstone with presence of high-angle scours are common and reflect polymodal transport directions and are commonly normally graded;

Fan-delta Gravel/Conglomerate

Deposits are poorly sorted, commonly ungraded, clast-supported gavel/conglomerate beds; beds display massive to crude horizontal bedding in coarse fractions and planar or horizontal beds in sandy conglomerate and conglomeratic sandstone; fabric of gravel/conglomerate, particularly in fringe areas, is a(t) b(i) with imbrication landward; giant foresets in conglomerate, up to 30 m high, are reported from shallow standing body of water of shallow marine or lake; deposits in shallow marine or lake represent Gilbert-type delta deposits, reflecting situations where density of incoming flow and ambient water are the same (homopycnal flow); non-marine to marine transition is dotted with features of beach and braided fluvial channel deposits; the upper sub-aerial part of the fan-delta is characterized by debris-flow and mud-flow deposits; mudstone shows mud-cracks and are interspersed with caliche horizons with rootlets and other soil features; depending on climatic factors, outer fringe fan-delta deposits may have features of spits, tidal flats, shallow marine bars, delta foreset and/or carbonate reefs or algal mounds; in shallow shelf areas, mud may be interbedded with near-shore bar sandstone; fossil and bioturbation structures are abundant in marine sandstone.

Alluvial Fan Gravel/Conglomerate

Debris flow is common in the upper part of alluvial fan gravels; beds are ungraded to graded and lack internal stratification; inverse grading may be present at the basal part because of shearing effect; progradation, however, may also result in coarsening-upward sequences; beds are texturally polymodal to bimodal, and the fabric displays either a clast-supported or a matrix-supported, commonly with 'outsized' clasts; statistical correlation is observed between the thickness of the bed and the maximum clast size; mid-fan areas are dominated by sandstone and gravelly sandstone with basal erosional surface; Fluvially dominated part of the gravelly deposits may show a(t) b(i) fabric; paleocurrent pattern radiate outward downfan; red colouration and evaporitic paleosols are characteristic deposits of semi-arid environment.

Submarine Fan Gravel/Conglomerate

Gravel/conglomerate is associated with sandstone inter-bedded with hemi-pelagic mudstone; gravel fractions mostly are re-sedimented; gravel/conglomerate is matrix-supported in deposits of near-slope and proximal submarine feeder channels, otherwise it is clast-supported; geometry of gravel/conglomerate is discrete, lenticular-shaped, bounded by lower scour surface; sandstone with fining- and thinning-upward sequence is more common; thickness ranges between 5-10 m; coarse clasts show a(p) a(i) fabric having landward imbrication; coarser clasts in gravel/conglomerate are usually normally graded with less common inverse grading; beds with clast-supported texture may alternate with clast-dispersed texture; gravel/conglomerate is dominantly structureless, massive; slump and fold structures are found in proximal slope and inter-fan deposits; olistostromal deposit is common in upper reaches of submarine fans; abundant deep-water fauna and ichnofossils are commonly found in distal part of the fan.

2.2. Sandstone

Sandstone is the lithified product of mechanically deposited sediments ranging in particle size from 1/16 mm to 2 mm. Sandy sediments are driven by and deposited from flows involving movement of air, water and glacier or sediment gravity flows (debris flow, turbidity current, grain flow, liquified flow). Sand deposits are common in almost all the sedimentary environments of the earth.

Compositionally, sandstone is dominantly siliciclastic; however, calciclastic (e.g. oolitic limestone), bioclastic and volcaniclastic varieties (Fig. 2.2.1) are also common in the rock record. Texturally the following varieties of sandstone are recognizable in the field: pebbly or gritty sandstone, coarse sandstone, medium sandstone and fine sandstone.

In the field, sandstone may occur as thick (tens of meters or more) monolithic units of stratigraphic status or as thin (less than a few meters) units inter-bedded with other lithologies such as, shale, conglomerate or non-terrigenous rocks (e.g. limestone).

Internal Features of Sandstone Beds

Massive and Graded Beds

Massive as well as graded sandstone beds are products of rapid deposition from high concentration sandy suspension flows (sediment gravity flows or storm suspensions). The particles settling from suspension are prevented from being reworked after reaching the depositional surface due to quick burial.

Fig.2.2.1: Field photograph of Tertiary volacaniclastic sandstone of Central Massif, France. Note well-developed stratification and volcanic clasts in the deposit. Length of the section is around 2m.

Deposition of massive sand beds results from sudden failure of capacity and competence of the flow. High sediment concentration may result (i) in higher viscosity of the flow (Middleton and Southard, 1978), (ii) suppression of turbulence and (iii) reduced settling velocity of sediment particles, hindering ripple/dune formation (Harms et al., 1982). Deposits from such flows appear structureless, massive in nature (Martin and Turner, 1998). Such sandy deposits, however, remain prone to the development of water escape structures because of loose primary packing (Lowe, 1982). A sandstone bed, however, may also appear massive due to post-depositional obliteration due to burrowing, fluidization or pedogenesis.

If deposition from sandy suspension flows takes place due to gradual failure of capacity and competence the particle size in the resultant sand bed decreases from bottom to top, i.e. normal grading develops (Fig. 2.2.2 a, b). Normal grading may have different expressions: 1) all particle size fractions fine upward, 2) only the coarsest fraction fines upward (Allen, 1982) and 3) the proportion of the coarsest fraction decreases upward (Fig. 2.2.3)

Sand beds characterized by a vertical increase in particle size, a feature termed reverse grading, are also products of rapid deposition from sandy suspension flows; however, in such flows the sand particles are held in suspension due to the dispersive force arising from grain collisions (grain flow); in grain flows, due to the operation of a process called kinetic sieving, finer particles tend to be disposed at the bottom of the flow giving rise to reverse grading in the resultant deposit (Harms et al., 1975). Hand (1997), on the contrary, suggested that inverse grading might be a consequence of streamwise size

Fig. 2.2.2 a, b: Different types of normal grading in sandstones.

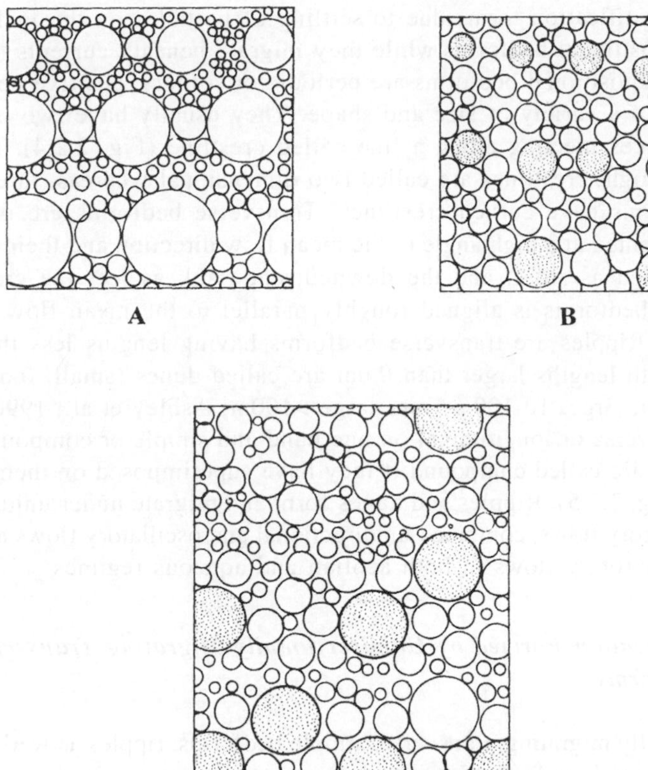

Fig. 2.2.3: Schematic illustration of different types of normal grading in sandstones. **(A)** All the grain population decreases in size vertically. **(B)** Only the coarsest fraction (stippled) decreases in size vertically. **(C)** The proportion of the coarsest fraction (stippled) decreases vertically.

segregation in the suspension flow containing heterogeneous mixture of sediment sizes. The streamwise size segregation arises as the coarse sediment fractions are transported more slowly than fine fractions. If deposition begins at a site before the arrival of the coarse fraction at the site. the resulting deposit may develop with an inverse grading.

Cross-stratification

Cross-strata are primary sediment layers inclined to the principal surface of accumulation at angles less than or equal to the angle of repose of the constituent sediment particles. They are the most ubiquitous internal structures of sandstone beds.

Cross-stratification forms due to settling of particles on the inclined surfaces of depositional bedforms while they migrate beneath currents (Harms et al., 1982). Depositional bedforms are periodic or solitary, positive bed surface undulations of a variety of size and shape. They usually have two oppositely sloping surfaces meeting along a line called crestline (Fig. 2.2.4). Bedforms that have straight crestlines are called two-dimensional whereas, three-dimensional bedforms have curved crestlines. Transverse bedforms are those with crestlines oriented at a high angle to the mean flow direction and their upcurrent flank is termed as stoss, and the downcurrent flank as lee; the crestline of longitudinal bedforms is aligned roughly parallel to the mean flow direction (Fig. 2.2.4). Ripples are transverse bedforms having lengths less than 0.6m. Bedforms with lengths larger than 0.6m are called dunes (small: 0.6-5m, medium: 5-10 m, large: 10-100m, very large >100m; Ashley et al., 1990). Dunes may be transverse or longitudinal on one hand and simple or compound on the other. Dunes are called compound if they have superimposed on them, smaller bedforms (Fig. 2.2.5). Ripples and dunes form and migrate under unidirectional flows, oscillatory flows, combined unidirectional and oscillatory flows and multidirectional or rotary flows in both aeolian and aqueous regimes.

Cross-stratification Formed by Unidirectionally Migrating Transverse, Simple Bedforms

Unidirectionally migrating, transverse, simple bedforms, ripples as well as dunes, produce unidirectionally dipping simple cross-strata as a result of deposition on the lee surface of the bedforms. Cross-strata may occur as a solitary set or multiple sets stacked one above another (Fig. 2.2.6). Form-sets are formed when the bedforms remain in the growing stage or migrate under the condition of zero net deposition (Fig. 2.2.6 b); Non-periodic or solitary bedforms also produce single sets with or without preservation of the bedform morphology (Fig. 2.2.6 a). Stacked-sets are, on the other hand, formed under the condition

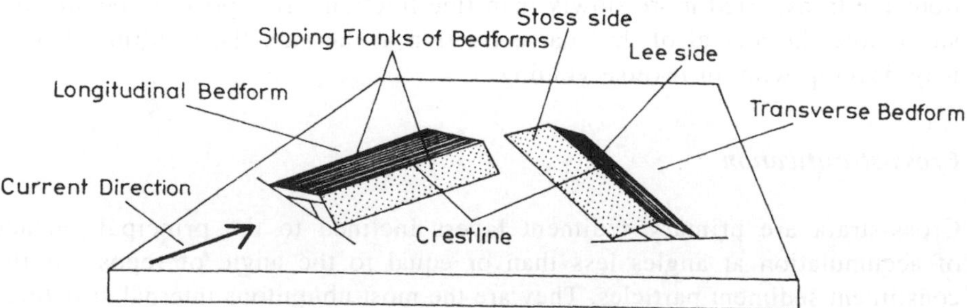

Fig. 2.2.4: Diagram illustrating definitions of longitudinal and transverse bedforms, crestline, stoss and lee surfaces with respect to the mean current direction.

Fig. 2.2.5: Photograph of a compound bedform. Note superimposed ripples on the foresets of the larger dune and orthogonal orientations of the ripple and dune crests.

of positive net deposition from a train of migrating bedforms (Allen, 1986; Rubin, 1987; Fig. 2.2.6 c). In a stack of cross-sets, individual sets may have a variety of shapes and the sets can be grouped into cosets (Fig. 2.2.7). A coset of cross-strata is produced due to migration and climbing of periodic bedforms under condition of net deposition (Figs. 2.2.8 and 2.2.9). In such a coset the set boundaries represent the paths of climb of the successive trough points of the bedforms and are essentially inclined opposite to the migration direction of the bedforms (Figs. 2.2.8 and 2.2.9). The amount of their inclination is directly proportional to the rate of net deposition (Fig. 2.2.9) As a result, the thickness of the sets (i.e., the perpendicular distance between two successive set bounding surfaces) also increases with increasing rate of net deposition. The inclination of the set bounding surfaces may be very close to zero and in many cases they appear sub-horizontal in the field (Fig. 2.2.6 c). A coset of cross-strata, formed by migration of ripples, is termed climbing-ripple cross-lamination (Fig. 2.2.8 a) The inclination of a set bounding surface may remain constant along the length if the rate of deposition does not vary temporally; it acquires a convex-up shape if the rate of deposition decreases steadily with time and a concave-up shape if the rate of deposition increases steadily with time; the surface assumes an undulatory shape if the rate of deposition fluctuates randomly with time (Rubin, 1987; Fig. 2.2.9 c). The shape of a cross-set is controlled by the relative inclination of the surfaces that bound it (Fig. 2.2.9 c).

Besides finding the paleocurrent from the azimuths of cross-strata, the other primary task of studying cross-stratification in the field is to reconstruct the type of bedform that generated it. Where form sets are preserved, they provide a direct clue to the geometry and size (wavelength and amplitude) of the bedform. Where the set bounding surfaces are inclined, the horizontal spacing between

Fig. 2.2.6: Different modes of occurrences of cross-stratification. **(a)** Solitary set of cross-strata without showing any bedform morphology. **(b)** Form set of cross-strata preserved within a ripple. **(c)** Vertically stacked sets of cross-strata.

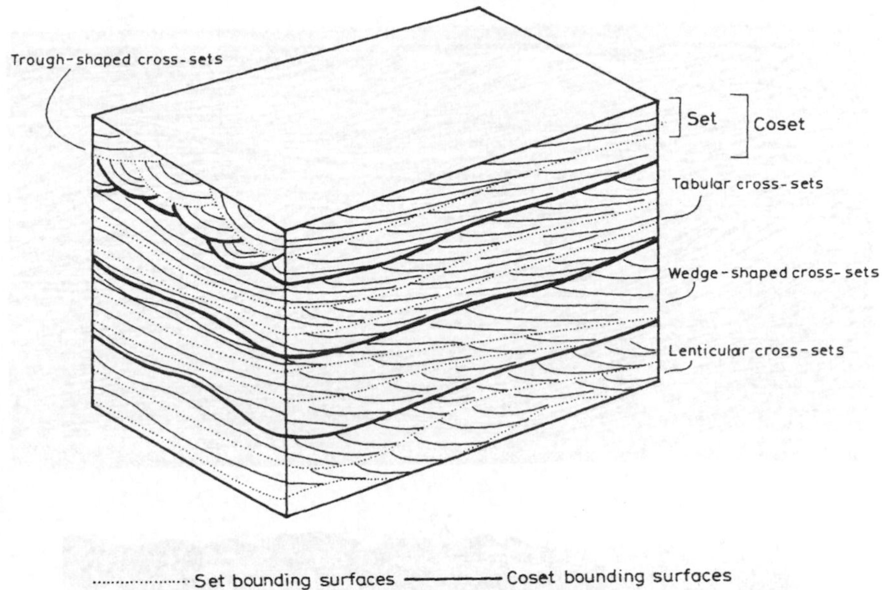

Fig. 2.2.7: Schematic diagram showing definitions of set, coset and different types of cross-set geometries.

two successive set bounding surfaces approximates the wavelength of the bedform (Fig. 2.2.10). Where the set bounding surfaces appear sub-horizontal in the field, the set thickness provides the least estimate of the amplitude of the bedforms and a rough estimate of the wavelength would be a few times the estimated amplitude.

The geometry of cross-strata represents the shape of the lee surface of the mother bedforms. In order to reconstruct the morphology of the lee surface, the cross-strata are to be observed in horizontal (bedding plane) as well as in longitudinal-vertical (parallel to the migration direction of the bedforms) sections (Figs. 2.2.11, 2.2.12 and 2.2.13).

The cross-set boundaries produced by 3-D bedforms characteristically appear concave-up in transverse (perpendicular to the bedform migration direction) and that is why such cross-stratification is often termed trough cross-stratification (Figs. 2.2.14 a, b, c). This should not be confused with channel fills as described later.

In addition to the morphology, the dip of the cross-strata is an important indicator of the type of bedforms. Angle-of-repose dip indicates lower flow regime bedforms, whereas low angle of inclination (less than 10^0) suggests transitional bedforms. In general, the lee surfaces of bedforms change from high angle, planar to low angle concave-up to sigmoidal with increasing flow strength (Chakraborty and Bose, 1992; Fig. 2.2.15). Sigmoidal cross-stratifica-

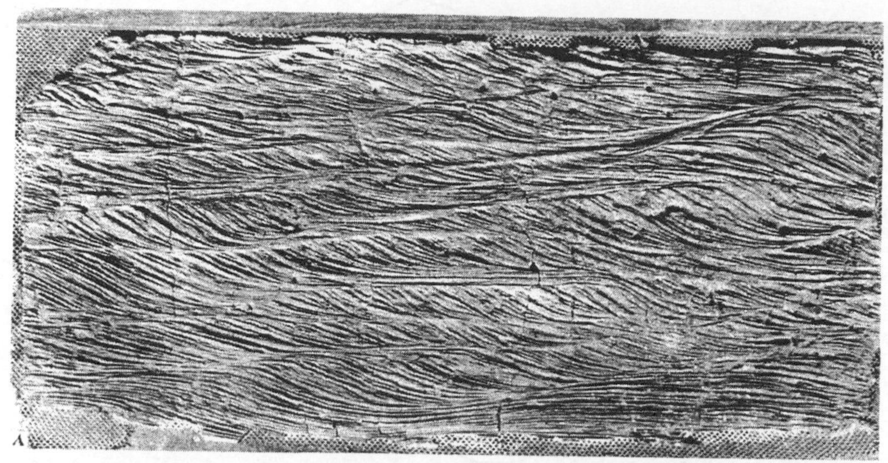

Fig. 2.2.8: Photograph of stacked sets of cross-strata produced due to migration and climbing of periodic ripples **(a)** and dunes **(b)**. Note upcurrent inclination of the set boundaries in both cases. **(a)** is a peel from recent River Ganges sediments. In **(a)** length of the peel is 38 cm; in **(b)** the diagonal scale (circled) is 15 cm long.

tion may eventually turn into inclined or horizontal parallel stratification (Fig. 2.2.16).

The internal features of individulal cross-stratum provide clues to the mechanics of deposition. Massive or inversely graded layers indicates deposition from grain flows on the lee side of the bedforms (Hunter, 1977); such layers are generally high-angle, planar, abut sharply against the lower set-bounding surface and may be tongue- or wedge-shaped (Fig.2.2.17). The layers formed due to deposition from suspension (grain fall) may be normally graded or massive,

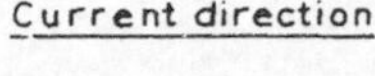

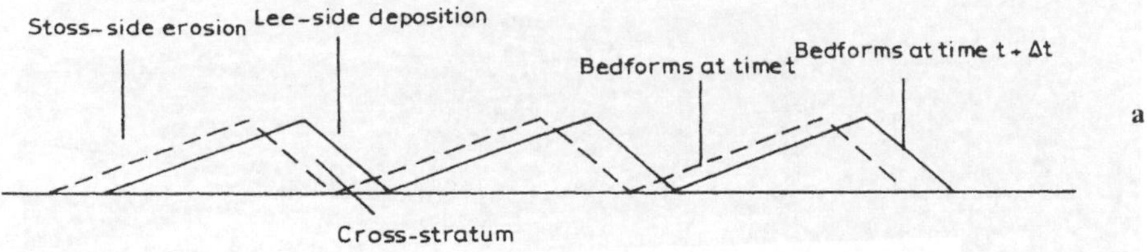

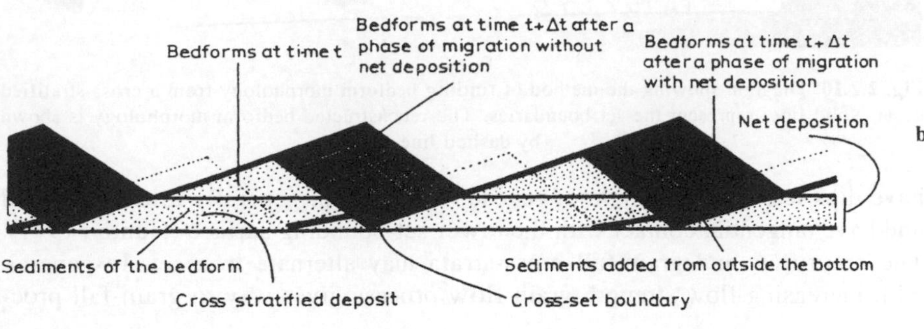

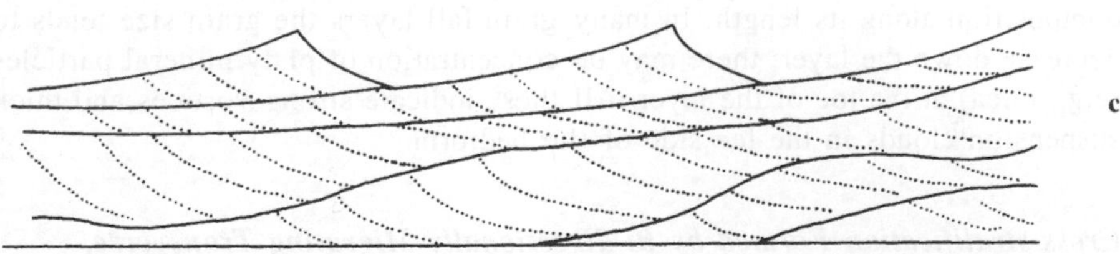

Fig. 2.2.9: Diagram showing the kinematics of formation of cross-stratified coset. Cross-stratification results from successive deposition on the inclined lee faces of the bedforms as they migrate by a process of stoss-side erosion and lee-side deposition. Note that in **(a)** only form sets are produced because the bedforms migrate without net deposition (i.e., without receiving extra sediments from the flow) - the volume of sediments deposited on the lee exactly balances the volume eroded from the stoss. In **(b)** the bedforms receive additional sediments from the flow which are deposited on the lee faces along with the sediments derived from the erosion of the stoss side; as a result, the bedforms climb and leave a cross-stratified deposit underneath (i.e., net deposition takes place); the set boundaries in the cross-stratified deposit represent the loci of migration of the trough points of the bedforms and are inclined opposite to the migration direction of the bedforms; note that the inclination of set boundaries would change if the amount of added volume of sediments varies. In **(c)** a cross-stratified deposit is shown wherein the cross-sets assume variable geometries and inclinations due to temporal fluctuation of the angle of climb of individual trough points and non-uniformity in the angle of climb of different trough points.

Fig. 2.2.10: Diagram showing the method of finding bedform morphology from a cross-stratified coset. Solid lines represent the set boundaries. The reconstructed bedform morphology is shown by dashed line.

have dips less than the angle-of-repose, concave-up, generally thicken upward and have tangential contact with the lower set-bounding surface (Hunter, 1977). The grain-flow and grain-fall cross-strata may alternate in a set. In general, with increasing flow strength grain-flow process gives way to grain-fall process.

Within an individual cross-stratum, there may be variation in texture and composition along its length. In many grain-fall layers the grain size tends to decrease down the layer; there may be concentration of platy mineral particles (e.g., mica) at the toe of the layer. All these indicate strong vortices and thick suspension clouds in the lee side of the bedform.

Cross-stratification Formed by Bi-directionally Migrating Transverse, Simple Bedforms

Among the bi-directionally migrating transverse simple bedforms the most commonly recognized variety is the wave ripple. Bi-directionally dipping cross-stratification arises due to deposition on the lee surface of the ripples that alternately reverses its orientation (Fig. 2.2.18 a). Successive sets (coset) of cross-strata are generated as the bedforms in a train migrate beneath flows under condition of positive net deposition. Form-sets are formed when the bedforms remain in the growing stage or migrate under the condition of zero net deposition. The different diagnostic features of wave-ripple stratification are shown in figure 2.2.18 h (see also De Raaf et al., 1977).

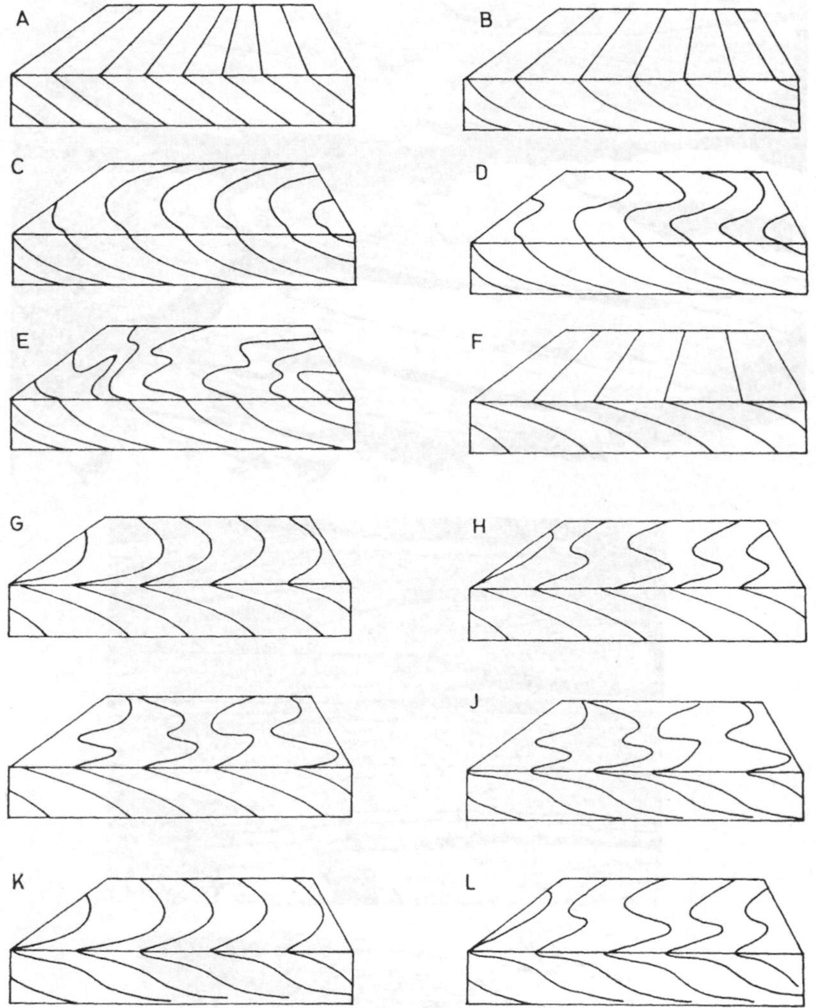

Fig. 2.2.11: Schematic diagram showing bedding plane and longitudinal cross-sections of different types of cross-stratification.

Cross-stratification Formed by Unidirectionally Migrating Transverse, Compound Bedforms

Transverse compound bedforms are large- mostly solitary- dunes with superimposed ripples (or smaller dunes) migrating across the crest of the dune. Unidirectionally dipping cross-strata is formed due to accretion on the lee surface of the dunes. However, in contrast to the cross-stratification formed by simple, transverse dunes, the cross-strata in this situation are internally made up

Fig. 2.2.12: **(a)** Photographs showing planar (lower set) and concave-up (middle set) cross-stratal geometries; **(b)** a set of convex-up cross-strata; **(c)** sigmoidal cross-stratification.

Fig. 2.2.13: Photographs of concave-downcurrent (a) sinuous (b) random (c) bedding plane geometries of cross-strata.

Fig. 2.2.14: Photographs showing longitudinal (a) and transverse (b, c) sections of trough cross-stratification.

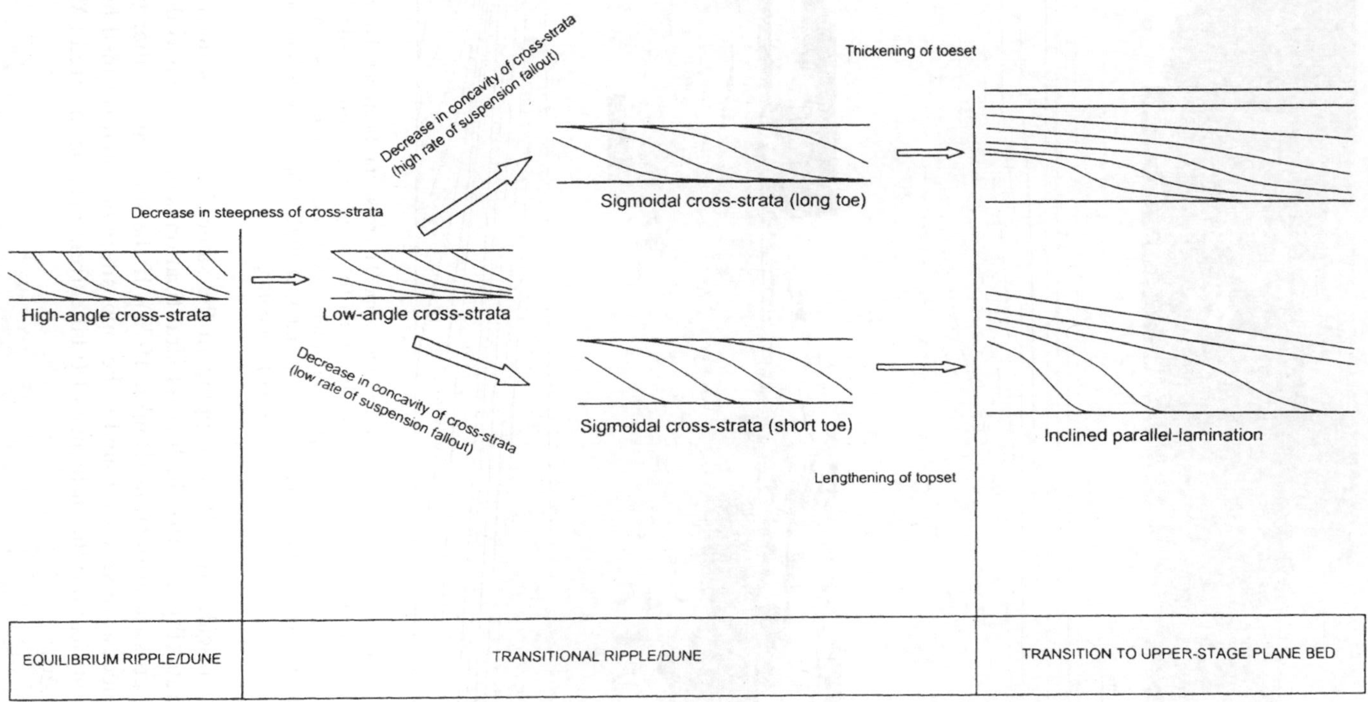

Fig. 2.2.15: Diagram illustrating the kinematics of the transition from planar cross-stratification to upper-stage parallel stratification.

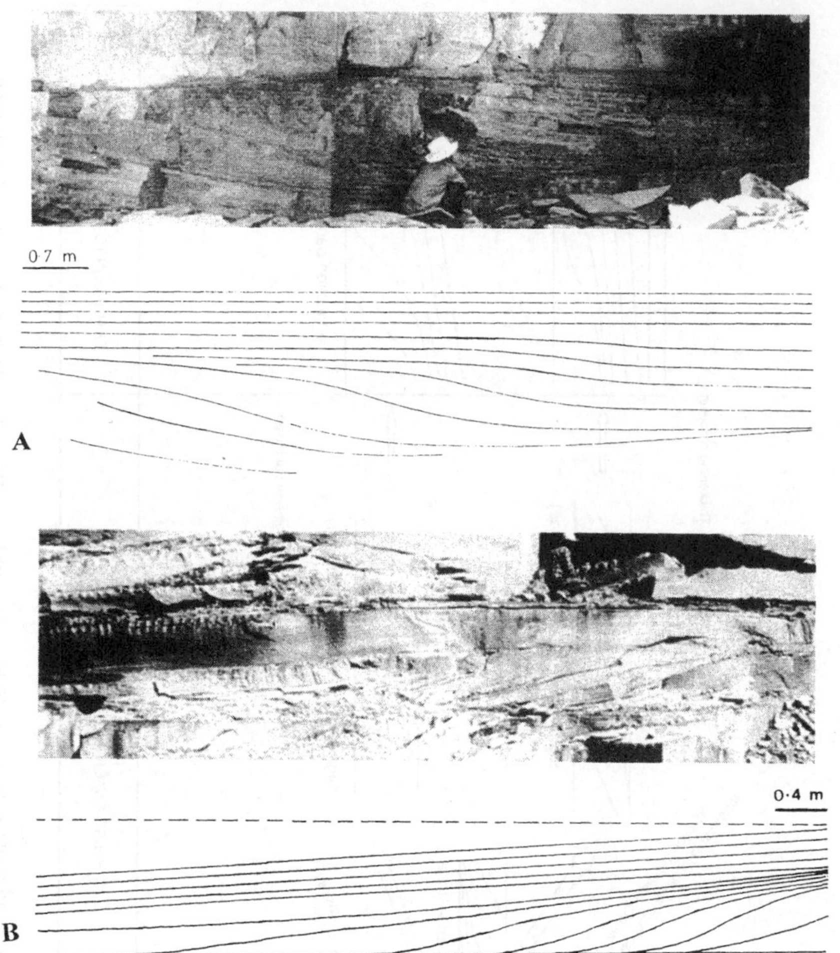

Fig. 2.2.16: Photographs showing transition from sigmoidal cross-stratification to horizontal parallel stratification (**a**) and inclined parallel stratification (**b**). (Reproduced from Chakraborty, C. and Bose, P.K., 1992, Ripple/dune to upper stage plane bed transition: some observations from the ancient record. Geol. Journal, Vol. 27, Figs. 3 and 4, © 1992, John Wiley & Sons, Ltd., with the kind permission).

of smaller scale cross-strata dipping in the same direction as the larger scale cross-strata (Fig. 2.2.19). Such cross-stratification is also known as downcurrent dipping cross-stratification (Banks, 1973). If the internal, smaller cross-strata in the successive sets are found to be oppositely oriented (termed herringbone cross-stratification) the influence of tidal current may be inferred (Allen, 1980; Fig. 2.2.20).

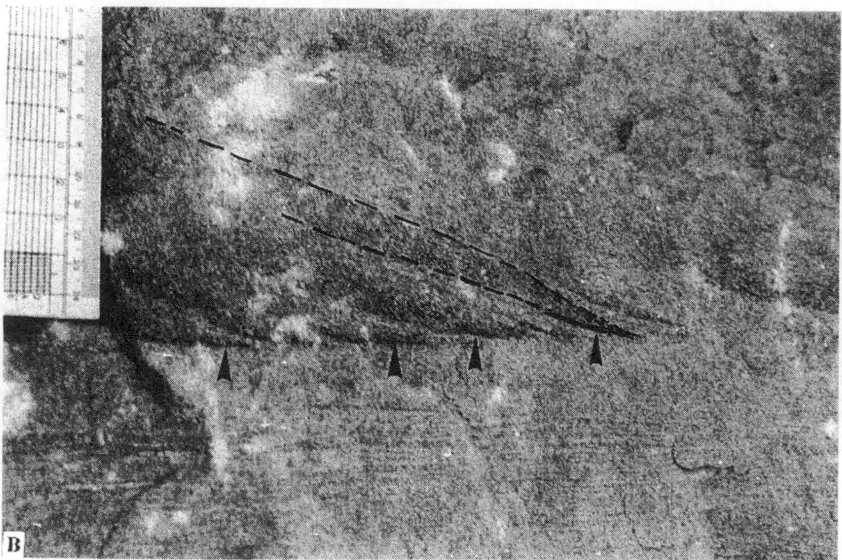

Fig. 2.2.17: Photographs showing grain-flow cross-strata. **(a)** Note high dip, planar shape and sharp abutment with the lower set boundary. **(b)** Tongue-shaped grain-flow cross-strata (arrows) interlayered with grain fall layers (Courtesy: Dr. T. Chakraborty, Indian Statistical Institute, Calcutta, India).

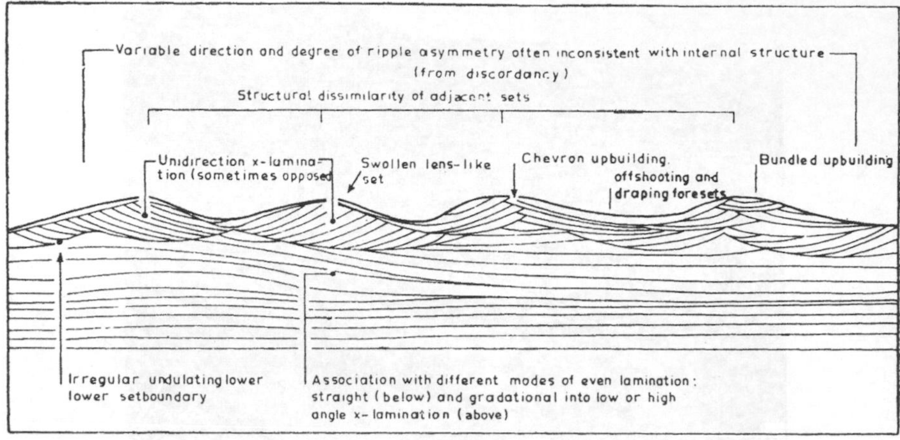

a

Fig. 2.2.18: Sketch **(a)** and photograph **(b)** showing principal diagnostic features of wave ripple stratification (a: reproduced from De Raaf, J. F. M., Boersma, J. R. and Gelder, A., 1977, Wave-generated structures and sequence from a shallow marine succession, Lower Carboniferous, County Cork, Ireland. Sedimentology, Vol.24, Fig. 7, p.459, with the kind permission from Blackwell Scientific Pub.; b: Geological photo, J. Sediment. Petrology, Vol. 51, No. 3, page 720, with the kind permission from Society for Sedimentary Geology (SEPM), Tulsa, Oklahoma, U.S.A.).

A

B

Fig. 2.2.19: Photographs of downcurrent dipping cross-stratification. (a) Note inclined set boundaries with internal smaller-scale cross-strata dipping in the same direction; (b) Close-up view of the same.

Fig. 2.2.20: Downcurrent-dipping cross-stratification with the smaller-scale cross-strata oriented towards, as well as, opposite to the inclination of the set boundaries (a) resulting in herringbone pattern of arrangement (b).

Cross-stratification Formed by Bi-directionally Migrating Transverse, Compound Bedforms

This type of cross-stratification is similar to that formed by unidirectionally migrating transverse, compound bedforms except that in this situation the larger scale cross-strata are oriented bi-directionally suggesting tidal current activity (Allen, 1980; Fig. 2.2.21).

Cross-stratification Formed by Unidirectionally Migrating Longitudinal, Compound Bedforms

Longitudinal dunes are compound bedforms with superimposed ripples (or smaller dunes) migrating parallel to the crest of the dune (Rubin and Hunter, 1985; Fig. 2.2.22 a) Unidirectionally dipping larger-scale cross-strata are formed due to accretion on one of the flanks of the dune (Fig. 2.2.22 b, c). The larger scale cross-strata is internally made up of smaller scale cross-strata dipping in the direction at a high angle to the dip direction of the larger scale cross-strata (Fig. 2.2.22). This type of cross-stratification is suggestive of an aeolian environment. Similar stratification develops also from laterally accreting point bars in fluvial environments and is commonly known as epsilon cross-stratification (see Chapter 4).

Cross-stratification Formed by Bi-directionally Migrating Longitudinal, Compound Bedforms

This type of cross-stratification is similar to that formed by unidirectionally migrating longitudinal dunes except that the larger scale cross-strata in this situation are oppositely oriented (Chakraborty, 1993; Fig. 2.2.23). The cross-stratification pattern is also suggestive of an aeolian environment.

Fig. 2.2.21: Schematic sketch showing the internal stratification style (in longitudinal section) of bi-directionally migrating compound, transverse bedforms.

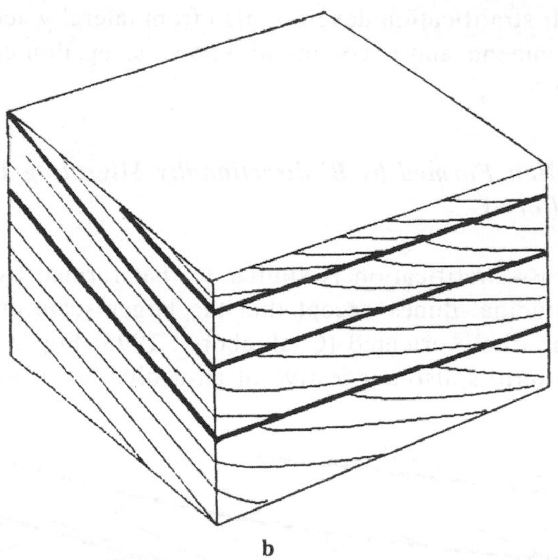

Fig. 2.2.22: **(a)** Photograph of a longitudinal dune. Note two oppositely dipping flanks one of which shows ripples oriented at right angle to the crest of the dune. **(b)** Schematic representation of the internal structure of longitudinal dunes. **(c)** A natural example of the internal structure of longitudinal dunes on a roughly crest-normal section. Note leftward dipping strata (translatent strata) produced due to migration of ripples across the plane of the section.

Fig. 2.2.23: An exposed longitudinal dune showing oppositely dipping strata that are internally made up of ripple cross-lamination with the laminae dipping along the crest of the dune.

Cross-stratification Formed under Multidirectional Flows

Under multidirectional aqueous flows bedforms tend to be domal in shape with an internal stratification termed hummocky cross-stratification described in Section 2.7. In an aeolian regime, multidirectional flows result into star dunes, the internal structure of which is poorly understood.

Other Features Associated with Cross-stratification

Within an individual set of cross-strata there may be several features indicative of flow fluctuations (Allen, 1973a):

(a) Within a cross-stratified set there may be erosional surfaces truncating the cross-strata (Figs. 2.2.24 a, 2.2.25 a, b). These surfaces, termed reactivation surfaces, indicate temporary cessation of bedform movement followed by an erosional event. If such surfaces are present periodically they divide the cross-set into bundles suggesting tidal current influence (Allen, 1980).

(b) Within a cross-stratified set some foreset surfaces may be characterized by burrows, rootlets or mud drapes (Fig. 2.2.25c). These surfaces indicate temporary cessation of bedform movement. If such surfaces are present periodically they divide the cross-set into bundles suggesting tidal current influence (Allen, 1980).

(c) Within a cross-stratified set there may be down-current variation in cross-strata characteristics. Within the set planar cross-strata may give way to low angle concave-up to sigmoidal cross-strata suggesting gradual transformation of dune to upper-stage plane bed (Chakraborty and Bose, 1992; Fig.2.2.15). In some cases thickness of cross-strata may change abruptly across a surface or there may be abrupt transition from simple cross-stratification to compound cross-stratification or vice versa.

(d) Within a cross-stratified set there may be convex-up or planar, sub-horizontal surfaces of limited lateral extent dividing the cross-set into two units locally (Fig. 2.2.24 b); the surface terminates both in the down-current and up-current directions within the cross-set and is termed as hanging reactivation surface.

(e) Within a cross-stratified set there may by down-current dipping surfaces that locally divide the cross-set into two units with the upper unit having cross-strata of greater inclination than that of the lower unit. The reactivation surface may be the last foreset of the underlying unit and terminate in down-current direction at the toe of the underlying foreset (Fig. 2.2.24 c). These surfaces record cessation of migration of a bedform followed by micro-delta-type filling of the space in front of it.

(f) The lower bounding surface of some cross-sets may be observed to have scooped down into the underlying set resulting in what is termed scalloped

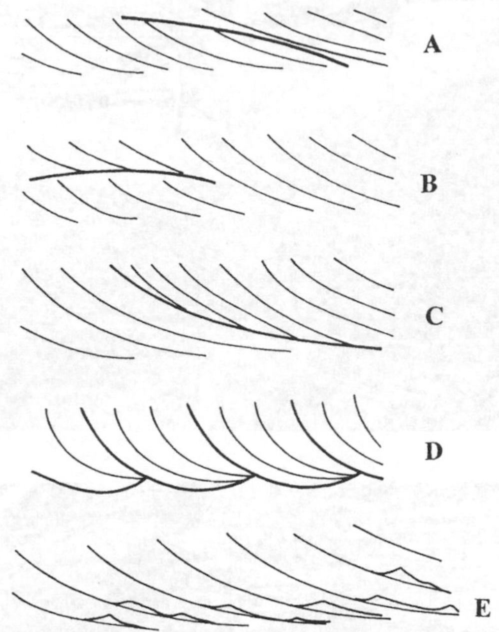

Fig. 2.2.24: Sketches showing different signatures of flow fluctuation with a set of cross-stratification. **(a), (b), (c)** show different types of reactivation surfaces; **(d)** shows scalloped lower set boundary; **(e)** shows oppositely facing ripples (back-flow ripples) at the toes of cross-strata; **(f)** is the photograph of back-flow ripples (arrowed). (Courtesy: Dr. Tapan Chakraborty, Indian Statistical Institute, Calcutta, India).

Fig. 2.2.25: **(a)** Erosion surface (arrowed) gently truncating cross-strata. **(b)** Erosion surface sharply truncating cross-strata. **(c)** Cross-strata (light coloured) punctuated by mud drapes (dark coloured). **(d)** Cross-stratified set with scalloped lower boundary. **(e)** Cross-stratified set with a scour on the upper boundary (arrowed); the scour-fill is also cross-stratified.

cross-stratification (Figs. 2.2.24 d and 2.2.25 d). Scallops indicate periodic fluctuation of bedform height under unsteady flows (Rubin, 1987).

(g) The upper bounding surface of a cross-set may be locally cut into scours filled with massive or stratified sediments indicating sudden and local fluctuations of the flow (Fig. 2.2.25 e).

(h) The toe of the foresets within a cross-set may show small asymmetric ripples directed opposite to the inclination of the foresets (Figs.2.2.24 e, f). These ripples are commonly termed as back-flow ripples if formed due to back-flow in the lee side vortex of the larger bedform. However, oppositely oriented ripples may also form due to reversal of the primary current, such as reversing tidal currents (Allen, 1980; Fig. 2.2.20).

Parallel Stratification

Parallel stratification is another common internal structure of sandstone. The term is reserved for sub-horizontal (less than 5^0) millimetre thick layers disposed parallel to each other (Paola et al., 1989; Fig. 2.2.26 a). If the parallel stratification is associated with parting lineation they may be recognized as products of upper flow regime (Harms et al., 1982). Under lower flow regime, parallel stratification is devoid of parting lineation. In an aeolian environment, migration and climbing of ripples at sub-horizontal angles often produce closely spaced cross-sets within which cross-stratification is hardly discernible and as a result appear as parallel stratification. Such stratification is commonly termed translatent strata (Hunter, 1977; Fig. 2.2.26 b). The distinguishing feature of translatent strata is the internal inverse grading (Kocurek and Dott, 1981).

Bedding Plane Features

Ripples

Ripples are common bedding plane features of sandstones (Fig. 2.2.27). Ripples may be symmetric and asymmetric in longitudinal profile. Symmetric ripples are usually diagnostic of oscillatory flows and are called wave ripples. Asymmetric ripples are commonly generated by both (a) unidirectional current and (b) velocity/time asymmetric oscillatory flow. The crest region of ripples may be sharp, rounded or flat (Fig. 2.2.27). Flat-topped geometry is considered not to be a primary feature of ripples but of erosional origin (Fig. 2.2.27 g).

The crest lines of ripples may be linear, curvilinear, sinuous, concave-downcurrent, convex-downcurrent, closed-polygonal, closed-curved (Fig. 2.2.27). Ripples with linear or curvilinear crest lines may locally show bifurcation of crest lines or double crest lines (Figs. 2.2.27 a, d, f,) suggesting their wave origin. There are also ripples that are domal in shape without any definite crest line (Fig. 2.2.27 c); these ripples are products of multidirectional oscillatory flows.

In some cases larger ripples often show smaller ripples on their troughs The smaller ripples may be aligned parallel to the larger ripples (Fig.2.2.27 h) or at a high angle to them. In the latter situation the term ladder-back ripple is used (Figs. 2.2.27 b, i).

On the bedding plane, ripples with diverse crest line orientations may be found superimposed on each other (Fig. 2.2.27 m). Such superimposition suggests multidirectional flows.

On some bedding planes of sandstone, ripple-like features are found arranged in a regular rhombic pattern (Fig. 2.2.27 l). These features, called rhombic ripples, are ubiquitous on modern beach surfaces.

Fig. 2.2.26: **(a)** Parallel stratification in sandstones. **(b)** Translatent strata in a low angle section showing reverse grading within each laminae (arrow). Note a ripple foreset preserved in the upper left part of the photograph. Light coloured layers are relatively coarse grained compared to dark coloured layers. Division of scale is in cm (Courtesy: Dr. Tapan Chakraborty, Indian Statistical Institute, Calcutta).

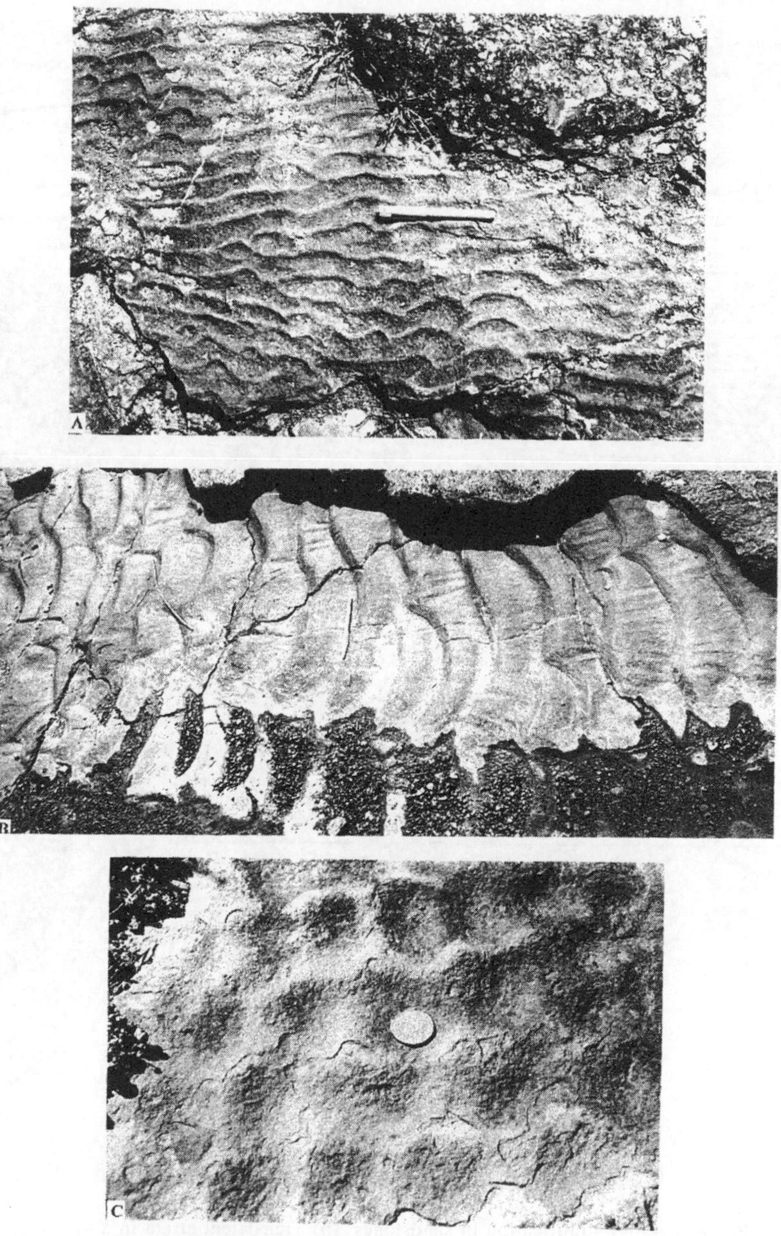

Fig. 2.2.27: Ripple marks on bedding plane. Note: **(a)** straight, sinuous and curved crestlines, **(b)** concave-downcurrent crestline, **(c)** domal crest, **(d)** crest bifurcation, **(e)** polygonal ripple, **(f)** double-crested ripple, **(g)** flat-topped ripple, **(h)** ripples with smaller, parallely oriented ripples in the trough, **(i)** ripples with smaller, transversely oriented ripples in the trough, **(j)** ripples with superimposed, smaller, transversely oriented ripples, **(k)** orthogonal superimposition of ripples of similar dimensions, **(l)** rhombic ripples, **(m)** ripples with diverse crestline orientations.

D
E
F

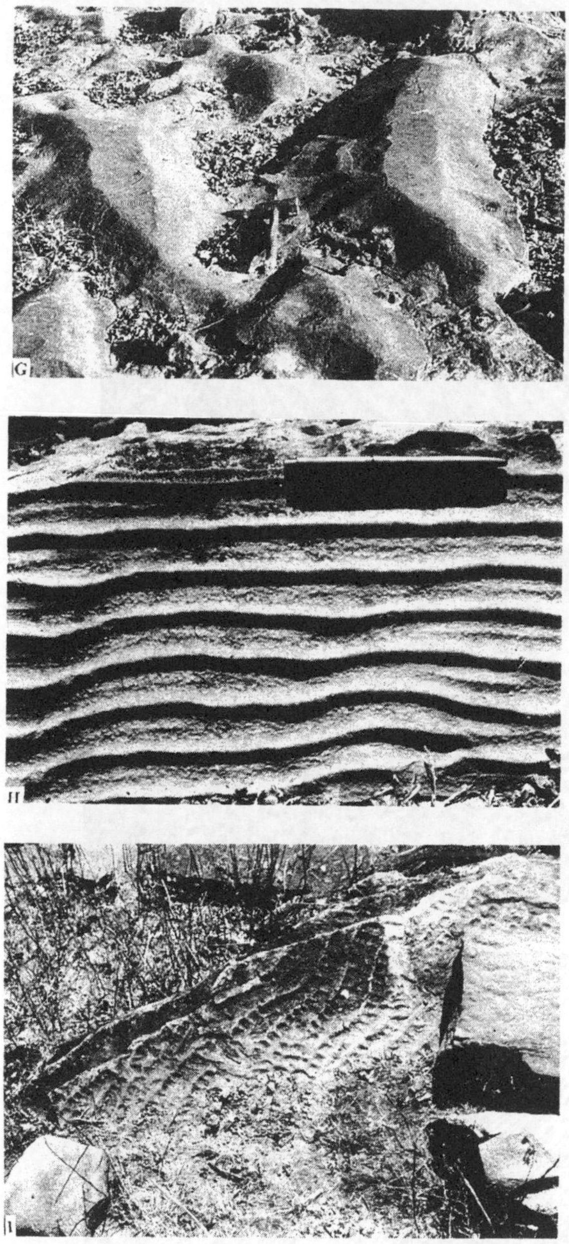
G
H
I

Parting Lineation

On the bedding planes of most parallel-stratified sandstone parallely-disposed lineations are observed (Fig. 2.2.28). Such lineation, called parting lineation or current lineation, are indicative of upper flow regime and are oriented parallel to the flow (Allen, 1986). Uniformly oriented parting lineation is characteristic of unidirectional flow and is rarely formed under oscillatory flows; if, however, they form under oscillatory flows their orientations are commonly diverse.

Current Crescent

These are scoured depressions with a U-shaped plan-geometry, formed by strong unidirectional currents around an obstacle (Figs. 2.2.29 a, b). The open end of the U indicates down-current direction. In rock records the depression may be filled up with sediments and the obstacle may be missing.

Rain Prints and Rain-impact Ripples

Exposed sandy surfaces are imprinted with circular depressions due to impact of raindrops (Fig. 2.2.30 a). If rain fall is associated with strong winds impact of rain drops reworks the sediment grains into irregular ripple-like forms termed rain-impact ripples (Kocurek and Fielder, 1980; Fig. 2.2.30 b). These features are common in sandstones of the rock record and are reliable indicators of sub-aerial condition.

Adhesion Structures

Adhesion structures form by the adhering of dry, wind-blown sand to a wet or damp surface (Kocurek and Fielder, 1980). Adhesion ripples are tiny, subparallel ridges aligned perpendicular to the wind direction (Fig. 2.2.30 c). They

Fig. 2.2.28: Parting lineation on the bedding planes of sandstone.

show a systematic, rhythmic pattern with even spacing, a few millimetres to 1 cm, and like height (a few millimetres). The ridges are asymmetric. Adhesion ripples accrete on their steeper upwind side by the adhering of saltating grains. Accretion is accompanied by upwind ridge migration and adhesion ripples climb over the deposits left by their upwind neighbours to generate climbing adhesion ripple stratification or adhesion cross-lamination defined by downwind dipping, thin (a few millimetres), crinkly laminae (Fig. 2.2.30 d). Adhesion warts are distinguished from adhesion ripples by their irregularity. They are dome-like features and tend to have a more random distribution than adhesion ripples. Small adhesion warts as well as adhesion ripples may resemble rain-impact ripples, but the latter are strictly a surface feature. Adhesion structures are common in the rock record and indicate aeolian environment.

Sole Structures

Several types of erosional scours are moulded as downward bulging features at the under-surfaces of many sandstone beds and are referred to as sole markings (Dzulynski, 1996). Erosional markings originate in a soft cohesive, muddy substratum whose essential property is that it allows the passage of a current without being washed away itself. A non-cohesive sandy substratum is not conducive to the formation and preservation of minor erosional markings, nor is it a hard bottom surface. However, cohesive sandy substrates may also preserve such markings. If the erosional event is followed by deposition of sandy sediments the impressions of the markings are preserved on the sole of the sand bed as relief features. The erosional markings preserved on the sole of sandstone beds are of two categories: (1) tool markings made by contact of

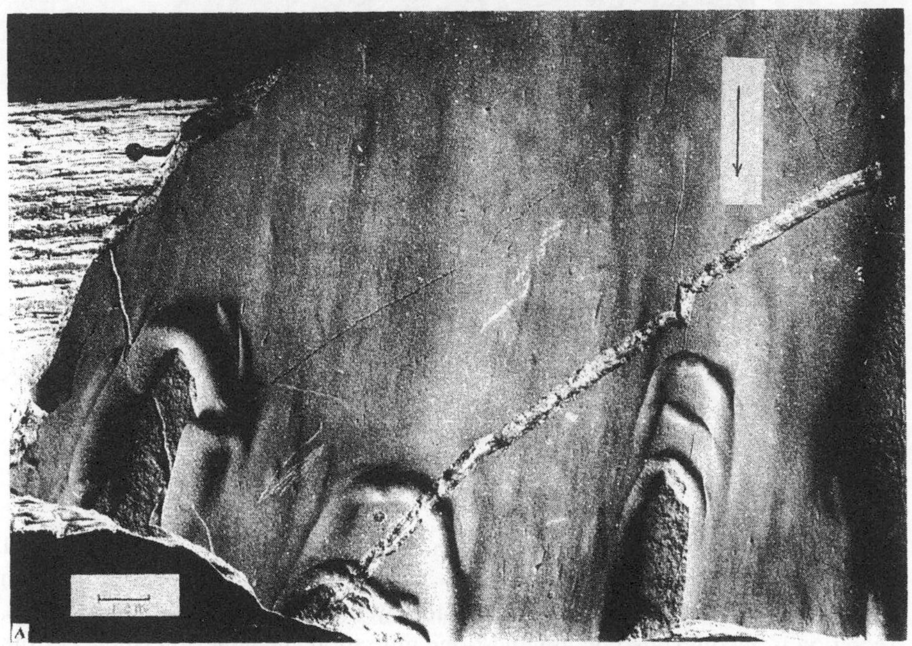

Fig. 2.2.29: **(a)** Moulds of current crescent; successive crescents indicate intermittent retreat of undercut edges of resistant patches. Oligocene Krosno Beds, Wernejowka (reproduced from S. Dzulynski, 1996, Erosional and deformational structures in single sedimentary beds: A genetic commentary. Annales Societatis Geologorum Poloniae, Vol. 66, Fig, 1a, p.141, with the kind permission of the author).

(b) Current crescent on bedding plane of Proterozoic Rewa Sandstone, Maihar, M.P., India

Fig. 2.2.30: **(a)** Rain prints. **(b)** Rain impact ripples. **(c)** Adhesion ripples, **(d)** Adhesion ripple stratification (Courtesy: Dr. Tapan Chakraborty, Indian Statistical Institute, Calcutta, India).

transported objects (the tools) and the bottom (see Section 2.3) and, (2) scour markings produced by the scouring action of the current itself.

Flute Marks

These are fluid-eddy generated erosional scours. They are elongate (length ranges upto a few centimetres) sub-conical hollows flaring out and merging with the bedding plane in the down-current direction. The up-current ends of such markings are bulbous, sharp-pointed or gently rounded. They may be present on the sole of sandstone beds as negative relief (Fig. 2.2.31).

It is worthwhile to draw attention to another set of primary structures—setulfs (flutes spelt in reverse order) and linear ridges—which resemble counterparts of flutes and grooves. While flutes and grooves are negative relief bedforms, setulfs and linear ridges are positive relief bedforms, reported from carbonate tidal flat at Abu Dhabi (Friedman and Sanders, 1974). The setulfs range in length from about 4 or 5 cm, width from 2 to 3 cm, and have positive relief of about 1 cm with respect to generally plane parts of the adjacent sediment surface (Friedman and Sanders, 1974). The maximum relief is at the

Fig. 2.2.31: Flute markings. Current direction from lower left to upper right (arrow). Oligocene Krosno Beds, Rudawka Rymanowska (Reproduced from S. Dzulynski, 1996, Erosional and deformational structures in single sedimentary beds: A genetic commentary. Annales Societatis Geologorum Poloniae, Vol. 66, Fig. 13a, p. 153, with the kind permission of the author).

pointed end, which is also the upcurrent direction. The relief of the setulfs gradually decreases downcurrent and flares out. Individual setulfs stand out alone and do not overlap one over another.

Linear round ridges that have been reported to be co-associated with setulfs in carbonate tidal flat at Abu Dhabi, are the other positive relief bed form, oriented parallel to the current direction and semi-elliptical in cross-section; their widths ranging between 5 and 6 cm, heights between 2 and 3 mm and are 5 to 30 cm long. These linear ridges die out gradually at both up- and down-current ends. Lateral spacing of the ridges range from about 1 to 25 cm.

Since setulfs and linear ridges are counterparts of negative-relief bedforms of flutes and grooves, extra caution is necessary while attempting to tell original top direction of strata particularly in tectonically deformed strata (Friedman and Sanders, 1974).

These positive-relief bedforms appear to have formed in response to complex patterns of flow lines in a shallow current, where secondary flows and flow separations are believed to have interacted (Friedman and Sanders, 1974).

Gutter Marks

Gutter marks are channel-like, fluid-eddy generated scours (Chakraborty, 1995; Myrow, 1992;). At the base of sandstone beds they occur as downward bulging sole structures (Fig. 2.2.32). However, isolated gutters are also

Fig. 2.2.32: Photograph showing an elongate, sand-filled gutter at the base of a sandstone bed interbedded with shale (reproduced from Chakraborty, C.: Gutter casts from the Proterozoic Bijaygarh Shale Formation, India: their implication for storm-induced circulation in shelf settings. Geol. Journal, Vol. 30, Fig. 2, p. 71, © 1995, John Wiley & Sons, Ltd., with the kind permission).

15cm
A

B

C

common (Fig. 2.2.33a). The gutters are elongate, linear to curvilinear features (Fig. 2.2.32). Their width and depth range from a few centimetres to a meter. In transverse cross-section gutters may be symmetrical or asymmetrical and have a variety of shapes and lateral relationships (Figs. 2.2.33, 2.2.34). The gutters may have, on their walls and floor, small-scale erosional features such as prod and groove marks (see Section 2.3).

The sandy fill of the gutters may be massive or horizontally/cross-stratified (Fig. 2.2.34). The alignment of the gutters provides the direction of the eroding current, whereas the nature of the in-fill gives the nature and direction of the in-filling current. Gutter casts, though common in shallow marine rocks, are also reported from non-marine rock records. In shallow marine setting, these structures are usually taken to be storm-generated features (Aigner, 1985; Bridges, 1972; Brenner and Davies, 1973; Hiscott, 1982; Kreisa, 1981; Leithold and Bourgeois, 1984). Under storm flows, gutters may be formed and filled by unidirectional as well as oscillatory currents. Accordingly, different types of gutters may be identified (Chakraborty, 1995).

Current-formed and wave-filled gutters: This type of gutter shows one or all of the following characteristics: 1. bipolar prod marks oriented perpendicular to the length of the gutter (Fig. 2.2.35), 2. wave ripples at the top trending parallel to the length of the gutter (Fig. 2.2.36), and 3. bidirectional or other wave-diagnostic cross-stratification striking parallel to the length of the gutter (Fig. 2.2.36).

Wave-formed and current-filled gutters: This type of gutters is typically filled with strata inclined in a direction at a high angle to the elongation direction, suggesting lateral accretion from a unidirectional current (Fig. 2.2.37). There may be bipolar prod marks aligned parallel to the length of the gutter suggesting oscillatory nature of the eroding current.

Wave-formed and wave-filled gutters: This type of gutter is filled with wave-diagnostic stratification and have bipolar prod marks oriented parallel to the elongation direction. The features evidently indicate the influence of wave orbital motion in the formation as well as filling of the gutters.

Current-formed and current-filled gutters: This type of gutter is filled with massive or horizontal, parallel-stratified sediments suggesting influence of unidirectional current in the formation as well as filling of the gutters.

Fig. 2.2.33: Photographs showing different types of gutters. **(a)** Isolated, sand-filled gutter protruding into the underlying shale; **(b)** Round, symmetrical gutter; **(c)** Hour-glass shaped gutter (Fig. 2.2.33a is reproduced from C. Chakraborty: Gutter casts from the Proterozoic Bijaygarh Shale Formation, India: their interpretation for storm-induced circulation in shelf settings. Geol. Journal, Vol. 30, Fig. 5, © 1995, John Wiley & Sons, Ltd., with the kind permission; Figs. 2.2.33 b, c reproduced from Paul M. Myrow: Pot and gutter casts from the Chapel Island Formation, Southeast Newfoundland. J. Sediment. Petrology, Vol. 62, No. 6, Figs. 2 and 3, © 1992, with the kind permission of Society for Sedimentary Geology (SEPM), Tulsa, Oklahoma, U.S.A.).

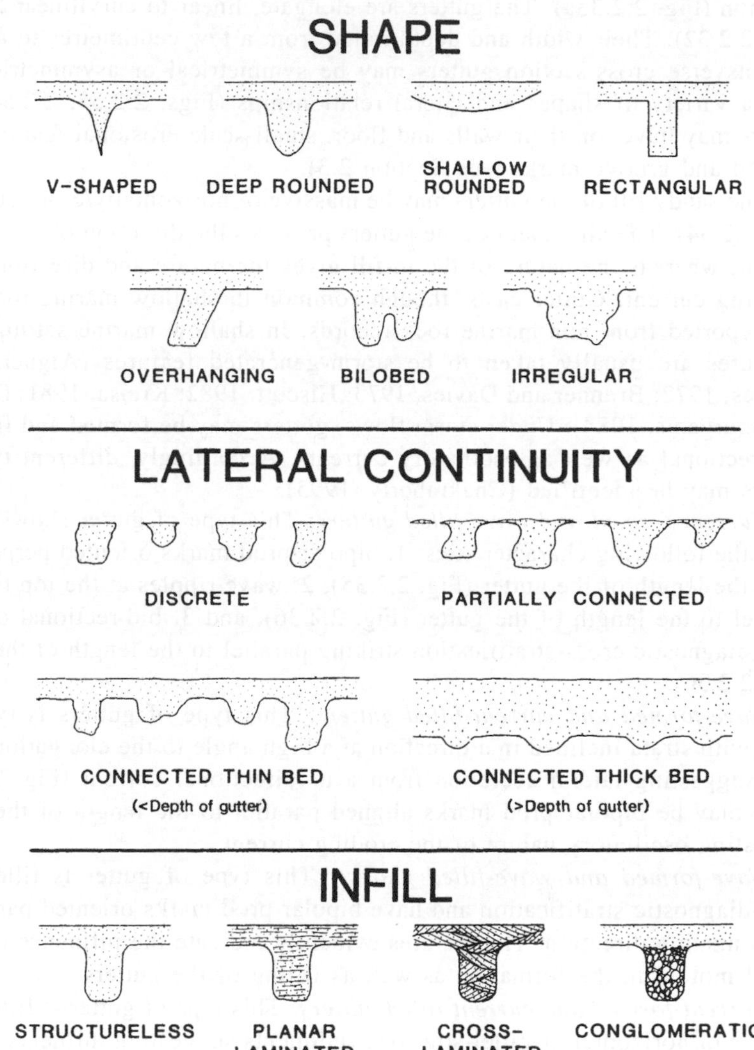

Fig. 2.2.34: Schematic representation of the cross-sectional shapes, modes of occurrence and patterns of infill of gutters (Reproduced from Paul M. Myrow; Pot and gutter casts from the Chapel Island Formation, Southeast Newfoundland. J. Sediment. Petrology, vol.62, No.6, Fig. 7, © 1992 Society for Sedimentary Geology (SEPM), Tulsa, Oklahoma, U.S.A. with the kind permission).

Pot Marks

These are rounded in plan and cylindrical in three-dimensional geometry, and are commonly associated with gutter casts (Myrow, 1992 b; Figs. 2.2.38). In fact, in many cases, gutter casts start from the down-current side of pot casts. The hollows of the potholes are usually filled-in by gravel or fine sand and

Fig. 2.2.35: Sole view of a gutter. Note bi-polar prods (arrowed) oriented at right angle to the orientation of the gutter (Reproduced from C. Chakraborty: Gutter casts from the Proterozoic Bijaygarh Shale Formation, India: their implication for storm-induced circulation in shelf settings. Geol. Journal, Vol. 30, Fig. 6, © 1995, John Wiley & Sons, Ltd., with the kind permission).

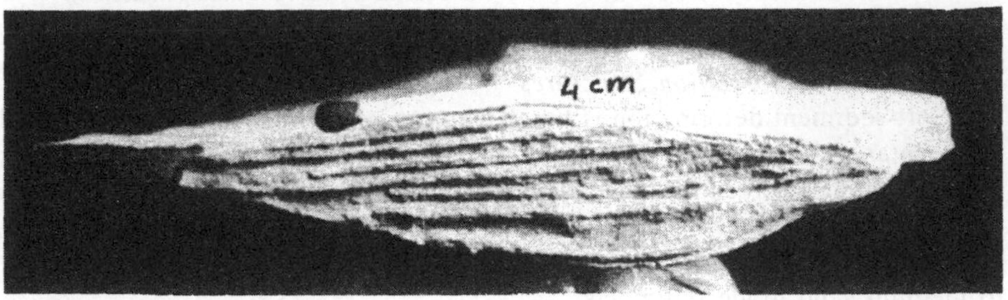

Fig. 2.2.36: Transverse profile of a gutter. Note wave ripples trending along the length of the gutter and wave-diagnostic lamination in the infill (Reproduced from C. Chakraborty: Gutter casts from the Proterozoic Bijaygarh Shale Formation, India: their implication for storm-induced circulation in shelf settings. Geol. Journal, Vol.30, Fig. 10, © 1995, John Wiley & Sons, Ltd., with the kind permission).

stand out prominently on the bedding plane surfaces on weathering. Pot casts range in shape from disc to rounded loaf-like form to tall pillows and in size range between 1 cm and 20 cm in diameter (Myrow, 1992 b). Pot casts are believed to be generated from spiral vortex that, under favourable and constant conditions, drills in to the rock in a "cork-screw" fashion (Alexander, 1932, cited in Myrow, 1992 b).

Fig. 2.2.37: Transverse profile of a gutter. Note downlapping lamination in the infill inclined towards left. (Reproduced from C. Chakraborty: Gutter casts from the Proterozoic Bijaygarh Shale Formation,India: their implication for storm-induced circulation in shelf settings. Geol. Journal, Vol. 30, Fig. 7, © 1995, John Wiley & Sons, Ltd., with the kind permission).

Soft-Sediment Deformation Structures

Soft-sediment deformation is the disruption of unlithified or semi-lithified sedimentary strata (Mills, 1983, Jones and Preston, 1987). Deformation occurs in the interval between the times of deposition and until the sediments remain water-saturated because water-saturation is an important pre-requisite for soft-sediment deformation to occur. As a result, it is common mostly in coarse silt to medium sand grade sediments.

The pore-water pressure in water-saturated sediments is greatly enhanced when subjected to external shocks such as seismic vibrations or cyclic wave loading or excess overburden. Excess pore-water pressure reduces the strength of the sediments making them easily deformable—a process called liquefaction. In liquefied state sediments may suffer gravitational collapse either vertically or along a slope and may be deformed even by the small shear stress exerted by the moving fluids above. In extreme situations, the pore-water may start expelling from the sediments and thereby deforming them—a process called fluidization. Deformation continues until the pore-water pressure regains equilibrium.

The process of fluidization produces a variety of fluid-escape structures, such as dish and pillar structures (Fig. 2.2.39), sand volcanoes and clastic dykes and sills. The role of air displaçement by influent water, combined with

Fig. 2.2.38: Bedding plane view of an in-filled, round pot mark (Reproduced from Paul M. Myrow, Pot and gutter casts from the Chapel island Formation, Southeast Newfoundland. J. Sediment. Petrology, Vol. 62, No. 6, Fig. 9, © 1992, Society for Sedimentary Geology (SEPM), Tulsa, Oklahoma, U.S.A. with the kind permission).

water escape has recently been highlighted in the formation of dish-and-pillar and tepee structures (McManus and Bajabaa, 1998). Extreme fluidization of sediments may lead to total churning of the sediments producing a structureless sediment body.

Intra-stratal vertical gravitational collapse of sediments during a liquified state results in plastic deformation producing convolute structures: regularly spaced, narrow sharp-crested anticlines separated by wider and rounded synclines (Fig. 2.2.40). De Boer (1979) has demonstrated, from the recent intertidal shoal in the Oosterschelde estuary, The Netherlands, that reversed density stratification resulting from trapped air in some sand layers may have played a role in the deformational process resulting in convolute lamination. On the other hand, in deposits where a denser, sandy stratum overlies a lighter, muddy stratum, liquefaction leads to inter-stratal gravitational collapse producing penetration of denser sandy stratum into the muddy stratum—a structure known as load structure (described in Section 2.3). When cross-stratified sands are liquefied, the fluid drag or down slope slide often leads to plastic deformation of the cross-strata producing structures known as overturned or recumbently folded cross-strata (Fig. 2.2.41).

Fig. 2.2.39: Dish structure in sandstone. Scale bar represents 2.25 cm.

Fig. 2.2.40: Convolute stratification in sandstone. Dot pen (14.5 cm) is for scale.

On liquefaction sediments may start moving down the slope in response to gravity—a phenomenon called slump. Dependent on the degree of disruption of the original stratification, slumps can be differentiated into either coherent or incoherent slumps. Coherent slump is distinguished by minimal mixing of sediments with general retention of the stratification. Incoherent slump reflects extensive mixing and with increased liquefaction, chaotic mixing and mobilization; slumped sediments may evolve into a variety of sediment gravity flows.

Fig. 2.2.41: Overturned cross-strata. Coin (3.5 cm) on the left is for scale.

In coherent slumps the moving sediment body deforms into a variety of structures depending upon the stress regime and sediment rheology. In the down-slope moving sediment body both tensile and compressive stresses develop in different parts of the body along with the basal shear stress. If the sediment body deforms plastically, folds (Figs. 2.2.42 a, b) and pinch-and-swell structures develop in areas of compression and extension respectively On the other hand, brittle response of the sediments produces slump breccias (Fig. 2.2.42 c), small-scale normal and reverse faults in areas of extension and compression respectively. During a slump movement the sense of the shear is oriented down the slope. The sense of shear determined from the deformation features indicates the direction of the slope (Figs. 2.2.42 a, b).

Slump structures are often difficult to distinguish from tectonically induced deformation features not related to down-slope movement. A commonly used criterion is confinement of the deformation features between un-deformed beds. However, in multi-layered units, tectonic deformation affects the incompetent or thin layers more than the competent or thick layers. As a result, in outcrops tectonic deformation may also be represented as being confined between apparently un-deformed beds. However, since slump takes place at the sediment-fluid interface, the slumped units are likely to be capped by master erosion surfaces underlying the un-deformed beds. In case of tectonic deformation the boundary between the deformed and the overlying un-deformed beds may not be erosional.

Intra-bed Vertical Variation

Sandstone beds may show several types of intra-bed vertical variation of different attributes such as grain size, sedimentary structure etc. Intra-bed vertical variation evidently indicates temporal velocity fluctuations in the flow during deposition. For example, most beds deposited from turbidity currents and storm flows characteristically record a vertical diminution of grain size as well as a

A

B

C

sequential arrangement of sedimentary structures demonstrating a gradual waning of the flow. The different patterns of vertical variations commonly observed in turbidite and tempestite beds are described in Section 2.7.

Structures Related to Microbial Mats

While the role of microbial mat in modern carbonate and siliciclastic depositional settings are well known and well documented (Schieber, 1999 and references therein; Walter, 1976), in ancient siliciclastic rocks its presence and structures mediated by it have rarely been documented. While positive proof of its presence and activities in ancient siliciclastic rocks are missing, comparison of sedimentologic and morphologic characteristics of modern and ancient siliciclatic deposits provides sufficient evidences of microbial mat activities. Recently, criteria have been suggested to identify the activities of microbial mat in ancient siliciclastic rocks (Gehling, 1999; Hagadorn and Bottjer, 1999; Pfluger, 1999; Schieber, 1998,1999; Simonson and Carney, 1999; Seilacher, 1999).

Features in siliciclastic rocks that may alert one to the possibility of microbial deposits are: i. domal build-ups, ii. cohesive behaviour during erosion, transport, deposition of thin sediment layers, iii. wavy and crinkly character of laminae, iv. irregular wrinkled bed surfaces, v. ripple patches, vi. laminae with mica enrichment, vii. irregular, curved-wrinkled impressions on bedding planes, and viii. lamina-specific distribution of early diagenetic minerals (dolomite, ferroan carbonate, pyrite). It should, however, be clearly understood that the above features are not positive proof of fossil microbial mat but are suggestive enough of its presence in ancient siliciclastic rocks.

i. *Domal build-ups, dolomite laminae and random micas*: In siliciclastic rock, this appears as low amplitude hummocks or domes or small hemispherical features in mudstone/shale; over-steepened laminae (greater than angle of repose) on one side of the domal feature suggest binding by microbial mats. Internally, the hummocky build-ups alternate with thin laminae of dolomite, silt and clay. These dolomite laminae in the build-ups are taken to be products of activities of microbial mats (Chafetz and Buczynski, 1992; Gebelein and Hoffman, 1973). Bacterial decay beneath mat surfaces enhances carbonate precipitation. Randomly oriented micas characterize the dolomite laminae, whereas mica in silt/shale laminae remain bedding parallel. Schieber (1999) suggests that the random orientation of mica flakes in dolomite laminae possibly reflects early cementation preventing align-

Fig. 2.2.42: Slump folds **(a, b)** Vergence of folds indicate a slope towards left. **(c)** carbonate slump breccia (Fig. 2.2.42 a is reproduced from Geological Photo: Richard L. Squires. J. Sediment. Petrology, Vol. 51, No. 3, p. 822, with the kind permission of Society for Sedimentary Geology (SEPM), Tulsa, Oklahoma, U.S.A.).

ment during subsequent burial. The above features (domal structures, over-steepened laminae, texture and composition) are found similar to those in recent microbial mat dominated siliciclastic sediment.

ii. *Cohesive behaviour*: Cohesive behaviour during erosion, transport and deposition of thin sediment layers has been suggested to be a useful indicator of microbial mat colonization (Schieber, 1999; Simonson and Carney, 1999). Presence of rolled-up layers, stretching and over-folding of thin layers, found in ancient siliciclastic deposits, are only possible in case of microbial-bounded layers; this is in stark contrast to those that are not microbial-bounded (dried-up and mud-cracked chips).

iii. *Wavy-crinkly laminae*: These features are associated with both sandstone and mudstone; in mudstone, wavy crinkly laminae are often found to be associated with carbonaceous layers; in non-mat carbonaceous shales, laminations are quite even and parallel. However, one has to be sure that wavy laminae are not due to differential compaction around micro-concretions, fecal pellets, silt lenses etc. (Schieber, 1999).

iv. *Irregular wrinkled bed surfaces*: Wrinkled structures, in general, are characterized by oddly contorted, wrinkled, irregularly pustulose, quasi-polygonal, commonly over-steepened surface morphologies that occur patchily on bed tops and bottoms (Hagadorn and Bottjer, 1999). Wrinkle structures have been described in literature as runzelmarken, wrinkle-marks, micro-ripples and variations of kinneyia. Wrinkle surface relief ranges from ca. 0.03mm to 3mm (from wrinkle crest to trough) and inter-crest distances range between 1-5mm. Crest shape, crest length, and ridge steepness vary from mosaic, polygonal, bifurcating, and elongate sub-parallel forms (Hagadorn and Bottjer, 1999; their Fig. 4). Based on close similarities with morphologic and sedimentologic characteristics of modern microbially dominated communities, it appears that ancient wrinkle structures are suspect microbial structures, and that associated laminated sediments are likely to be microbially dominated (Hagadorn and Bottjer, 1999).

v. *Ripple patches*: The presence of patches of ripples in an otherwise smooth surface gives an indication of microbial surface colonization. Partial removal of the mat surface due to erosion (storm activity) may expose the underlying sandy and silty surfaces to the formation of wave- and current-ripples (Schieber, 1999). Features identical to this may be produced when a sandy/silty layer, overlying a rippled surface, is partially removed due to erosion. Such a rippled surface is nothing but a 'window' to the underlying rippled surface. These two features can be discriminated by looking at the nature of contact at the bedding normal face; in the former the contact will be smooth between the rippled area and the mat-covered surface, while in the latter situation a sharp break will be observed between the two.

vi. *Lamina-specific diagenetic minerals*: Presence of lamina-parallel diagenetic minerals, e.g., dolomite, ferroan dolomite, pyrite siderite, in ancient siliciclastic rocks points to the existence of microbial mat. Anaerobic decay beneath a microbial mat surface creates a chemical condition that is strongly reducing (Gerdes et al., 1985). The reducing chemical condition favours precipitation of dolomite, pyrite, siderite, ferroan carbonate etc. (Gebelein and Hoffman, 1975; Schieber, 1999) in the pore spaces of siliciclastic sediments. Thus, as Schieber(1999) has suggested, "well-defined, thin, stratiform horizons of these minerals in a shallow water sandstone may be a 'tip-off' of the former presence of microbial mats".

Why is it necessary to look for evidences of activities of microbial mats in siliciclatic sediments? It helps one to establish (a) paleoecology and paleobiology of the earliest metazoans, (b) mat surfaces may be primary component in the preservation of soft-bodied early metazoans, e.g., Ediacaran biota, (c) this helps in pinpointing redox facies, independent of body fossils, or trace fossils. Microbial mat surfaces in ancient siliciclastic rocks are temporally restricted to the Vendian-early Cambrian times, since Ordovician and thereafter saw the rise of metazoan grazers that thrived on microbial mats and destroyed them.

Paleocurrent

Many of the different sedimentary structures found in sandstone are useful in determining the direction of the depositing flow (Allen, 1966). Among them some indicates the line as well as the direction of the flow, whereas there are some that indicate only the line of the flow.

Cross-strata characteristically dip in the direction of the depositing flow. Therefore, their azimuths are reliable indicators of the paleocurrent direction. However, the azimuths of a trough cross-stratum vary due to its curvature and it is advisable to measure paleocurrent directions from such cross-strata on bedding plane sections (Fig. 2.2.43; Rubin, 1987).

The azimuths of cross-strata produced by transverse, compound dunes represent the long term sediment transport direction and the azimuths of the smaller cross-strata present within the dune strata provide the directions of short term, local flows.

Bedforms occurring on the flanks of longitudinal dunes and migrating parallel to the dune crest arise when the regional flow reaches the dune crest and deflects at the crest to become crest-parallel on the flank of the dune (Chakraborty, 1993). Therefore, the azimuths of longitudinal, compound dune strata would not indicate the direction of the regional flow; on the other hand, the azimuths of the smaller cross-strata, occurring within the dune cross-strata, would only indicate the directions of the deflected components of the regional flow. In order to find the regional paleo-flow direction measurements should be made on other features that have seemingly formed as a direct response to the regional flow (see Chakraborty, 1993).

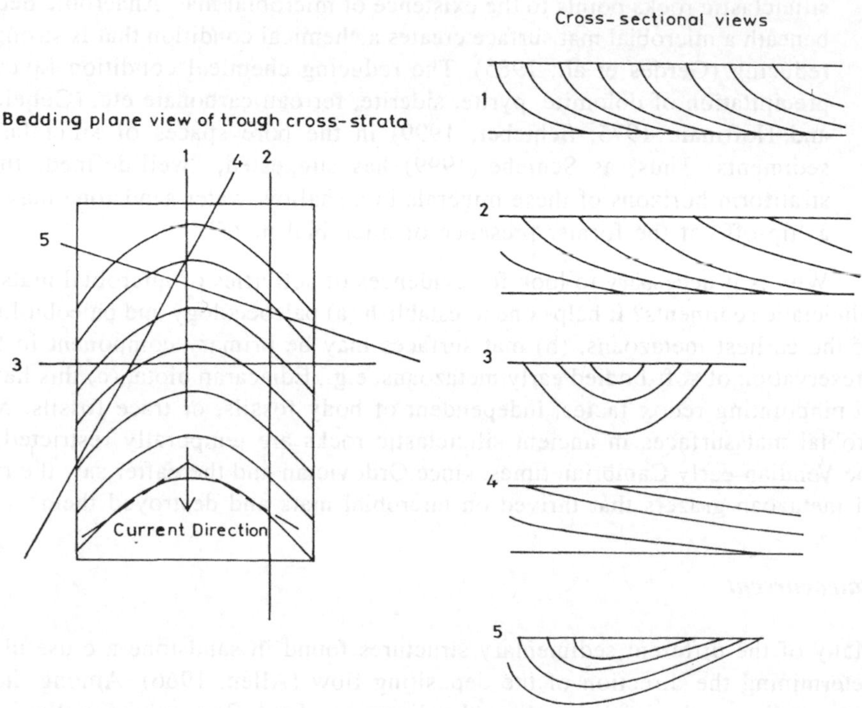

Fig. 2.2.43: Diagram illustrating the likelihood of erroneous paleocurrent measurements from trough cross-stratification. The reliable paleocurrent direction is given by (i) the bisectrix of the tangents to the traces of the cross-strata on bedding plane (shown by dashed lines), (ii) the azimuths of cross-strata appearing in section 1 and (iii) the orthogonal direction of the cross-strata appearing in section 3. The azimuths of the cross-strata appearing in sections 2, 4 and 5 would provide incorrect paleocurrent directions.

Ripples are generally asymmetric in the direction of the associated current. Since asymmetric ripples may form under combined oscillatory and unidirectional flow with the asymmetricity of ripples indicating the direction of the unidirectional current component, it is advisable that while determining paleocurrents from ripples, the crestal trends of the ripples also be measured along with the sense of asymmetricity. The line of the flow will be perpendicular to the crestal trends.

Among the sole structures, flutes and prods are reliable indicators of the paleocurrent directions since they are aligned parallel to the current and strongly length-asymmetric. The bulbous nose of the flutes indicates the up-current direction whereas that of the prods indicates the down-current direction. The trends of gutters and grooves, on the other hand, indicate only the line of the flow.

Parting lineations are good indicators of the line of flow. However, if they are associated with current crescents the direction of the current can also be inferred with the open end of the crescent indicating the downcurrent direction.

In order to arrive at a comprehensive understanding of the implications of the paleocurrent data they are represented in rose diagrams. The distribution pattern of the data in the rose diagram is suggestive of nature of the flow from which deposition took place (see Collinson and Thompson, 1982; Tucker, 1982a, 1991).

2.3 Shale

Shale is a terrigenous sedimentary rock composed dominantly of mud-sized particles and is the most abundant sedimentary rock type in the stratigraphic record, constituting nearly 50 percent of the total rock record. This rock type is variously known as lutite, psammite, argillaceous sediment, mudstone, argillite; phyllite and slate are its metamorphosed equivalents. While some workers want to retain the term 'shale' and 'mudstone' for showing fissile and non-fissile features respectively, following Tourtelot's (1960) review, the term 'shale' is retained here irrespective of presence or absence of fissility.

Classification and Field Description of Shale

While mud-size particles dominate shale, silt- and sand-size particles are found associated in various proportions. Mud of present environments averages approximately 15 per cent sand, 45 per cent silt and 40 per cent clay (Picard, 1971). Potter, Maynard and Pryor (1980) have proposed a classification for shale (Tables 2.3.T1 and 2.3.T2). This classification emphasizes on (i) clay-size constituents and (ii) bedding thickness. The terminology can be refined further by adding colour, type of cement, organic matter content, fossil content and nature of bedding in shale.

If shale is fossiliferous, special care should be taken in noting their type, abundance and relative positions in a vertical litholog of the shale section. It is necessary to note the mode of occurrence of microfossils while collecting them (flat, horizontal, inclined or vertical). This helps in understanding the paleocurrent direction (Seilacher and Meischner 1964; Jones and Dennison 1970). While noting the orientation of the macrofossils, it is also necessary to note whether the orientation is an effect of post-mortem transport or living *in situ* position. It is also important to note the geometry and position of the burrows. This has been discussed in detail in a sub-chapter on ichnofossils. These observations on the orientation of the macrofossils and the burrows help in understanding the depositional environment, its salinity, depth of light penetration,

Table 2.3.T1: Classification of shale (Reproduced from 'Sedimentology of Shales: Study Guide and Reference Source': P. E. Potter, J. B. Maynard and W. A. Pryor, Table 1.2, page 14, ©1980, Springer-Verlag GmbH & Co. KG. , with the kind permission)

<table>
<tr><td colspan="3">Percentage of clay-size constituents</td><td>0 - 32</td><td>33- 65</td><td>66 – 100</td></tr>
<tr><td colspan="3">Field Adjective</td><td>Gritty</td><td>Loamy</td><td>Fat or slick</td></tr>
<tr><td rowspan="2">NON-INDURATED</td><td>Beds</td><td>>10mm</td><td>BEDDED SILT</td><td>BEDDED MUD</td><td>BEDDED CLAYMUD</td></tr>
<tr><td>Laminae</td><td><10mm</td><td>LAMINATED SILT</td><td>LAMINATED MUD</td><td>LAMINATED CLAYMUD</td></tr>
<tr><td rowspan="2">INDURATED</td><td>Beds</td><td>>10mm</td><td>BEDDED SILTSTONE</td><td>MUDSTONE</td><td>CLAYSTONE</td></tr>
<tr><td>Laminae</td><td><10mm</td><td>LAMINATED SILTSTONE</td><td>MUDSHALE</td><td>CLAYSHALE</td></tr>
<tr><td rowspan="3">METAMORPHOSED</td><td rowspan="3">Degree of Metamorphism</td><td rowspan="3">Low ↓ High</td><td>QUARTZ ARGILLITE</td><td colspan="2">ARGILLITE</td></tr>
<tr><td>QUARTZ SLATE</td><td colspan="2">SLATE</td></tr>
<tr><td colspan="3">PHYLLITE AND/OR MICA SCHIST</td></tr>
</table>

temperature, turbidity of the water column, wave and current energy, and rate of sedimentation.

Primary Sedimentary Structures

Primary sedimentary structures in pure shale are i) horizontal stratification and ii) parting. The stratification is classed into two types: bed and lamina. Parting in shale has been classed as slabby, flaggy, platy, fissile and papery (Table 2.3.T2). But when shale is interstratified with silt and fine sand, host of primary sedimentary structures occur in them (Table 2.3.T3). To study these structures, it may be necessary to adopt certain processes to enhance them. Since shale is prone to weathering, it is necessary to dig back into the shale to open up a fresh surface. The prepared vertical surface may then be treated with dilute HCl,

Table 2.3.T2: Stratification and parting of shale (Reproduced from 'Sedimentology of Shale: Study Guide and Reference Source': P. E. Potter, J. B. Maynard and W. A. Pryor, Table 1.3, page 16, ©1980, Springer-Verlag GmbH & Co. KG., with the kind permission)

THICKNESS	STRATIFICATION		PARTING	COMPOSITION
30 cm	Thin	Bedding	Slabby	Clay and organic content ↓ ↑ sand, silt and carbonate content
3 cm				
	Very thin			
10 mm	Thick	Lamination	Flaggy	
5 mm	Medium		Platy	
1 mm	Thin		Fissile	
0.5 mm	Very thin		Papery	

water, mineral oil, or acrylic spray (such as, krylon) in order to reduce light scatter. This procedure enhances the internal features for sketching and photography.

Bedding and Laminae

These two are most prominent structures that are associated with any shale sequence. The distinction between bed and lamination is kept at a thickness of 10 cm—above is the bed and below is the lamination (Potter et al., 1980). As shown in Table 2.3.T2, lamination can be further divided into flaggy, platy, fissile and papery depending upon its thickness. The shale becomes more fissile with increasing abundance of clay minerals and organic matter. Massive shale does not necessarily indicate absence of stratification. The bioturbation may have destroyed all primary depositional features, depending upon intensity of bioturbation, giving a deceptive massive look to shale. It becomes evident when a thin slab is exposed to x-ray radiography (for details, consult Bouma 1969).

In addition to study of bedding and lamination in shale and their thickness distribution, it is also important to note the stratification style. The stratification can be (i) continuous and parallel, (ii) discontinuous but even, parallel, (iii) wavy or lenticular or parallel, non-parallel or discontinuous. The boundaries can be sharp or gradational.

Table 2.3.T3: Sedimentary structures in shale and their origin (Reproduced from Sedimentology of Shales: Study Guide and Reference Source - P. E. Potter, J. B. Maynard and W. B. Pryor, Table 1.5, page 22, 1980, ©1980, Springer-Verlag GmbH & Co. KG., with kind permission)

PART– A: PRIMARY STRUCTURES		
TYPES	**SUB-DIVISIONS**	**ORIGIN**
STRATIFICATION	Parallel horizontal Massive	Episodic suspension in still water Continuous, rapid sedimentation from suspension or bioturbation
	Parallel discontinuous	Episodic suspension with some bottom currents
	Lenticular, wavy	Episodic traction transport with possibly some deposition from suspension
	Varves	Suspension grading with rapid sedimentation in spring and slow sedimentation in winter
RIPPLE MARKS, FLASER BEDDING	Ripple: Straight crested, Sinuous, Lunate, Linguoid, Cuspate. Flasers: Simple,Wavy, Bifurcated, & Bifurcated wavy	Traction transport of silt, sand and mud aggregates as ripples with some deposition from suspension
CROSS BEDDING	Planar tabular, Wedge shaped and Trough cross-bedded	Traction transport of sand and silt as large ripples and dunes
PARTING LINEATION		Traction transport of silt and sand in the 'flat bed' mode
SOLE MARKS	Flutes, Grooves, Prod casts, Skip marks, Brush marks, Longitudinal furrows and ridges, Chevron marks, Gutter and Pot casts, Bounce marks, Rill casts	Bottom scour followed by deposition
GRADED BEDS	Normal and Reverse grading	Deposition by turbidity currents
MASSIVE SAND BEDS		Deposition by grain flow
CONVOLUTE LAMINATION, DISH STRUCTURES, FLUID ESCAPE PIPES		Formed by fluidized sediment flow
PEBBLY MUDSTONE BEDS, CONGLOMERATE BEDS		Deposition by debris flow
CLAY CLASTS		Local erosion and deposition of cohesive clay layers
RAINDROP PRINTS, IMPACT RIPPLES		Sub-aerial impact by rain drops

contd.

Table 2.3.Ts. contd.

PART B : COMPACTIONAL AND DEFORMATIONAL STRUCTURES		
MUD CRACKS, SLICKENSIDES		Desiccation and shrinkage either by sub-aqueous syneresis or sub-aerial drying
LOAD CASTS	Ball and Pillow structures Flame structures	Soft sediment displacement of sands and silts in to underlying mud Soft sediment displacement of sands and silts in to underlying mud
MUD LUMPS, DIAPIRS		Large-scale upward displacement of plastic mud and shale
PART – C : DIAGENETIC STRUCTURES		
CONCRETIONS	Nodules, Septaria, Geodes, Spherulites	Local cementation, commonly early, without major displacement of mud matrix; commonly form around organic nucleus
CONE-IN-CONE, CRYSTAL CASTS, COLOUR BANDING		Crystal growth. Colour banding probably due to diffusion or may be related to weathering

Continuous and parallel even laminae signify deposition from suspension in a still water body with little turbulence. Interplay of silt/sand and mud, in response to fluctuating flow regime, produce a set of distinctive primary sedimentary structures, broadly known as **flaser bedding**. When mud stringers drape ripple trough and crest or only trough, this texture is commonly known as flaser bedding (Reineck and Wunderlich, 1968). This feature indicates alternate periods of current activity (when silt/sand is thrown into ripple form) with periods of quiescence when mud settles down from suspension, draping ripple trough and crest or ripple trough only. Three distinct types of flaser bedding are recognized: **simple flaser, bifurcated flaser** and **wavy flaser** (Reineck and Wunderlich, 1968).

Simple flaser beds are those where mud drapes only the trough regions of ripples and mud flasers are not interconnected (Figs. 2.3.1, 2.3.2 a, b and c). Bifurcated flasers are those where mud flasers are seen to bifurcate (Fig. 2.3.1). This is the effect of contact of partially exposed flasers of an earlier generation with the later-formed flasers and forms due to intermittent current reworking. In wavy flaser bedding, mud flasers cover ripple troughs and crests but fail to form a continuous bed (Fig. 2.3.1); this is possibly the result of partial erosion of the ripple crests.

In wavy bedding, mud and sand form continuous layers with mud flasers completely covering the ripple troughs and crests (Figs. 2.3.1, 2.3.3 a, b and c). In other words, mud flasers follow the concavity and convexity of the

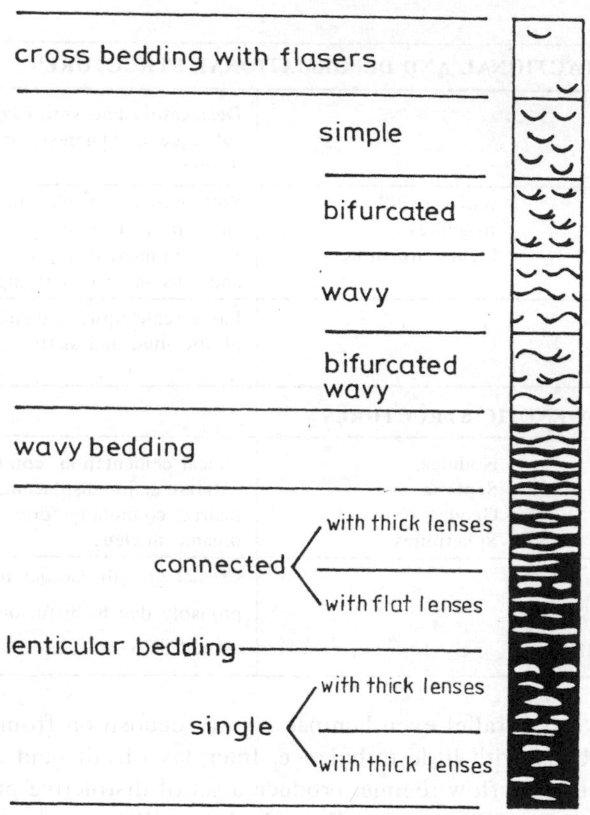

Fig. 2.3.1: The division of flaser and lenticular bedding (Black = mud or shale; White = sand or sandstone / silt or siltstone) (Reproduced from H.-E. Reineck and F. Wunderlich: Classification and origin of flaser and lenticular bedding. Sedimentology, Vol. 11, Fig. 1, © 1968, Blackwell Science with the kind permission).

underlying rippled surface. This is repeated vertically forming a sequence of wavy mud layers that alternate with ripple-bedded sandy / silty layers. This type of bedding indicates alternate periods of current activity followed by a period of quiescence (Reineck and Wunderlich, 1968). The flaser bedding of different styles, as described above, is usually obtained in sediments of tidal-flat environment. While this interpretation is generally true, non-tidal association of this structure is also reported. Such features can also form in environments where currents fluctuate, as in large river systems. Bhattacharyya (1997), in a recent paper, has reported one such occurrence from recent fluvial environment of the Ajay River, NE India. He has demonstrated that the thin mud layers that are deposited over sandy ripples and dunes, associated with point bars, during waning stages of floods, are curled up due to desiccation on being sub-aerially

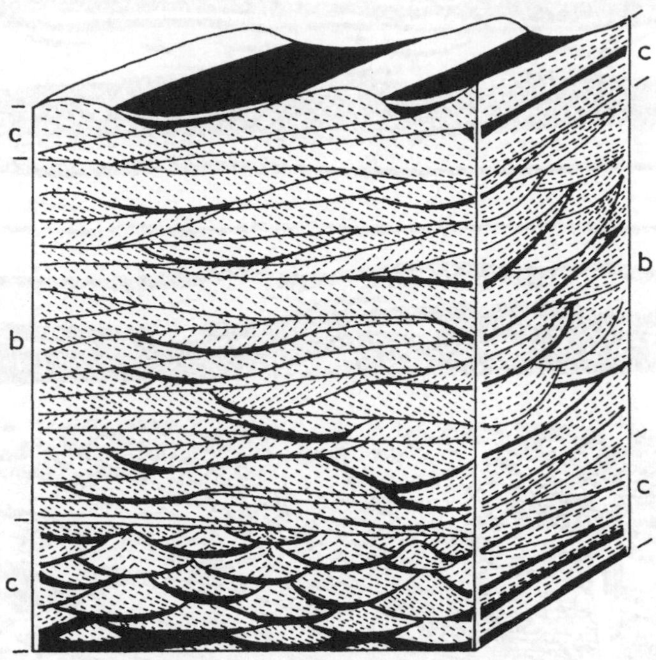

Fig.2.3.2 a: Flaser bedding on the surface, the mud is lying in the troughs of the ripples. This mud will be called "flaser" when preserved by overlying sand. In the top, flaser bedding formed from current ripples with straight crests; b: flaser bedding formed from current ripples with curved crests; c: flaser bedding formed from oscillation ripples (Reproduced from H. -E. Reineck and F. Wunderlich: Classification and origin of flaser and lenticular bedding. Sedimentology, Vol. 11, Fig. 2, © 1968, Blackwell Science with the kind permission).

exposed, and roll down into the ripple troughs. These mud curls, when submerged by the next sedimentation phase, are compacted into mud lenses, and look like mud flasers. However, flaser bedding so generated in fluvial environment is "generally less systematic in pattern compared with those formed in tidal settings" (Bhattacharya, 1997). While flaser bedding, flat-topped ripples, mudcracks, reactivation surfaces, herringbone cross-stratification are commonly associated with tidal-flat sediments, it may not always be so. One has to be careful in this respect. Alam et al. (1985) have reported occurrence of herringbone cross-stratification in fluvial sediment, and so also in longshore troughs of large lakes (Friedman et al., 1992). What these represent are alternations in current directions and velocity that may occur in many sedimentary environments. The interpretations of these structures must be considered in the background of the entire package of sedimentary structures and processes.

In lenticular bedding, ripples are found isolated both laterally and vertically by a thick mud layer (Figs.2.3.4 a, b and 2.3.5). This indicates that sand/

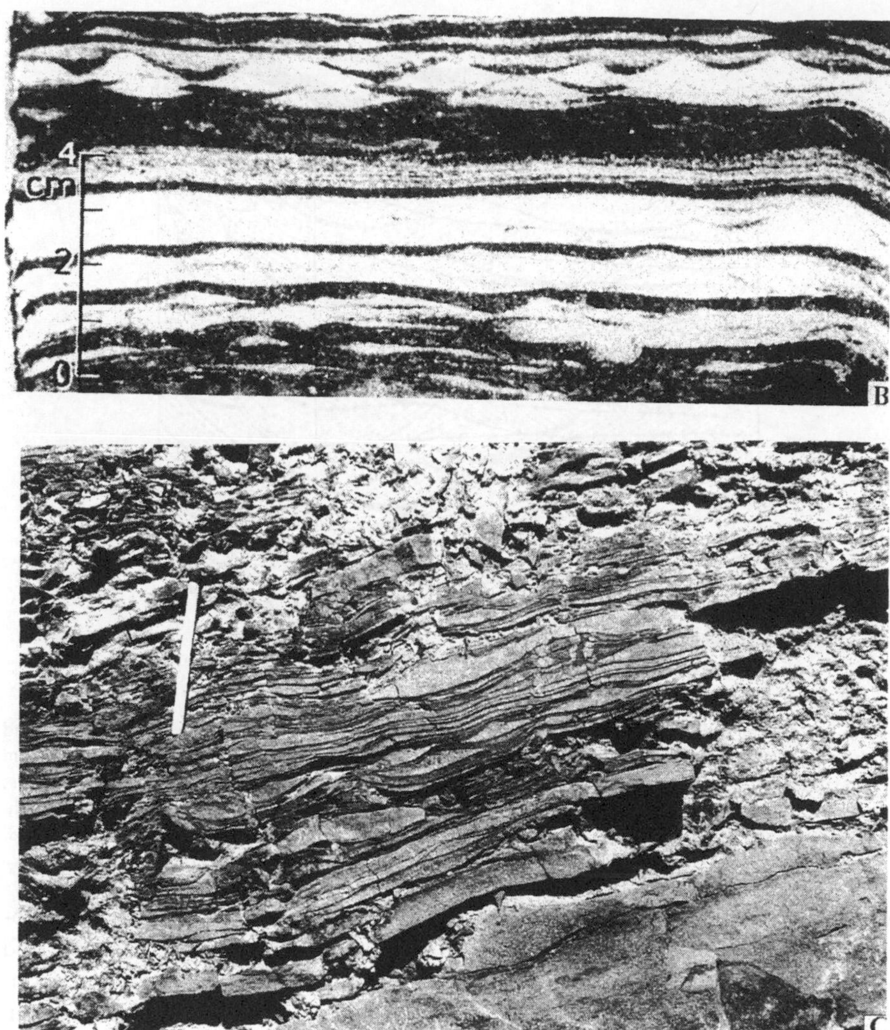

Fig. 2.3.2 b: A box-core from sediments of intertidal flats showing various types of bedding. In the top zone, lenticular and flaser bedding are of symmetrical wave ripples origin. North Sea tidal flats (Reproduced from Reineck, H. -E. and Singh, I. B: Depositional Sedimentary Environments, Fig. 193, © 1980, Elsevier Science-NL, Sara Burgerhartstraat 25, 1055 KV Amsterdam, The Netherlands with the kind permission).

c: Photograph showing lenticular bedding. Note isolated, symmetrical ripple forms (white) encased within mud (dark). Chorhat Formation, Proterozoic Vindhyan Supergroup, Chorhat, M.P., India.

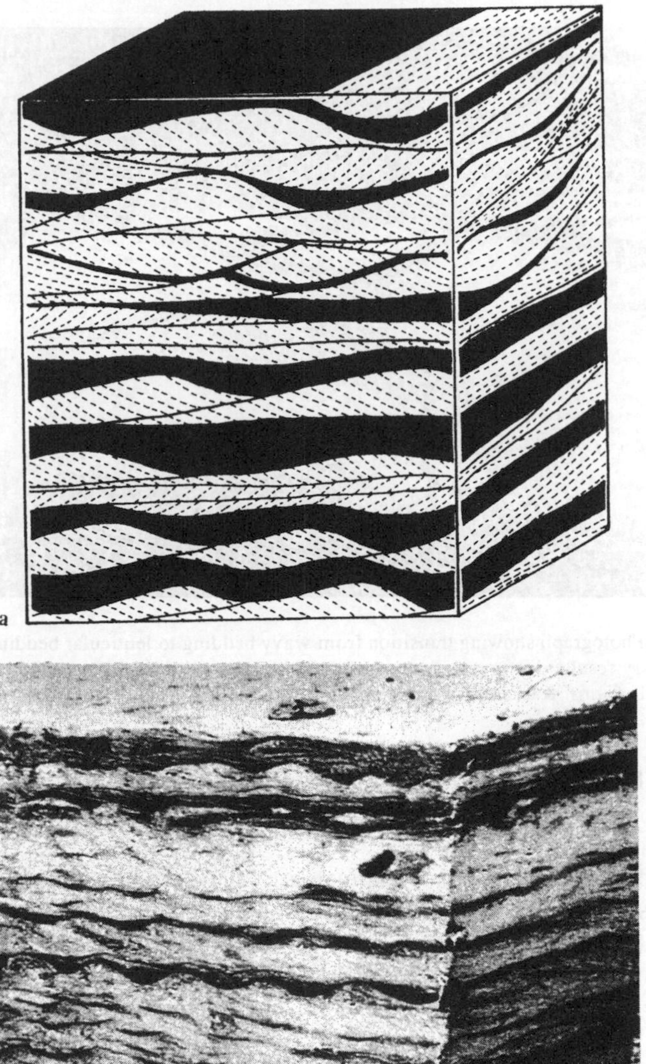

Fig. 2.3.3 a: Wavy bedding when flaser bedding covers both ripple troughs and crests but fail to form continuous layers (Reproduced form H. -E. Reineck and F. Wunderlich: Classification and origin of flaser and lenticular bedding. Sedimentology, Vol. 11, Fig. 3, © 1968, Blackwell Science with the kind permission).

b: Photograph of wavy bedding in the sediments of intertidal flats, North Sea (Reproduced from Reineck, H. -E. and Singh, I. B: Depositional Sedimentary Environments, 1973, 1980, Fig. 196, with the kind permission of Elsevier Science-NL, Sara Burgerhartstraat 25, 1055 KV Amsterdam, The Netherlands).

Fig. 2.3.3 c: Photograph showing transition from wavy bedding to lenticular bedding. Note that the wavy geometry results due to preservation of ripple forms and the thin mud drape (dark) between the ripple forms. Karharbari Formation, Gondwana Supergroup, Ramgarh, Bihar, India.

silt supply was meager, forming incomplete ripple trains being covered totally by mud. This situation points to a basically mud depositing environment with occasional supply of silt/sand forming isolated and disconnected ripples (starved ripple).

Tidal bundles and rhythmic laminae are fairly good diagnostic structures of tidal deposits in the rock record. Tidal bundles display sets of cross-laminae formed by the stronger set of current in an asymmetric tidal cycle and show subtle changes, from base to top, in response to changes in the current in each tidal cycle and bundles are set-off from one another by reactivation surfaces (Friedman et al., 1992). Thickness of tidal bundles changes in response to variation in amplitude on a monthly cycle (neap-tide, spring-tide). Tidal bundles may superficially resemble fluvial cross-strata sets, but can easily be separated from the association of other tidal signatures. Rhythmic laminae, on the other hand, are stacked sequences of horizontally to sub-horizontally laminated sediments, where millimeter-thick laminae thicken and thin systematically (Fig. 2.3.6). Individual laminae are interpreted as the products of semidiurnal tides, and the cycles of varying lamina thickness reflect fortnightly (neap-spring) fluctuation in tidal currents.

Fig. 2.3.4 a: Schematic cross-section of lenticular bedding with thick connected lenses. Upper part with current ripples; lower part with oscillation ripples (Reproduced from H. -E. Reineck and F. Wunderlich: Classification and origin of flaser and lenticular bedding. Sedimentology, Vol. 11, Fig. 4, © 1968, Blackwell Science with the kind permission).

b: Field photograph of lenticular bedding. Sand lenticles are asymmetrical wave ripples. In the lower part, convolute bedding is visible, North Sea tidalflats (Reproduced from Reineck, H. -E. and Singh, I.B.: Depositional Sedimentary Environments, 1973, 1980, Fig. 195, with the kind permission of Elsevier Science-NL, Sara Burgerhartstraat 25, 1055 KV Amsterdam, The Netherlands)

Fig.2.3.5: Lenticular bedding. Upper part with single flat lenses, lower part with single thick lenses. Some of them are oscillation ripples (Reproduced from H. -E. Reineck and F. Wunderlich: Classification and origin of flaser and lenticular bedding. Sedimentology, Vol. 11, Fig. 5, © 1968, Blackwell Science, U.K. with the kind permission).

Cross bedding

Cross-bedding in shale is not a common primary sedimentary structure but may develop where silt/sand fractions dominate (e.g., along basin margins). Their types and hydrodynamic implications are discussed separately (Section 2.2).

Parting lineation

This is another common structure found on bedding planes of thin-bedded siltstone/sandstone and its description and geological implications are discussed in the section on sandstone (Section 2.2).

Sole Structures

These structures are so named because they occur at the undersides of siltstone/sandstone that overlie shale. There are various forms of sole structures, e.g., flute and groove casts, load casts and various other hieroglyphs, and are seen as positive relief on the underside of siltstone/sandstone beds, formed as filling of the scoured surfaces on the shale (Fig. 2.3.7). Hence, these are casts. **Flutes** are raised elongated ridges with a bulbous up-current nose that flare down current and are discussed in greater detail in the section on sandstone (Section 2.2). **Groove casts** are raised and elongated ridges of sand that

Fig. 2.3.6: Polished section of a sequence of laminations. Spring tidal deposits are characterized by alternating light laminae and dark drapes, whereas neap tidal deposit is defined by darker organic-rich zone. Some laminae have been disrupted by bioturbation (Reproduced from M. A. Brown, A. W. Archer and E. P. Kvale: Neap-spring tidal cyclicity in laminated carbonate channel-fill deposits and its implications: Salem Limestone (Mississippian), South-Central Indiana, U.S.A. Vol. 60, No. 1, Fig. 6, p. 156, © 1990 Society for Sedimentary Geology (SEPM), Tulsa, Oklahoma, U.S.A. with the kind permission)

are remarkably straight with a relief of 1-2 mm (Figs. 2.3.8 and 2.3.9). Occasionally, two such sets may occur on the same surface cross-cutting at an acute angle. Groove and flute casts are mutually exclusive structures never occurring together on the same surface. Flute is a scour structure formed by the combined action of vertical and sub-horizontal eddies (Dzulynski, 1996), while groove casts are tool marks formed by the dragging of a tool (shell, bone of a vertebrate, pebble or any hard object) on a stiff mud substrate. Very rarely one may come across the tool at the down-current end that made the grooves. Since these structures are very helpful in determining paleocurrent direction, it is important to note their orientation. In case of groove casts, where tools are not found at the down current end, sense of current direction cannot be determined. Hence, one should measure and record the undirected lineation (e.g., 80^0–260^0).

Besides grooves, there are other diverse groups of tool marks formed by skipping and rolling of tools on soft mud substrate. These are **bounce marks** (Fig. 2.3.7), **prod marks** (Fig. 2.3.7) and **brush marks** (Fig. 2.3.7). Besides, several other structures are also reported from sandstone-shale sequences, e.g., **rill marks** (Figs. 2.3.10 a, b), **load structures** (Fig. 2.3.11a, b, c), **frondescent marks** (Figs. 2.3.7 and 2.3.12) and **skip casts** (Fig. 2.3.13).

Prod marks and bounce marks occur on the underside of many sandstone/siltstone beds as sharp, discontinuous and often elongated marks, produced by the impact of tools on a soft substrate. In prod marks, these are asymmetric in longitudinal section, in contrast to bounce marks, with gentle up-current and steep down-current sides (Figs. 2.3.14 B, D). In some examples, the tool, after hitting the substrate at high angle, may be dragged along the substrate producing groove marks; in such a situation, the steeper slope of the marking is on its upcurrent end (Dzulynski, 1996). Size in both types may vary from several centimeters wide and tens of centimeters long with depths roughly proportional to width. While recording such structures in the field, it is important to measure their size and direction and their longitudinal asymmetry, since these differentiate the prod marks from bounce marks.

Bounce marks, when arranged linearly with even spacing, are known as skip marks (Fig. 2.3.13). This suggests that the tool must have bounced on the soft substrate generating marks that are even spaced. Occasional differences in the shape of the skip marks may have resulted from the rotation of the tool (Dzulynski, 1996).

Brush marks (Figs. 2.3.7 and 2.3.14 C) are generated where tools impinging upon the bottom at a relatively low angle without being arrested on impact (Dzulynski, 1996). These markings show one or more moulds of overhanging mud bulges at their down-current end.

Roll marks (Fig. 2.3.14 F) are records of rolling objects on the soft mud substrate like the tread of an automobile wheel. Discoidal shale fragments, traveling in an upright position, produce narrow grooves that look similar to those produced by dragging of sharp-pointed tools.

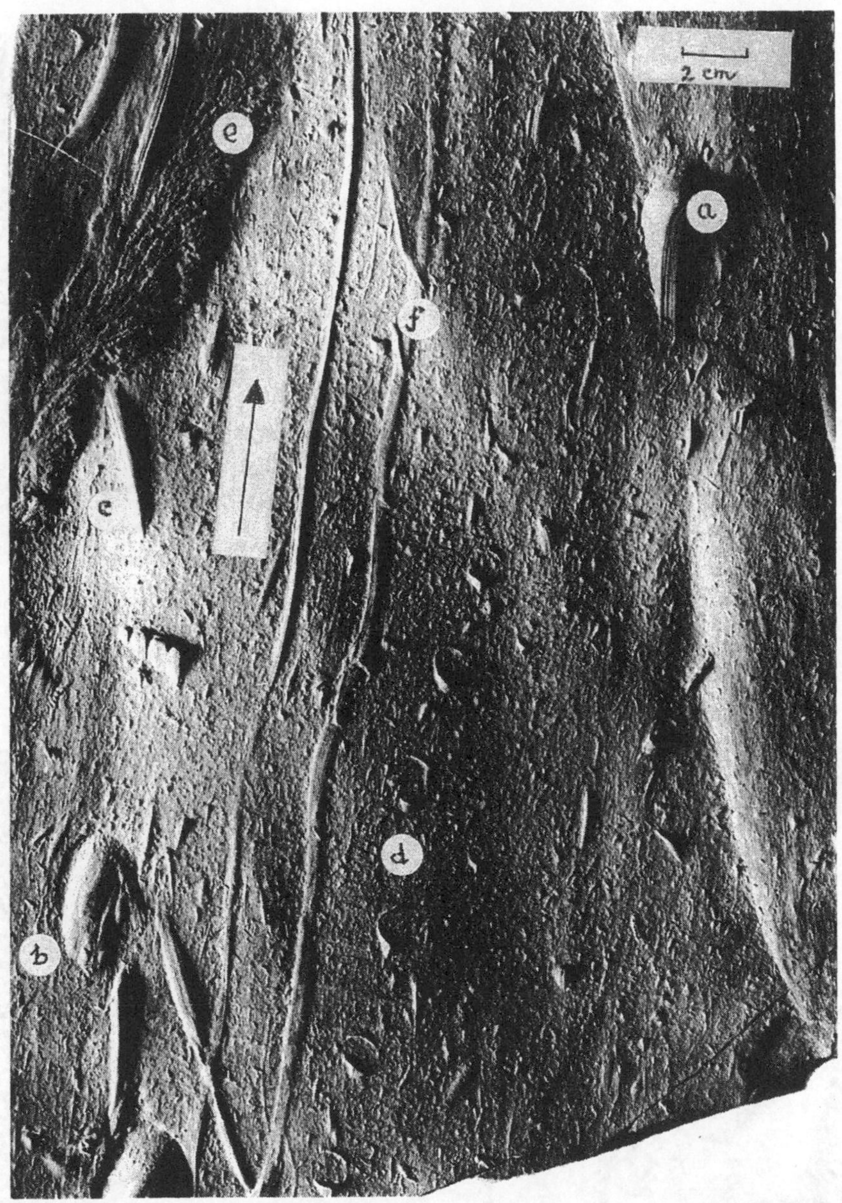

Fig. 2.3.7: Base of sandstone with moulds of: a, prod marks; b, brush marks, c, bounce marks; d, saltation marks of fish vertebrae; e, frondscent mark; f, double groove mould. Oligocene Krosno beds, Wetlina Polana (Reproduced from S. Dzulynski: Erosional and deformational structures in single sedimentary beds: A genetic commentary. Annales Societatis Geologorum Poloniae, Vol. 66, plate 4, 1996, with the kind permission of the author).

Fig.2.3.8: Broad groove mould, slightly curved; original mark made by flexible tool. Prod moulds on either side of groove mould, Oligocene Krosno beds, Rudawka Rymanowska, Carpathians, Poloniae. Length of the bar is 2 cm (Reproduced from S. Dzulynski and A. Slaczka: Directional structures and sedimentation of the Krosno beds (Carpathian flysch). Rocz. Pol. Tow. Geol., Vol. 28, Fig. 7a, 1958, with the kind permission of the author).

Fig. 2.3.9: Rounded groove moulds, Oligocene Krosno beds, Rudawka Rymanowska, Poloniae (Reproduced from S. Dzulynski: Erosional and deformational structures in single sedimentary beds: A genetic commentary. Annales Societatis Geologorum Poloniae, Vol. 66, plate 8a, 1996, with the kind permission of the author).

Fig. 2.3.10a: Slightly meandering moulds of rill markings, Oligocene Krosno beds, Wernejowka, Poloniae (Reproduced from S. Dzulynski: Erosional and deformational structures in single sedimentary beds: A genetic commentary. Annales Societatis Geologorum Poloniae, Vol. 66, Plate 16a, 1996, with the kind permission of the author).

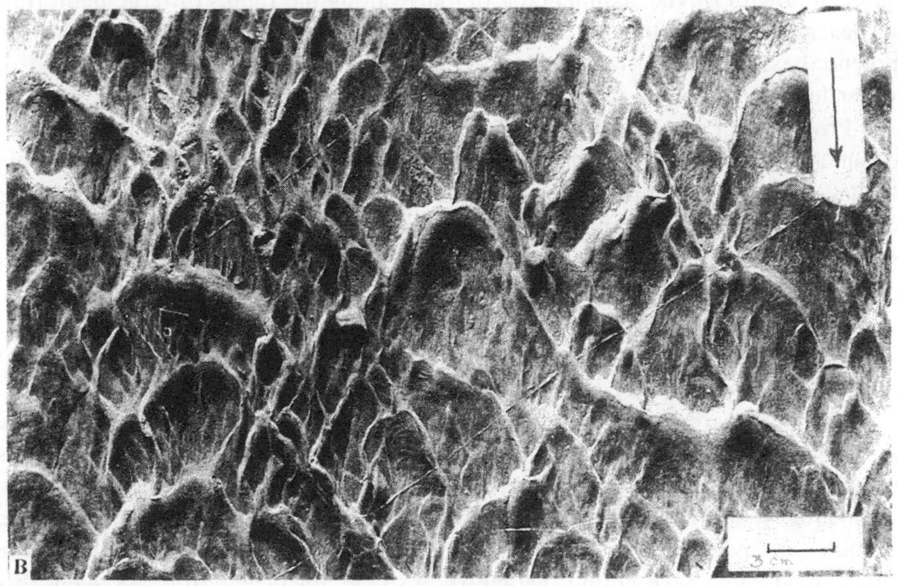

Fig. 2.3.10 b: Rill markings arranged in diagonal pattern, Oligocene Krosno beds, Wernejowka, Poloniae (Reproduced from S. Dzulynski : Ersosional and deformational structures in single sedimentary beds: A genetic commentary. Annales Societatis Geologorum Poloniae, Vol. 66, Plate 16b, 1996, with the kind permission of the author).

Fig. 2.3.11 a: Genetic model of formation of load casts. Load casts generate at the interface of sand bed overlying mud. Note how load cast is transformed to sand balls on passing from left to right. (Reproduced from J. R. L. Allen, 1970, Physical Processes of Sedimentation: An Introduction. Fig. 2.10, page 86, with the kind permission of the author).

Current crescents are narrow semi-circular grooves excavated in sand/silt substrate by current turbulence in the vicinity of small obstacles (pebble, mud fragment, shell) resting on the sediment and is discussed in detail in the section on sandstone

Rill marks (Figs. 2.3.10a, b) are sinuous, anastomosing shallow grooves (a few centimeter wide), arranged in rhomboid pattern on the surface of sand/silt laminae. Rill marks form at the exposed sediment surface following a fall in water stage (Collinson and Thompson, 1982). This structure is commonly associated with fluvial sediment, particularly along the flanks of large bedforms, in beaches, and on slopes of larger tidal bedforms. However, its preservation potential is very low, since this structure is easily washed out at the rise of water level.

Frondescent marks (Figs. 2.3.7 e and 2.3.12), resembling foliating leaves, occur as sole structures in many turbidite beds. The characteristic spreading 'foliage' of the structure is downcurrent oriented in most cases; they may also diverge radially upcurrent from point sources. Various mechanisms have been suggested for its origin. Summing up, frondescent marks appear to be the result of the injection of liquefied sediment from the base of settling turbidite onto the underlying clay layer (Dzulynski, 1996). This structure, as seen in vertical section, appears as bulbous or flat-bottomed. This structure has no paleocurrent implication, since it may be oriented either upcurrent or downcurrent.

Load casts are irregular bulbous structures formed at the interface of shale and overlying siltstone/sandstone bed or even between limestone beds (Figs. 2.3.11 a, b). This structure forms as a result of gravitational instability or unequal loading (Kuenen 1958). The dimensions may be the same as those of flutes or much longer, depending upon degree of load adjustment. Unlike flutes, load casts have neither geometrical shape nor orientation. As such, this structure has no value in paleocurrent determination. The term "load casts" is a misnomer, since these are not casts in the true sense of the word but represent

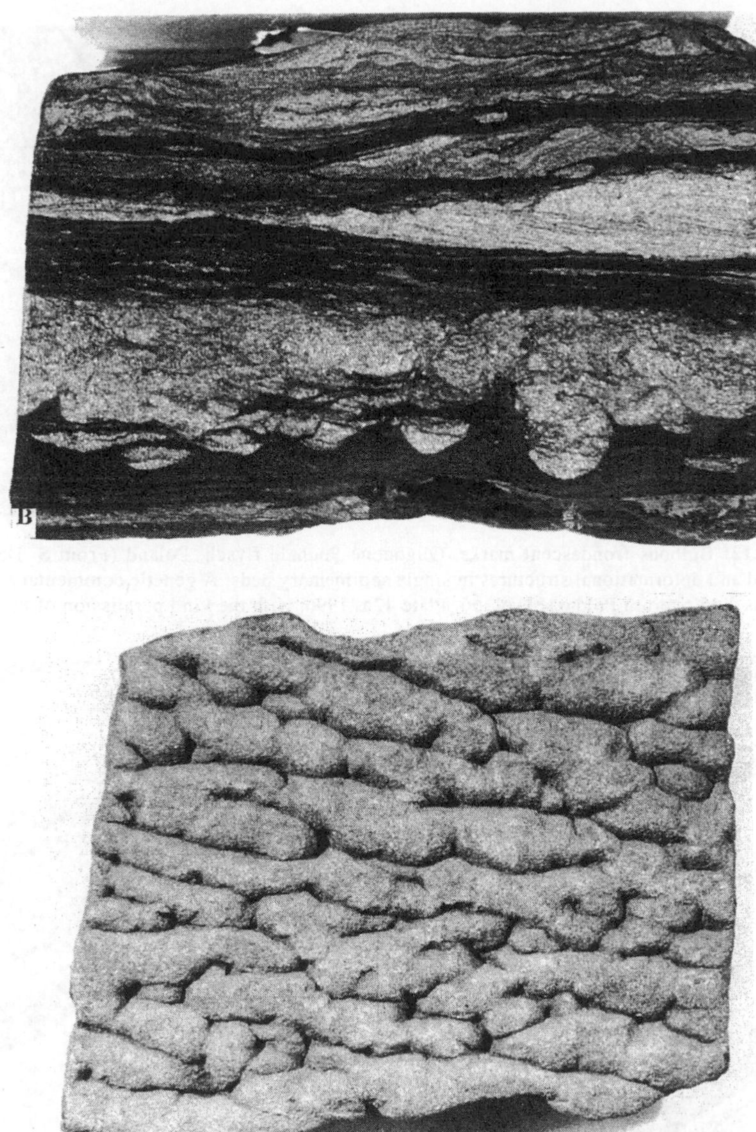

Fig. 2.3.11 b: Polished slab showing load casts at the junction of black carbonaceous shale and overlying gray fine siltstone, Permo-Carboniferous Barakar Formation, Gondwana Supergroup, Bihar, India. Horizontal distance is 11 cm.

c: Photograph of torose load casts, Proterozoic Bundi Hill Sandstone, Vindhyan Supergroup, Maihar, India. Horizontal length is 14.5 cm.

Fig. 2.3.12: Bulbous frondescent marks. Oligocene Podhale flysch, Poland (From S. Dzulynski: Erosional and deformational structures in single sedimentary beds: A genetic commentary. Annales Societatis Geologorum Poloniae, Vol. 66, Plate 42a, 1996, with the kind permission of the author).

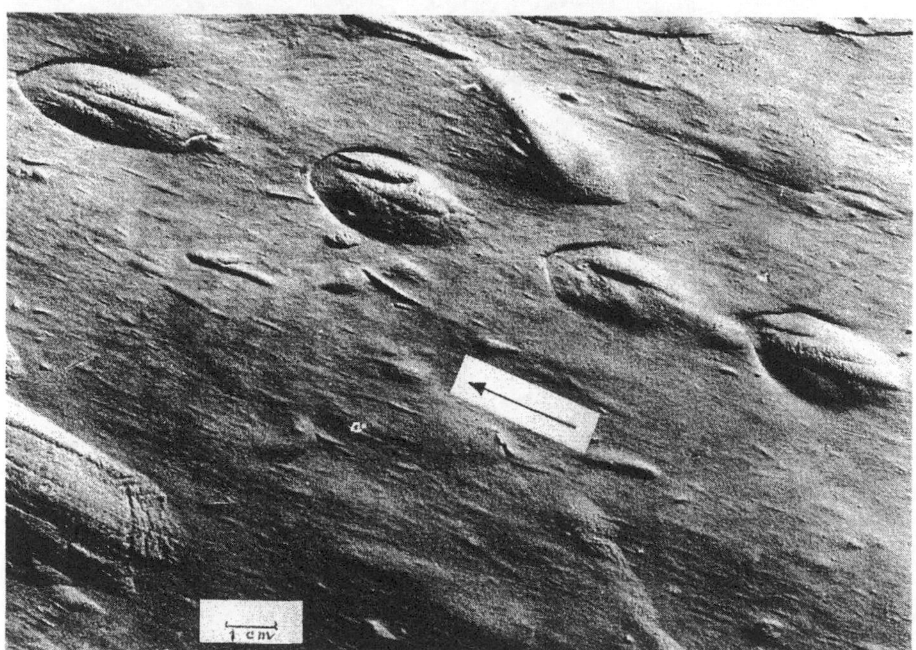

Fig. 2.3.13: Skip casts, Oligocene Krosno beds, Katy, Poland (Reproduced from S. Dzulynski : Erosional and deformational structures in single beds: A genetic commentary. Annales Societatis Geologorum Poloniae, Vol. 66, Plate 5A, 1996, with the kind permission of the author).

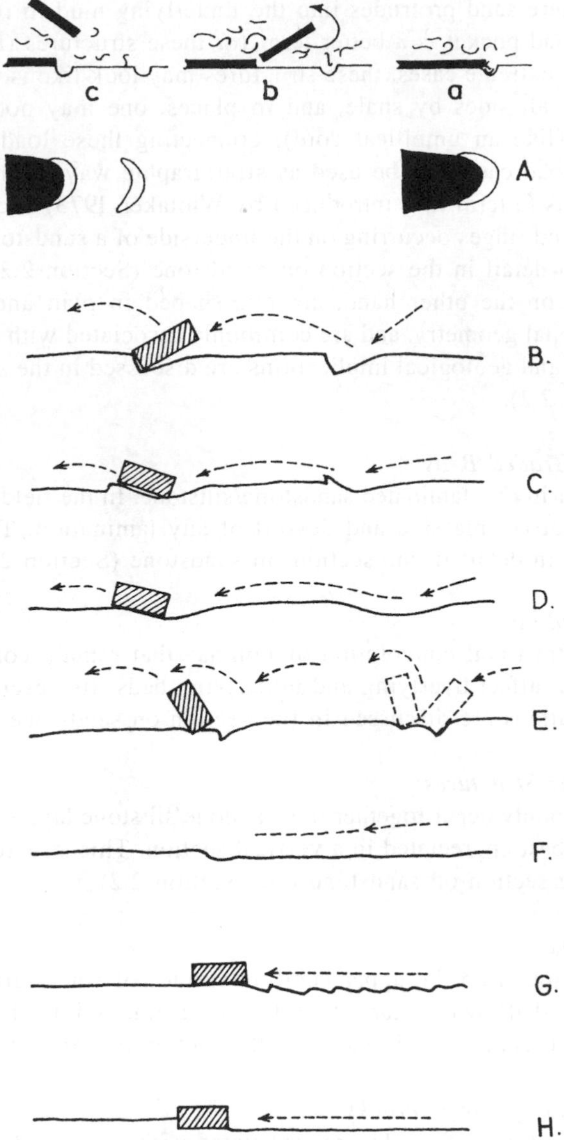

Fig.2.3.14: Diagram showing formation of tool marks. A, formation of successive crescent marks; B, prod marks; C, brush marks; D, bounce marks; E, saltation marks; F, roll marks; G, chevron marks; H, groove marks (Reproduced from S. Dzulynski: Erosional and deformational strucrtures in single beds: A genetic commentary. Annales Societatis Geologorum Poloniae, Vol. 66, Figs. 1 and 2, 1996, with the kind permission of the author).

a structure where sand protrudes into the underlying mud in response to load adjustment. Load pocket is a better term for these structures (Pettijohn, 1975, p. 120). Under extreme cases, these structures may look like isolated load balls surrounded on all sides by shale, and in places, one may notice a thin constricted neck (like an umbilical cord), connecting these load balls with the mother bed. Load casts can be used as stratigraphic way-up indicators.

Gutter casts (a term first introduced by Whitaker, 1973) occur as elongated and isolated sand ridges occurring on the underside of a sandstone/siltstone and is discussed in detail in the section on Sandstone (Section 2.2).

Pot casts, on the other hand, are cup-shaped in plan and cylindrical in three-dimensional geometry, and are commonly associated with gutter casts and its description and geological implications are discussed in the section on sandstone (Section 2.2).

Massive and Graded Beds

They occur in non-laminated sandstone/siltstone. In the field, one may come across beds that are massive and devoid of any lamination. This structure is also discussed in detail in the section on sandstone (Section 2.2).

Convolute bedding

It is an intrastratal convolution of laminae that remain confined within a bed; it does not affect overlying and underlying beds. Its description and geological implications are discussed in the section on sandstone (Section 2.2).

Dish and Pillar Structures

They commonly occur together in sandstone/siltstone layers associated with shale, and are best appreciated in a vertical section. This structure is discussed in detail in the section on sandstone (see Section 2.2).

Raindrop prints

These are found on the upper bedding surface of fine-grained terrigenous rocks as small, shallow circular pits with raised rim, and detail discussion and geological significance are discussed in the section on sandstone(Section 2.2).

Desiccation and Syneresis cracks

Broadly two types of cracks are associated with recent sediments and sedimentary rocks: 1) desiccation cracks and 2) syneresis cracks. In modern sediments, they occur at the top of the sediment surfaces, and both at the lower and upper bedding surfaces when it occurs in rocks.

Desiccation cracks develop on the dried-up sediment floors and tend to occur as polygons having more or less hexagonal patterns; many polygons may, however, be quadrangles or triangles on bedding plane surfaces. The polygons range in size from centimeters to meters in diameter, having crack width that may range up to several centimeters. The cracks are usually parallel sided in plan view and are found to taper downwards in vertical sections. The depth of the cracks depends upon degree of dehydration. Since the spacing of cracks is proportional to the thickness of the layer undergoing brittle failure, the earlier generation cracks tend to be more closely spaced than the later generation cracks (Fig. 2.3.15 a). The desiccation cracks, as the name implies, form due to dehydration that give rise to isotropic, horizontal, tensional stress within the sediment. Vertical cracks form to release the horizontal stress. These cracks taper downward, since the tensional stress diminishes downward to a point of zero stress. Desiccation cracks are not found in pure sands since there is no volume decrease on drying and hence no horizontal tensional stress.

The cracks are commonly gravity-filled by sand; in some cases, cracks may be filled-in by wind-blown sand. Mud chips, derived from desiccation of thin mud layers may be reworked and deposited within the cracks. Where desiccation cracks are well-preserved, mud curls may be found lying above the cracked surfaces. Some cracks may show repeated episodes of desiccation and infilling. The preservation potential of sub-aerial desiccation cracks is, however, low, and where preserved, may be in places bounded by erosional surfaces.

Syneresis cracks are believed to develop sub-aqueously at the sediment-water interface or intrastratally. In plan view, individual syneresis cracks are lenticular, 'eye-shaped' and about 1-10 mm wide and 1-30 cm long. They may be parallel or randomly oriented and occur either as discrete or interconnected in plan view, resulting in configurations either as rectangular to polygonal with variable degrees of completeness (Pratt, 1998). Syneresis cracks are filled-in by sands or silts injected from above or below. In bedding normal sections, syneresis cracks are found to taper downward and commonly upward or may appear as parallel-sided. Margins are straight, curved, lobate or ragged. In-fills may often appear ptygmatically folded due to variably oriented shear stresses and the host beds show a reduction in bed thickness.

Subaerial desiccation cracks may be distinguished from syneresis cracks by their in-fill; in-fills in desiccation cracks are deposited geopetally, whereas in syneresis cracks in-fills are injected forcefully from below and/or the top. Besides, host beds betray other features (particularly the facies association) that provide unequivocal evidence of subaerial exposure.

As Pratt (1998) has pointed out, 'syneresis' in colloid chemistry denotes slow contraction of gels without any external physical stimulation and this process was invoked to explain these sub-aqueously formed cracks. Various theories have been suggested to explain this structure: (i) trace fossils (Harding and Risk, 1986; Retallack, 1994), (ii) sub-aerial desiccation cracks with the lenticular geometry being explained as molds of evaporite minerals, particularly gypsum (Astin and Rogers, 1991; Kidder, 1990), and (iii) the most commonly held view is the spontaneous clay deflocculation and lattice contrac-

Fig.2.3.15a: Photographs showing two orders of dessication cracks on top of a mudstone, Proterozoic Ganurgarh Shale, Vindhyan Supergroup, Maihar.

b: Spindle-shaped dessication cracks on top of Proterozoic Sasaram Formation, Vindhyan Supergroup, Churk, U.P, India.

tion due to salinity changes in pore water (Burst, 1965; Plummer and Gostin, 1981).

Two recent publications (Tanner, 1998; Pratt, 1998) have discussed these structures in more detail and deserve attention. Pratt (1998) has suggested that syneresis cracks, under shallow burial depths, are generated under conditions of strong ground movements due to syn-sedimentary earthquakes; such strong ground movements result in dewatering of argillaceous sediments and liquefaction and injection of inter-bedded sands and silts in to the fissures so created. Nature of in-fill, upward gradation of cracks into micro-faults, cracking of numerous layers at once and deflection of cracks around overlying sole structures are points that indicate intrastratal origin of the syneresis cracks (Pratt, 1998). Moreover, shale fragments, commonly associated with the in-fills, are derived from layers brecciated intrastratally, and not from fragmentation of surface sediment such as might occur during sub-aerial desiccation.

On the other hand, Tanner (1998) has suggested that the sand-filled cracks develop through "intrusion of water-saturated sand into shrinkage cracks in mud or muddy-sand, not, as previously thought, as a result of sub-aerial desiccation, or sub-aqueous cracking of the sediment surface (syneresis). These cracks likely resulted from layer-parallel contraction caused by compaction of mudstone layers during burial. Seismic shock may have provided the required trigger for the preferential development of polygonal crack patterns in these layers Thus "sand-filled cracks, forming polygonal patterns on the bases of beds should not be automatically interpreted as having been caused by sub-aerial desiccation or some form of syneresis". Tanner (1998) concludes "interstratal cracking is a mechanism which rivals sub-aerial desiccation in importance and is more common in geological record than is currently realized".

Diagenetic Structures in Shale

Shale commonly contains bodies that are either spherical, elliptical or irregular in shape and are distinct in chemistry compared to host rock. These are different diagenetic structures variously known as **nodule, concretion**; **cement** and **vein** (Selles-Martinez, 1996). These commonly occur in distinct vertical profiles and stand out prominently because of their shape.

Nodule does not incorporate clastic material during growth and acquire their shape that may display great variations in size and shape. Where nodules occur in profusion along the same plane, they may tend to converge, forming a discontinuous sheet.

Concretion, on the other hand, refers to that portion of sedimentary rock which is cemented differentially from its host (Selles-Martinez, 1996) and is usually spherical or ellipsoidal in shape. Concretions result "from precipitation of authigenic material incorporating clastic framework", and thus make them different from nodules. Warping of internal laminae toward the border of the concretion and gradual change in texture and fabric are taken as evidence of syncompactional origin. Occasionally, uncrushed and undeformed fossils (both invertebrate and vertebrate) may occur in the core of these concretions. Sometimes, the central part of the concretions may display irregular, lenticular cracks

(Fig. 2.3.16 a) the width of which is independent of their position in the nodule; cracks near the margin of these concretions wedge out and do not extend outside. These cracks, known as **septarian structure,** are variously filled-in by coarse calcite crystals (Fig. 2.3.16 b). Septarian cracks display a double set of fracture systems: (i) one is radial and characterized by polygonal intersections of fracture plane, and (ii) the other displays a wedge-like geometry that thins outwards and ends before reaching the exterior. The infillings of septarian cracks are similar to those in veins in their morphology, but are very complex in their mineralogy and isotopic composition, reflecting the continuous evolution of parent solutions. In some other concretions, curious-looking angular pattern of fractures are found around the margins involving stacked conical fractured surfaces. These structures in nodules are known as **cone-in-cone structures** (Fig. 2.3.17).

In a recent paper, Hounslow (1997) has analyzed the stress conditions and crack morphology within the septarian nodules and observed that "the cracks in septarian concretions result from tensional failure (sub-critical crack growth), as a consequence of localized excess pore pressure". He has further suggested that cracks generate at depths of less than 10m.

The nodules and concretions, when examined chemically, are variously made up of calcite, siderite, pyrite, chert, evaporite, hematite, barite or manganese. The various compositions of these nodular bodies represent a "frozen" record of the chemistry of pore water at the time of precipitation. Nodules and concretions may have formed either syngenetically (manganese nodules in deep seafloor), penecontemporaneously or epigenetically.

Laminae into and around these nodules and concretions are found to conformably wrap around and the laminations within are found to converge toward the major axis of the nodule. These features indicate penecontemporaneous nature of the nodules and concretions. Presence of uncrushed fossils or ooids inside, in contrast to crushed and deformed fossils or ooids outside the nodules and concretions, are other signatures of penecontemporaneous origin of these bodies. Absence of all these features signal epigenetic origin of the nodules and concretions.

When nodules and concretions occur in shale, the following features are to be noted: (i) the geometry and their dimensions, (ii) the nature of the laminae in the host sediment immediately surrounding the nodules and concretions, (iii) whether these nodules are bedding parallel or perpendicular; if found perpendicular, try to establish if these are burrow filling or tree root filling, (iv) determination of the composition of these nodules in the field, if possible, and finally (v) collection of oriented samples for further laboratory studies.

In a recent paper, Raiswell and Fisher (2000) have presented a major review of mudrock-hosted carbonate concretions. They have emphasized that existing interpretations of cement textures and isotopic compositions may significantly under-estimate the depth and duration of growth of concretions. This

Fig. 2.3.16 a: Radial and concentric cracks within septarian nodules in-filled with calcite 'veins' which radiate out from the centre. These often form within concretions and each segment is known as septa (meaning dividing walls or partitions), hence the name as septarian nodules (Courtesy J. Forsyth, Institute of Geological and Nuclear Sciences Ltd., New Zealand).

b: Septarian nodules, known locally as Moeraki boulders in New Zealand, are found scattered along the shore between Moeraki and Hampden on the east coast of New Zealand's South Island. These concretions, ranging in size from small pellets to huge spherical bodies, spanning three meters across, are enclosed in shales of Paleozoic age (Courtesy Dr. D. Weston, Geology Department, University of Otago, Dunedin, New Zealand).

Fig. 2.3.17: Cone-in-cone structures (Courtesy Prof. I. B. Singh, Lucknow, India)

arises partly from the difficulty of recognizing and evaluating the extent of concretion deformation during burial. The strongest evidence in favour of shallow burial growth of concretions is the preservation of soft-bodied fossils or biogenic structures within concretions. The growth of concretions may have started at shallow burial depth but available petrographic, chemical and isotopic evidences indicate a prolonged growth history, possibly involving deep burial process to the tune of tens to hundreds of meters. Furthermore, Raiswell and Fisher (2000), indicate that cements within concretions were neither infilled passively nor minus-cement porosity need indicate initial porosity of the host sediment. Regarding growth model, Raiswell and Fisher (2000) recognize two growth modes: i) concentric where concretions progressively increase in size outwards and ii) pervasive where cement crystals grow throughout the concretion simultaneously.

Why is it necessary to study these nodules and concretions? This is because (i) nodules and concretions contain "frozen" records of the chemistry of pore water at the time of precipitation, (ii) it helps, in some cases, to determine the degree of compaction of the host sediment subsequent to deposition. The following publications on nodules and concretions may be helpful (Collinson and Thompson, 1982; Feistner 1989; Hennessy and Knauth, 1985; Huggett, 1994; Johnson, 1989; Mozley, 1989,1996; Mozley and Burns, 1993; Pettijohn, 1975; Raiswell, 1971, 1976; Raiswell and Fisher, 2000; Scotchman, 1991; Selles-Martinez,1996.

2.4 Limestones

2.4.1 Stromatolite

Krumbein (1983) has redefined stromatolite as "laminated rocks, the origin of which can clearly be related to the activity of microbial communities that by their morphology, physiology and arrangement in space and time interact with the physical and chemical environment to produce a laminated pattern that is retained in the final rock structure". As will be noticed from the above definition, the role of cyanophytes as principal contributor to stromatolite building has been negated because examples of stromatolite exist which are formed by fungi, chemo-organotrophic and chemo-lithotrophic bacteria, and by obligate anoxygenic phototrophic bacteria. Again, the term 'algae' has been removed from the earlier definitions, since algae (green algae), being obligate oxygenic and obligate photolithotrophs, are poor candidates for stromatolite formation.

Field Description Of Stromatolites

Many Archaean and Proterozoic limestone and dolomite are replete with occurrences of stromatolites (Fig. 2.4.1.1); this is also true for the Phanerozoic rock record. For any serious study of stromatolites, they need to be systematically described accompanied by scaled field sketches, photographs, and collection of good samples of various forms of stromatolites for detail laboratory studies.

Following are the steps, suggested by Preiss (1976), to be followed in an attempt to extract maximum information from a stromatolite-bearing limestone/dolomite.

1. Determination of the thickness of the stromatolite beds.
2. Determining if there is a single or several stromatolite morphologies in the bed. If there are several morphologies, it is useful to delineate them in the bed properly with the help of chalk/felt pen.
3. Determination of the geometry of the bed containing stromatolites, and its lateral and vertical relationships with the associated lithologies.

Fig. 2.4.1.1: A photograph showing columnar stromatolite, Proterozoic Bhander Limestone, Vindhyan Supergroup, Maihar, Madhya Pradesh, India

4. Determination of the shape and the relief of successive growth interface of the stromatolite. This information is important for paleoecology, since it provides an idea of minimum depth of water for the growth of stromatolite. The above features help to categorize the mode of occurrence of stromatolite horizons as either bioherm (tabular-, domed-, subspherical-, tonguing-bioherm) or biostrome (tabular-,and domal-biostrome) (Fig. 2.4.1.2).
5. Determination of the nature of constituent stromatolites: are they stratiform stromatolites (flat laminated, laterally linked, undulatory, pseudocolumnar or cumulate) or columnar-layered. It is important to note their lateral and vertical relationships (Fig. 2.4.1.2).
6. If the stromatolites are columnar, are they branching? If so, it is necessary to note their style of branching (parallel, slightly divergent, markedly divergent, coalescing columns) (Fig. 2.4.1.2).
7. On good bedding plane section, it is necessary to determine the nature of outline of the columnar stromatolites—are they round, elliptical, oblong, or polygonal. Fig. 2.4.1.3 shows bedding plane view of composite stromatolite.
8. Determination of the nature of the column shape and surface ornamentation of columnar stromatolites (smooth, bumps, cornices, peaks, projections,

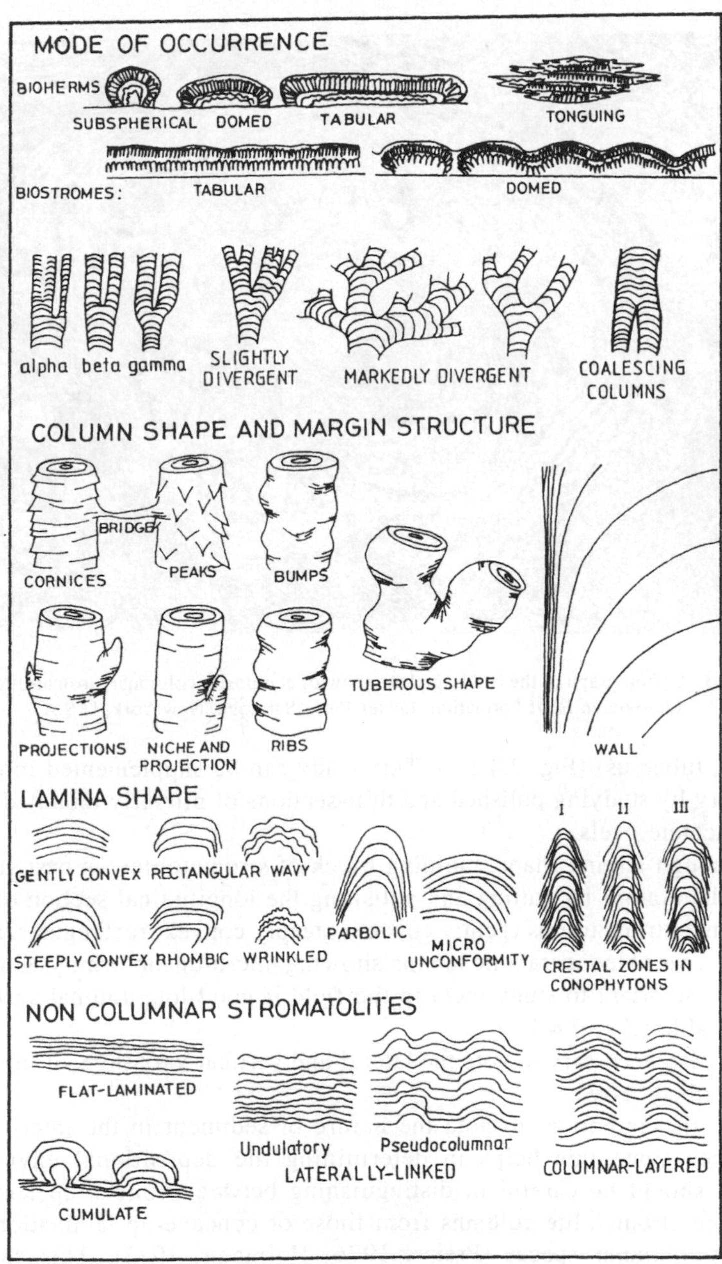

Fig. 2.4.1.2: Diagrammatic illustrations of terms used in the description of diagnostic characters of stromatolites (Reproduced from W. V. Preiss: The systematics of South Australian Precambrian and Cambrian stromatolites. Part 1. Trans. Royal Soc. of South Australia, Vol. 96, © 1972, Fig. 1, Royal Society of South Australia, Inc., Adelaide, Australia with the kind permission).

Fig. 2.4.1.3: A photograph of the bedding plane view of composite columnar stromatolite, Cambro-Ordovician Hoyt Formation, Lester Park, Saratoga, New York, U.S.A.

ribs, tuberous) (Fig. 2.4.1.2). This study can be supplemented in the laboratory by studying polished and thin-sections of properly mounted samples or acetate peels.

9. Although lamina shape, building block of stromatolites, is best studied in the laboratory by cutting and polishing the longitudinal section of the columnar stromatolites (gently convex, steeply convex, rectangular, rhombic, wavy, wrinkled, parabolic lamina showing micro-unconformity within), it is also important to study them in the field if good longitudinal sections are available (Fig. 2.4.1.2).
10. It is also necessary to note the lateral and vertical variations of the features described above.
11. It is also important to note the nature of sediment in the inter-columnar spaces since this helps in determining the depositional environment. One should be careful in distinguishing between convex-up laminations within stromatolite columns from those of concave-up laminations in the inter-columnar spaces (Preiss, 1976; Hofmann, 1973). This can easily be discriminated if one notices the direction of branching of columnar stromatolites, where laminae is always convex-up. Thus concave-up laminae indicates inter-columnar filling.

12. Sampling of stromatolite is another important step. A sample that shows a clear cross-section of several complete columns of stromatolite, representing different forms, should be collected to prepare serial sections in the laboratory; this helps in three-dimensional reconstruction of stromatolite columns following techniques as outlined by Preiss (1976) and Hofmann (1969, 1976). When stromatolite columns are too large to sample, careful field study and collection of small but representative samples to study the geometry of the stromatolite lamellae and the nature of the margin of the stromatolite columns is important.

The parameters, as noted above, e.g., i) mode of occurrence of stromatolites, ii) column shape and margin, iii) branching style, iv) laminae shape, and v) the microstructure of the layers are the very basis of identifing forms. Natural assemblages of forms help one to identify groups. There are at least 70 groups (form-genera) of stromatolites identified so far.

Stromatoloids

The term "stromatoloid" is used for structures that are morphologically similar to stromatolite but are of uncertain origin: these are geyserite, speleothem and calcrete.

Geyserite is made up of opaline silica formed non-biologically within and around hot springs and geysers (e.g., Yellowstone National Park, U.S.A.), and develops a variety of forms that are strikingly similar in structure to stromatolites. Walter (1976) has suggested criteria to distinguish them from stromatolites: (i) geyserites occur locally around vents and fissures discharging water, (ii) micro-cross-lamination is only found in geyserites, and (iii) extremely thin regular laminations present in geyserites (whose regularity is a striking feature).

A variety of crystalline speleothems occur, resembling stromatolites in its gross morphology and internal structure. Thrailkill (1976) has suggested criteria to distinguish them. They are: (i) association with other speleothems such as stalactites, (ii) scale of the structure, (iii) presence of initially deposited or early diagenetic magnesium minerals, such as dolomite, huntite, or hydromagnesite, (iv) indication that most of the deposition was due to precipitation from solution rather than entrapment of detritus, (v) non-vertical orientation, and (vi) lack of filaments or other biogenic structures.

Calcrete soil profiles may sometime superficially resemble wrinkly laminated stratiform stromatolites. Following features may help to discriminate: (i) calcrete laminae commonly thicken into depressions (rather than over crests of the depressions as in stromatolites), (ii) presence of sinkholes and collapse stucture in calcretes, (iii) presence of micro-unconformities or solution unconformities in calcretes (iv) ooid and pisolite beds (though resembling cryptalgal oncolites) are devoid of any mechanical structures, (v) common in situ brecciation and fracture in calcretes, (vi) polygonal fitted structure and perched inclusions in pisolites, and (vii) presence of solution cavities followed by cementation and internal sedimentation in calcretes. While fenestral fabric has been suggested by Read (1976) as one of the identifying criteria, it needs to be mentioned that such fabric is not universally associated with laminated stromatolites; moreover, fenestral fabric has been reported from the soil profiles. Again, vertical occurrence of many generations of stylolites occurring subparallel to bedding may produce irregularly laminated rocks mimicking wrinkly laminated stratiform stromatolites. This situation is further complicated if stylolites develop in chemically pure rocks where stylolamination may not develop at all or may be very indistinct. The only definitive criterion, in such a situation, is provided by the microstructure of the laminations.

Lastly, structures simulating the morphology of pseudo-columnar or even columnar stromatolites may also develop by flexural-slip deformation of thinly laminated sediments, particularly when associated with pressure-solution features (Buick et al., 1981). As a rule of thumb it can be said that structures that show diversity and branching habit are unlikely to be of inorganic origin.

It is pertinent to point out here that recently the biogenicity of stromatolites has been questioned (Grotzinger and Rothman, 1996). They have used the principles of fractal geometry to analyze the morphology of one form of stromatolite collected from the Proterozoic of Canada and have concluded that the morphogenesis of these stromatolites, and perhaps many others, could be a reflection of abiotic processes. The processes involved in the morphogenesis of stromatolite, they point out, are four abiotic processes: i) fallout of suspended sediments, ii) down-slope movement of that sediment, iii) surface normal precipitation, and iv) random effects. Grotzinger and Rothman (1996) have suggested that "future studies should concentrate on the surfaces upon which stromatolite growth is initiated and the way that lamina shape changes upward".

Utility of Stromatolite Study

1) Taking the age of the earth as 4.5 billion years and placing the beginning of the Phanerozoic at 560 m.y., it is found that nearly 88 percent of the rock record falls below the Cambrian. Because of abiogenic nature of this huge pile of rocks, proper stratigraphic correlation is a challenge to geologists. Stromatolites provide a good element for such stratigraphic correlation not only intrabasinally but also on an intercontinental scale. It is now established that distinctive assemblages of columnar stromatolites succeed themselves over roughly the same time interval in geographically widely separated areas. The precision of correlations based on stromatolites has been checked by classical methods (position of unconformities, marker horizons etc.), and geochronological results (Bertrand-Sarfati and Raaben

1971; Bertrand-Sarfati and Trompette 1976; Cloud and Semikhatov 1969; Glaessner et al. 1969; Preiss 1972; Walter 1972). The precision and reliability of such correlations depend upon good definition and identification of stromatolites; in all such correlations, the microstructure of the laminations is of prime importance.

2) Anisotropic growth of stromatolites has been successfully used in paleocurrent and paleogeographic analyses. Statistical study by Hoffman (1967) has convincingly demonstrated the feasibility of using the shape and orientation of the stromatolites in the reconstruction of the paleocurrent system.
3) Stromatolites can also be used as a way-up indicator in a structurally disturbed terrain by taking note of the fact that convexity of the laminae within stromatolites always point upward. Direction of branching of columnar stromatolites can also be used for this purpose.
4) Stromatolites have also been used as a) paleo-tide gauge, b) paleolatitude reckoner and as a, c) as geochronometer. Hofmann (1973) has evaluated their various utilities and applicability, and summarized in Table 2.4.1.T1.

2.4.2 Diagenetic Bedding

The science of stratigraphy basically deals with strata where it is not only sufficient to look into the composition, internal primary sedimentary structures, or its biotic content but at the same time it is important to ensure that the planes separating the beds represent primary depositional planes. This is important since a bed is the basic building block of facies, paleoenvironmental and basin analyses (Simpson, 1985). Definition of a bed, as proposed by Campbell (1967), is "a layer of sedimentary rock or sediment bounded above and below by bedding surfaces. Beds are deposited under essentially constant physical conditions and their internal composition and texture can be uniform or heterogeneous, rhythmically variable or systematically gradational. Bedding planes are essentially deposition surfaces that reveal the rock layering or bedding".

Cyclic or rhythmic alternations of limestone and shale or marl (impure limestone) are common in marine carbonate platform, pelagic and hemi-pelagic carbonates of various ages. This rhythmic alternation has been explained as due to the earth's orbital cycles of precession, obliquity and eccentricity with periods of approximately 21,000, 41,000, and 100,000 years respectively. This periodic movement of the earth forces a change of climate in the form of ocean current systems, fluctuations of sea level etc. Rhythmic bedding is thus believed to reflect climatic forcing.

However, the bounding surfaces of many of the apparently rhythmic platform, pelagic and hemi-pelagic limestone, when examined in the context of above, do not appear to be primary but diagenetically enhanced bedding

Table 2.4.1.T1: Summary of information on utility of fossil stromatolites (Reproduced from H. J. Hofmann: Stromatolites: Characteristics and Utility. Earth-Science Reviews, Vol. 9, Table 1, 1973, Elsevier Science - NL, Sara Burgerhartstraat 25, 1055 KV Amsterdam, The Netherlands with the kind permission)

Evaluation of reliability (combination of criteria)	**Ordinal scale suggesting degree of confidence (as of end of 1972)**				
	None		low		high
	0	1	2	3	4
Biogenicity: identification of times and places of past biotic activity				3	
Evolutionary markers: data on biosphere, type of life, photosynthesis			2		
Paleoenvironments, Paleoecology				3	
Paleogeography: ancient shorelines				3	
Paleocurrents: circulation patterns					4
Paleotide gauges	0				
Paleolatitude: phototropism, polar shift	0				
Paleoclimatology: carbonate lithotopes in tropical and sub-tropical zones			2		
Correlation: physical facies equivalence					4
Chronostratigraphy: biostratigraphy				3	
Rhythmometry:astronomic,geologic, biologic and climate cycles			2		
Geochronometry: geologic calendar based on rhythmometry		1			
Lunar orbital evolution: evolution in the tidal ranges	0				
Sediment gauges: rates of sediment accumulation		1			
Geopetal use					4
Fossil search: prospecting ground microfossils			2		
Economics: prospects for economically interesting materials		Variable			
Tourist attraction: scientific, aesthetic, recreational, economic		Variable			

characterized by swarms of dissolution seams (Fig. 2.4.2.1), fitted fabric (framework of interpenetrant grains) and stylolites (Figs. 2.4.2.2, 2.4.2.3 and 2.4.2.4) which are horizontal to sub-horizontal to the primary depositional surfaces. The

Fig. 2.4.2.1: Diagenetically enhanced bedding. Quarry face, Cefn Mawr Limestone, Brigantian, Dinantian, Carboniferous. Graig Quarry, Llanarmon-yn-Ial, North Wales, U.K. Length of the bar is 50 cm. (Reproduced from R. G. C. Bathurst: Pressure-dissolution and limestone bedding; Fig. 1A, page 451, in, Cycles and Events in Stratigraphy: G. Einsele, Werner Ricken and Adolf Seilacher, eds., © 1991, Springer-Verlag GmbH & Co. KG., with the kind permission).

dissolution seams are found to be mainly parallel to the primary depositional surfaces with few exceptions, and can be laterally traced for hundreds of kilometers (Fig. 2.4.2.2) (Bathurst, 1987, 1991; Ricken, 1986). These secondary surfaces, being parallel to the primary depositional surfaces, are very difficult to identify in the field, unless enhanced by weathering. Many horizons previously regarded, in the field, as shale or marls derived from primary non-carbonate mud, are now known, through petrographic study, to be only concentrations of dissolution seams in limestone (Fig. 2.4.2.5). Pressure-dissolution appears to have played a significant role in the development of pseudo-bedding in limestone (Bathurst, 1991).

Again, dolomitization may destroy the primary bedding and accentuates secondary planar surfaces that are likely to be mistaken as primary bedding. While penecontemporaneous dolomitization can be parallel/sub-parallel to bedding, late dolomitization may obliterate primary sedimentary structures and crosscut primary bedding (Bathurst, 1991). It is possible that some primary depositional surfaces of limestone beds are sharpened by the development of dissolution seams and fitted fabric (Fig. 2.4.2.4), in many other centimeter-meter scale layers of limestone (otherwise lithologically and structurally

Fig. 2.4.2.2: A concentration of dissolution seams which has been followed for 20 km using gamma ray logs. Leasingham core, Lincolnshire Limestone, Bajocian, Jurassic, Lincolnshire, England. Length of the bar is 20 cm. (Reproduced from R. G. C. Bathurst: Diagenetically enhanced bedding in argillaceous platform limestones: stratified cementation and selective compaction. Sedimentology, Vol. 34, Fig. 23, p. 767, © 1987, Blackwell Science, U.K. with the kind permission).

homogeneous) represent pseudo-beds. Wanless (1979) sounds prophetic when he observed "a major challenge of the future will be to re-examine our assumptions about the primary character of limestone units and sequences". During the last decade, considerable work has been done on this topic (Bathurst

Fig. 2.4.2.3: Stylolites as bedding planes in coccolithic chalk, Upper Chalk, Upper Cretaceous, Flan Borough Head, Humberside (Old Yorkshire), England, overlooking the North Sea. Length of the bar is 50 cm. (Reproduced from R. G. C. Bathurst: Pressure-dissolution and limestone bedding, Fig. 2, p. 452, in, Cycles and Events in Stratigraphy: G. Einsele, Werner Ricken and Adolf Seilacher, eds., © 1991 Springer-Verlag GmbH & Co. KG, with the kind permission).

1987, 1991; Eder 1982; Ricken 1986; Ricken et al., 1982; Ricken and Eder, 1991; Simpson, 1985). Following are the features that help in distinguishing primary beds from that of pseudo-beds:

Field Characteristics

a. Primary bedding traces on the vertical faces of limestone in a quarry or natural exposure are usually faint without any weathered recessive grooves unlike the pseudo-bedding surfaces. Such faint traces of bedding surfaces always occur between a pair of prominent pseudo-bedding surfaces and always demarcate changes in lithology and/or primary sedimentary structures. On tracing such surfaces laterally, one may find them nearer to small stylolites or dissolution seams.
b. Prominent weathered surfaces showing shaly recessive layers or just a perceptible groove in the vertical quarry face usually define pseudo-beds in limestone. Such surfaces, when examined carefully, are found to be undulose to planar, and are usually regularly spaced on vertical section that mimick primary beds. Such surfaces do neither define lithological change nor change in primary sedimentary structures across the boundary. Again, when such

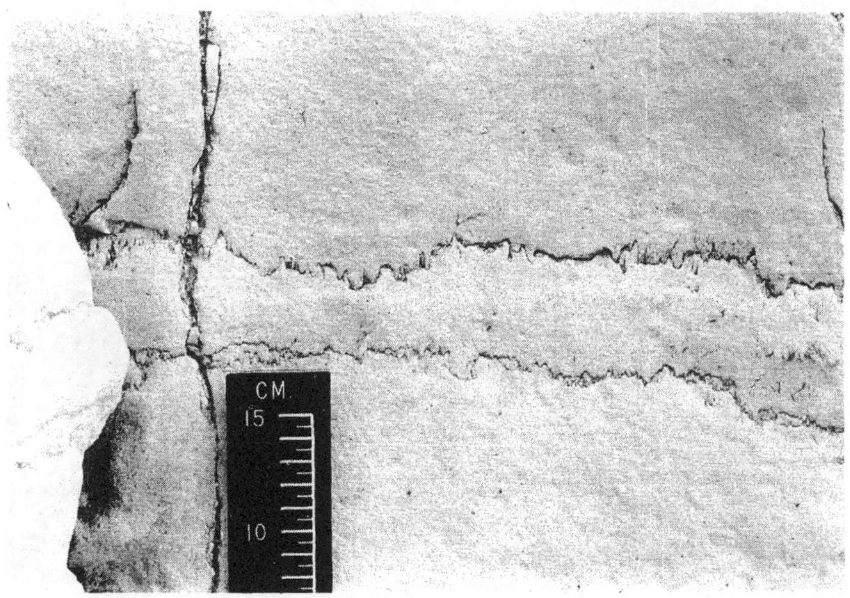

Fig. 2.4.2.4: Two of the bedding planes defined by stylolites as in Fig. 2.4.2.3. Details as in Fig.2.5.2.3. (Reproduced from R. G. C. Bathurst: Pressure-dissolution and limestone bedding, Fig. 2, p.452, in, Cycles and Events in Stratigraphy: G. Einsele, Werner Ricken and Adolf Seilacher, eds, © 1991, Springer-Verlag GmbH & Co. KG, with the kind permission).

surfaces are traced laterally, one will invariably find them passing into swarms of dissolution seams and/or stylolites (Bathurst, 1987; Simpson, 1985). The above features cumulatively suggest that these weathered and recessed surfaces do not represent primary depositional surfaces but clearly are diagenetic. This view is further strengthened from the study of thin-sections and peels of samples collected across the planes revealing several millimeters of insoluble residue (a product of pressure-dissolution) (Bathurst, 1987; Simpson, 1985).

c. Difference in faunal association in primary beds may also be used as a criterion, but caution is necessary here. Some of the fauna might have been obliterated earlier through dissolution; siliceous organisms are known to be notoriously susceptible to dissolution (Einsele, 1982).
d. Bioturbation texture is also a very good criterion in this respect, if, of course, rhythmic sequence is not totally homogenized through bioturbation. Burrows in one bed may be subsequently filled by sediment of other beds. Change in the nature and intensity of burrowing may also help in this respect (Einsele, 1982).
e. Change in primary sedimentary structures is one of the best evidences in distinguishing primary beds from the diagenetic one.

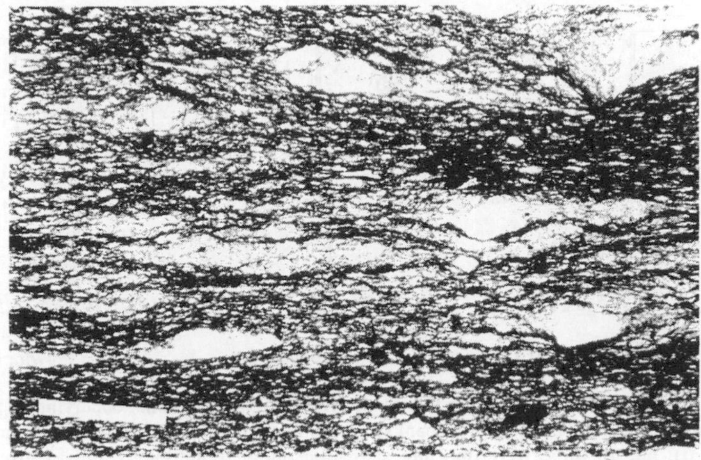

Fig. 2.4.2.5: Thin-section of a wackestone in a fissile limestone showing dissolution seams. Concentration of such dissolution seams marks the weathered-back zones that separate the beds in Fig. 2.5.2.1. Details as in Fig. 2.4.2.1. Length of the bar is 500 m. (Reproduced from R. G. C. Bathurst: Diagenetically enhanced bedding in argillaceous platform limestones: stratified cementation and selective compaction. Sedimentology, Vol. 34, Fig. 8, p. 751, © 1987, Blackwell Science, U.K. with the kind permission).

2.5 Trace Fossils

Ichnology (derived from the Greek word *ichnos* meaning a track or trace) deals with the traces of animals (invertebrate and vertebrate) and plants, made within the soft or consolidated sedimentary rocks, their systematic description, classification, and interpretation. It constitutes a field of study that encompasses the disciplines of stratigraphy, paleontology and sedimentology (Frey, 1973).

Since these traces are inseparable from the rock, unlike body fossils, trace fossils have to be studied in the field. Geologists are interested in the study of trace fossils as they provide information on depositional environment, paleoecology, facies pattern, substrate consistency, rate of sedimentation, storm activities, identification of erosional surfaces, and Precambrian-Cambrian correlation—a problem which still eludes geologists.

It must be clearly understood that the trace makers cannot be identified from the traces they create in soft sediments or hard rocks, unless body fossils are found alongside; it has been convincingly demonstrated that (i) a simple organism can produce traces that vary a great deal in morphology signaling more than one behavioral trait (ii) similar traces can be made by different organisms (Fig.2.5.1). Nevertheless such heterogenous types can be useful indicators as long as they express a similar response to the same environmental conditions (Seilacher, 1967). Hence, trace fossils have a good potential in sup-

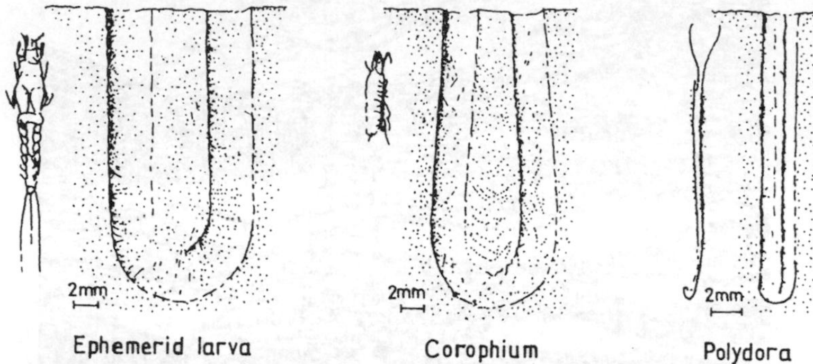

Fig. 2.5.1: Similar looking spreit burrows made by different organisms occurring in a variety of environments (Ephemerid larvae in river banks; Corophium volutator and Polydora ciliata in intertidal environment), but all three were dug into semi-consolidated sediments and exposed to highly turbulent water (Reproduced from Adolf Seilacher: Bathymetry of trace fossils, Marine Geology, vol. 5, p. 413-428, Fig. 1, © 1967, Elsevier Science - NL, Sara Burgerhartstraat 25, 1055 KV Amsterdam, The Netherlands, with the kind permission).

plementing other lines of environmental evidences. Trace fossils represent behavioral response of organisms, and are controlled by the environment, substrate consistency, rates of sedimentation, temperature, and salinity. Extremely few kinds of trace fossils can be linked with particular genera or species of organisms. Names of genera or species of trace fossils, which have little to do with discrete genera or species of organisms, are binomial, based totally upon morphology of the respective biogenic structures (Frey and Seilacher, 1980). While all fossil and recent traces are biogenic structures, the reverse is not true. Stromatolite, despite being a biogenic sedimentary structure, is not a trace fossil.

This section, emphasizes how to study trace fossils in the field and what information they convey in unraveling the geological history. This information, amalgamated with other geological information, is a powerful tool in understanding past geological history.

Many terms and concepts are involved in the study of trace fossils; these should be clearly defined at the outset before further discussion on the topic and is given in Table 2.5.T.1.

Observational Methods and Field Techniques

A close observation is called for in the study of trace fossils in the field and their ultimate three-dimensional reconstruction. Farrow (1966), following the suggestions of Ager (1963), has suggested quadrat and/or line method in the study of trace fossils to avoid (i) attention to more spectacular trace fossils at the expense of less spectacular but geologically more significant ones, (ii) to

Table 2.5.T1: Basic concepts in the study of biogenic structures (Reproduced from R. W. Frey: Behavioral and ecological implications of trace fossils, Table 1, page 45, in, Paul B. Basan, ed., Trace Fossils Concepts, SEPM Short Course No. 5, © 1978, Society for Sedimentary Geology (SEPM), Tulsa, Oklahoma U.S.A. with kind permission)

A: Definition and differentiation of biogenic structures

Biogenic structure: in ichnology, tangible evidence of activity by an organism, fossil or recent, other than the production of body parts. Embraces the entire spectrum of substrate traces or structures that reflect a behavioral function: biogenic sedimentary structures, bio-erosion structures, and other miscellaneous features representing biological activity. Excludes molds of body fossils that result from passive contact between body parts and the host substrate, but not imprints made by the body parts of active organisms.

Biogenic sedimentary structure: biogenic structure produced by the activity of an organism upon or within an unconsolidated particulate (detrital) substrate: bioturbation structures and biostratification structures.

Bioturbation structure: biogenic sedimentary structure that reflects the disruption of biogenic and physical stratification features or sediment fabrics by the activity of organisms: tracks, trails, burrows, feeding traces, and similar structures.

Biostratification structure: biogenic sedimentary structure consisting of stratification features imparted by the activity of an organism: biogenic graded bedding, certain stromatolite, and others.

Bioerosion structure: biogenic structure sculpted mechanically or bio-chemically by an organism into a rigid substrate: boring, gnawing, scraping, biting, and related traces.

B: Some fundamental terms

Bioturbation: reworking of sediments by an organism.

Bioerosion: degradation of consolidated sediments or other rigid substrate by an organism.

Ethology: in ichnology, the study or interpretation of the behavior of organisms as reflected by their traces.

Trace: in ichnology, an individually distinctive biogenic structure, fossil or recent, especially one that is related more or less directly to the morphologic part(s) of the organism that made it: tracks, trails, burrows, boring, gnawings, coprolite, fecal castings, and similar features. Excludes bio-stratification structures and other features lacking diagnostic anatomical characteristics.

Trace fossil: a fossil trace (= ichnofossil).

Lebensspur: synonymous with trace (plural = lebensspuren).

help in the study of mutual relations between various trace fossils, and (iii) comparison between different trace fossil populations and their relative numerical abundance. Quadrat method is more suitable for study on bedding planes and line method for vertical sections of rocks. The size and shape of a quadrat can be 1 square meter for higher concentration of trace fossils; larger quadrat is desirable for sparser assemblages and smaller ones for exceptionally higher concentration.

In practice, "the most convenient material form for a quadrat is probably just four skewers or spikes with cords between" for the study of trace fossils in recent sediments and "a folding wooden frame is more usable on hard rock surfaces" (Ager, 1963). "The folding wooden 1m quadrat in conjunction with a frame of 1m by 10cm, having the frame marked-off at 10cm intervals along the longer sides may be used. A line of standard length is stretched along a section parallel to the bedding planes, and every specimen which it crosses is counted" (Ager, 1963). The line transect method can be fruitfully utilized when trace fossils are abundant; it looses its practicality when the concentration decreases below a certain density (Ager, 1963). A net of 20cm mesh can be hung over the vertical rock face for the purpose (Wurster, 1958, quoted in Farrow, 1966).

Peeling can be done when trace fossils are studied in modern sediments for better results. Bouma (1969) has described different peeling and casting techniques.

Photography and sketching are two important field procedures for the study of trace fossils. The photographic style is slightly different from normal photography; strong side light is necessary to bring out the relief of the structures.

Where trace fossils are not very prominent on the exposure surfaces, one has to adopt certain indigenous techniques in order to highlight them and following are the tips:

i) Wetting of the burrowed faces with water,
ii) Smearing of ink over and then washing it off from the face,
iii) Whitening of the face is sometime necessary to bring out the finer details, i.e., claw scratches, and photographing the face with strong side light,
iv) Outlining of inconspicuous traces with the help of felt-tipped pen,
v) When photography does not produce the desired result, it is better to take out latex peels of critical areas (see, Bouma 1969, p. 42-43 for various techniques and material required) to bring out the finer details of the structures and photograph the latex peel in the laboratory with strong side light.
vi) Application of organic stains (Alizarin Red S, Methylene blue) may help in bringing out the finer details of the structure. These dyes are preferentially absorbed by clay mineral and highlight them. This is best employed in fine-grained sediments (since they are rich in clay minerals) than in coarse-grained one.
vii) Spraying carbonate-cemented rocks with a solution of dilute HCl, or siliceous rocks with KOH may enhance the relief of trace fossils facilitating sketching and photography.
viii) If the surface of exposure is more or less smooth, it is better to prepare acetate peels on the etched surfaces in the same way as practiced in the laboratory. The most inexpensive method is to dissolve discarded celluloid film (collected from different photographic studios) and dissolve them in

acetone in such a way that a viscous solution is prepared. Various other laquers can be used for peeling (see, Bouma, 1969).

After adopting any of the techniques, as suited and outlined above, sketching of the trace fossils is the next important step. While sketching, it is important to measure the length, width and diameter of the trace fossils and their lateral and vertical variations, if any. Study of trace fossils, both on bedding plane and their vertical sections, are necessary for three-dimensional reconstruction. It is also important to note and sketch their branching pattern - whether simple straight tubes, simple but curved, branching or U-tube. If these are branching, note if these are regular/irregular, branching patterns and note also if the diameter remains constant or variable.

It is important to examine the burrow wall to find out if these are lined with mud or plastered by fecal pellets. Note also if the adjacent sedimentary laminae show any deflection and if so, note the direction of deflection. Study of the burrow-fill and their nature are also necessary. Are the burrows filled with coarser or finer sediment compared to the adjacent sediment? Are the burrow-fills made up of skeletal debris or fecal pellets? Note also the curvature of the burrow fill. Most are consistently concave in the direction the creator organisms move. Spreiten, representing the composite traces resulting from multiple feeding forays of the organisms, are usually associated with U-shaped burrows (Fig. 2.5.2).

While examining trails on bedding plane surfaces, it is important to study if these trails show regular or irregular pattern and whether these are straight, curved, coiled, meandering or radial. It is necessary to note also if these trails form continuous ridge/furrow, and if there is any central division and ornamentation, e.g., chevron pattern. If footprints are found, e.g., dinosaurs, it becomes necessary to measure their size and spacing of impressions to understand their gait. Of all these features photography is most essential for the purpose of documentation.

Classification of Trace Fossils

There are various classifications of trace fossils proposed by ichnologists. For descriptive purpose, toponomic classification of trace fossils proposed by Martinsson (1970) is useful and has gained popularity amongst ichnologists (Fig. 2.5.3). If the traces occur as concave grooves or convex ridges on the top surface of stratum, the term epichnia or epirelief structure is used. If the traces occur as concave groove or convex ridge on the underside of the stratum, the term hypichnia or hyporelief structure is used. Endichnia is a term used for full relief structure occurring within a stratum when the burrow is filled-in with the same material as surrounding, and exichnia, (also a full relief structure),

Fig. 2.5.2: Zoophycos from Miocene marlstone heteropic with turbidites of Marnosoarenacea Formation, northern Appenines (Courtesy Dr. Ricci Luchi, Italy).

for those burrows that are filled-in with different material compared to their surrounding.

Classification of trace fossils, based on ethology/behavior of trace makers, has been proposed and is the direct outcome of elementary descriptive-genetic concepts (Table 2.5.T2). Amongst ethological classifications, the one suggested by Frey and Pemberton (1985) and given in Table 2.5.T4, and Fig. 2.5.4 are found to be more compact and elegant. This scheme of classification reduces the vast majority of traces to six major categories. The advantages of ethological classification are (a) emphases on similarities in animal behavior rather than on geometric or toponomic similarities of trace fossils, (b) morphology and function of traces remain constant despite evolutionary changes of trace makers from Cambrian to Recent (Frey and Seilacher, 1980).

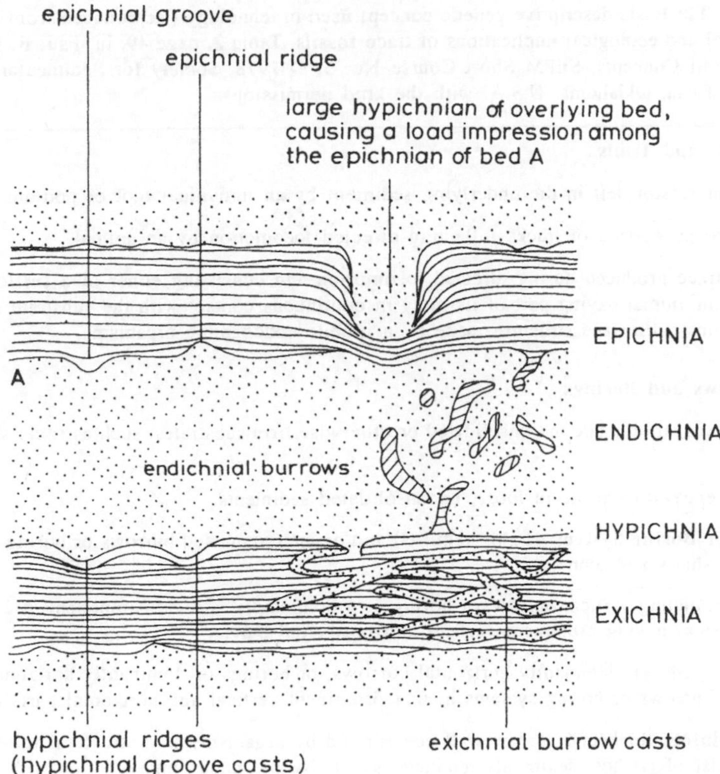

Fig. 2.5.3: Diagramatic representation of a toponomic terminology based on the main medium of preservation (the 'casting' medium). Siltstone beds (stippled) are seen in cross-section, embeded in shales (ruled). The four key terms to the right of the diagram refer to bed A. The endichnia and exichnia are drawn as mud-filled (or mineral filled) burrows and silt-filled burrow casts respectively (Reproduced from A. Martinsson: Toponomy of trace fossils, *in*, T. P. Crimes and J. C. Harper, eds., Trace Fossils, Special Issue no. 3, Fig. 1, © 1970, John Wiley & Sons Ltd. with the kind permission).

Interpretations of Trace Fossil Assemblages

Trace fossils, when studied carefully in the field and the laboratory, can provide information on various aspects of sedimentology, e.g., on nature and rate of sedimentation, water depth, salinity, energy level at the depositional site. The information when clubbed together and analyzed with other physical features of sedimentary rocks, help in understanding the depositional environment.

Rate of sedimentation: Dynamic response of sedimentary environment can be understood from the study of trace fossils. The following Table 2.5.T3 illustrates the relationships of bioturbation vis-a-vis deposition and erosion.

In this respect, two examples are cited to highlight how the nature of sedimentation or erosion can be deciphered from the pattern of trace fossil. First

Table 2.5.T2: Basic descriptive-genetic concept used in ichnology (Reproduced from R.W. Frey: Behavioral and ecological implications of trace fossils, Table 2, page 49, in, Paul B. Basan, ed., Trace Fossil Concepts, SEPM Short Course No. 5, © 1978, Society for Sedimentary Geology (SEPM), Tulsa, Oklahoma, U.S.A. with the kind permission)

A. Tracks and Trails

Track: impression left in the underlying sediment by an individual foot or podium.

Trackway: succession of tracks reflecting directed locomotion by an animal.

Trail: a trace produced during directed locomotion and consisting either of a surficial groove made by an animal having part of its body in continuous contact with the substrate surface, or a continuous subsurface structure made by a mobile endobenthic organism.

B. Burrows and Borings

Boring: excavation made in consolidated or otherwise firm substrates, such as rock, shell, bone, or wood.

Burrow: excavation made in loose, unconsolidated sediments.

Burrow or Boring System: Highly ramified and/or interconnected burrows or borings, typically involving shafts and tunnels.

Shaft: dominating vertical burrow, or boring or a dominantly vertical component of a burrow or boring system having prominent vertical and horizontal parts.

Tunnel (= gallery): dominantly horizontal burrows, or borings or dominantly horizontal component of a burrows or boring systems having prominent vertical and horizontal parts (=gallery).

Burrow lining: thickened burrow wall constructed by organisms as a structural reinforcement. May consist of (i) host sediments retained essentially by mucus impregnation, (ii) pelletoidal aggregates of sediment shoved into the wall, like mud-daubed chimneys, (iii) detrital particles selected and cemented like masonry, or (iv) leathery or felted tubes consisting mostly of chitinophosphatic secretions or organisms. Burrow linings of types (iii) and (iv) are commonly called “dwelling tubes”.

Burrow cast: sediments infilling a burrow (= burrow fill). Sediment fill may be either “active”, if done by animals, or “passive”, if done by gravity or physical processes. Active fill is termed “back-fill” wherever U-in-U laminae, etc., show that the animal packed sediment behind itself as it moved through the substrate.

C: Bioturbation

Bioturbate texture: gross texture or fabric imparted to sediments by extensive bioturbation. Typically consists of dense, contorted, truncated or interpenetrating burrows or other traces, few of which remain distinct morphologically. Where burrows are somewhat less crowded and thus are more distinct individually, the sediment is said to be “burrow mottled”.

D: Miscellaneous

Configuration: in ichnology, the spatial relationships of traces, including the disposition of component parts and their orientation with respect to bedding and/or azimuth.

Spreite: blade-like to sinuous, U-shaped, or spiralled structure (plural = spreiten) consisting of sets or cosets of closely juxtaposed, repetitious, parallel or concentric feeding or dwelling burrows or grazing traces. Individual burrows or grooves comprising the spreit commonly anastomose into a single trunk or stem (as in *Daedalus*) or are strung between peripheral "support" stems (as in *Rhizocorallium*). "Retrusive" spreiten are extended upward, or proximal to the initial point of entry by the animal, and protrusive spreiten are extended downward, or distal to the point of entry.

Table 2.5.T3: General relationships between bioturbation and rates of deposition and erosion (Reproduced from James D. Howard: Sedimentology of trace fossils, Fig. 5, page 22, in, Paul B. Basan, ed., Trace Fossil Concepts, SEPM Short Course No. 5, © 1978, Society for Sedimentary Geology (SEPM), Tulsa, Oklahoma, U.S.A. with kind permission).

A. Continuous Deposition

1. Slow rate of deposition—complete biogenic reworking. Examples: protected parts of estuaries; offshore areas below storm wave-base; the deep sea.

2. Fast rate of deposition—sparse burrows; escape structures; little biogenic reworking except near top of beds. Examples: turbidite, splays, laterally accreting bars, spits and beaches; delta foresets.

B. Discontinuous Deposition

1. Without erosion
   - i. Slow rate of deposition—very similar to Part A1.
   - ii. Fast rate of deposition—repetitive stacking of beds such as those in Part A2; thick beds separated by very thin layers of finer-grained sediments representing settlement from suspension.

2. With erosion
   - i. Slow rate of deposition—similar to Part A1; but sequence is punctuated by scour horizons and burrow truncations.
   - ii. Fast rate of deposition—similar to Part B1ii; bedding is replete with scour horizon, burrow truncations and, possibly, lag concentrates; individual bedding units may be graded; burrows very sparse near base, becoming somewhat common near top.

---

example is taken from Goldring (1964). The summary diagram from that paper is sufficient to explain points indicated in Table 2.5.T3. In the illustrations (Fig. 2.5.5 A-G), heights of solid arrows show amount of erosion/deposition.

Fig. 2.5.5 A depicts the movement pattern of pelecypod Mya that has a single siphon. With stationary sedimentary surface (1), growing organism gradually burrows deeper; bottom of structure is wider than top. With rapid sedimentation (2), organism migrates towards surface, leaving behind in-filled burrow the width of the shell. With degradation of the surface (3), organism migrates downward, leaving the burrow of same width but having different internal structure.

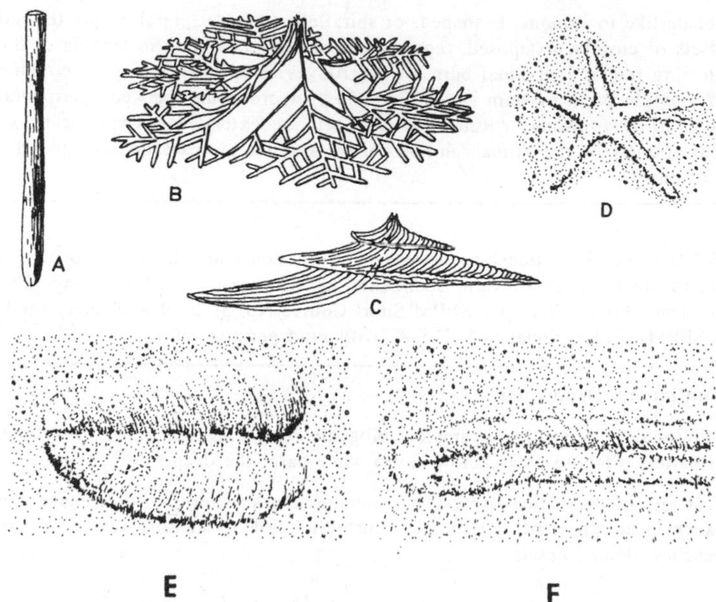

Fig. 2.5.4: Examples of behaviour represented by trace fossils. A, dwelling structure; B, feeding structure chondrites; C, grazing trace zoophycos; D, resting trace Asteriacites; E, resting trace Rusophycus; F, crawling trace Cruziana. Not to scale (Reproduced from R. W. Frey: The realm of ichnology, its strengths and limitations, *in*, R. W. Frey, ed., The Study of Trace Fossils, Fig. 2.4, page 23, © 1975, Springer-Verlag GmbH & Co. KG., with the kind permission).

Fig.2.5.5B depicts movement pattern of polychaete worm Nereis. Older colonized surface (1) is rapidly covered by sediment (2), and during deposition, paths of escape are directed upward. With stabilization, new colonization surface (3) has irregular "normal" burrows. Structures in (1) and (3) are generally mucus lined, whereas in (2) they are unlined.

Fig. 2.5.5 C shows movement pattern of anemone Cerianthus, an organism dwelling in a single tube. With sedimentation, animals move upward, leaving an unfilled burrow. A similar pattern may be expected in traces such as Skolithos and Monocraterion.

Fig. 2.5.5 D depicts movement represented by trace fossil Asteriacites lumbricalis, resting place of a stelleroid. With sedimentation, animals migrated upwards in stages a-c; combined (1) and separate (2) plan of all impressions.

Fig. 2.5.5 E depicts preservation patterns of trace fossil Chondrites. Tunnel system (1) is in-filled (2), following a change in type of sediment being deposited (bed-junction preservation). Slight degradation of surface (3) removes the proximal shafts before further sediments of different types accumulate. Renewed degradation of surface winnows away sediment, leaving mucus-lined infilled tunnels as burial preservation (4).

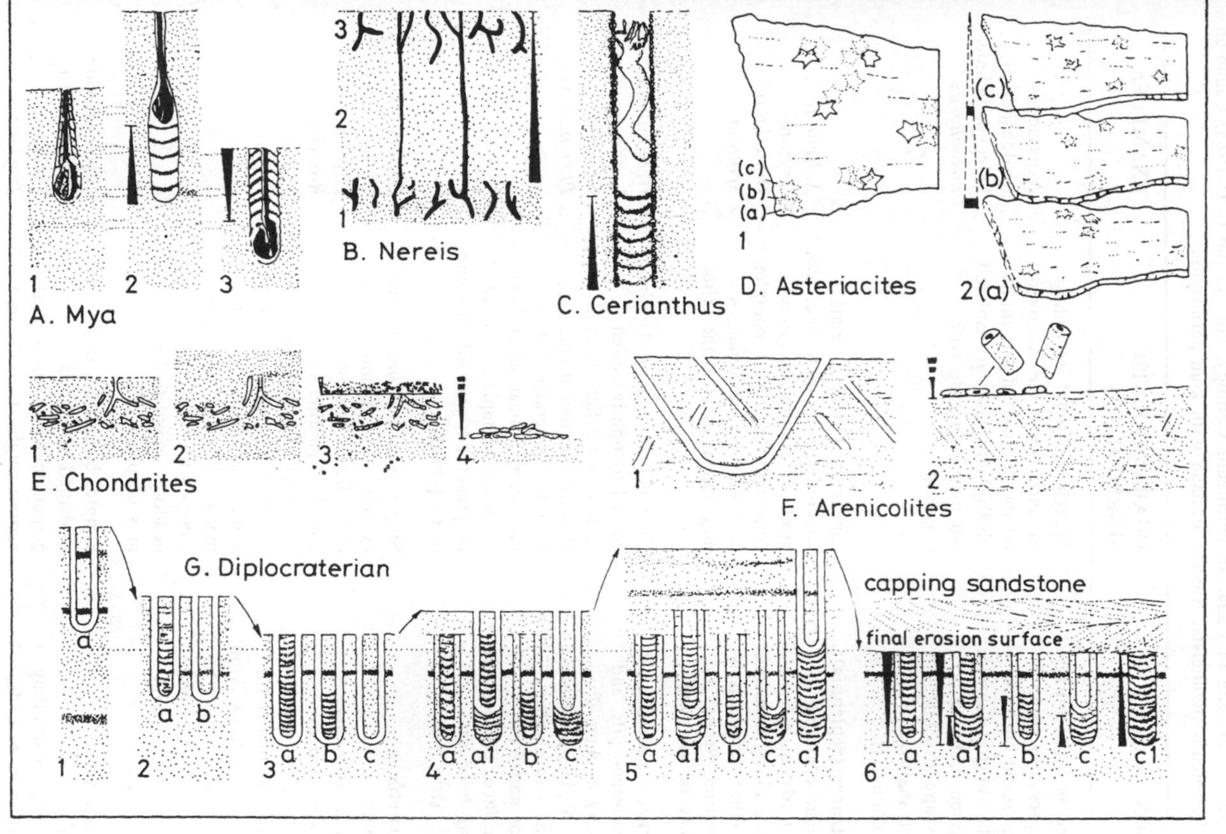

Fig. 2.5.5: Response of organisms to erosion and sedimentation. Detail explanation is given in text (Reproduced from R. Goldring : Trace fossils and the sedimentary surface in shallow marine sediments, in, L. M. J. U. Van Straaten, ed., Deltaic and Shallow Marine Deposits. Development in Sedimentology 1, Fig. 1, © 1964, Elsevier Science-NL, Sara Burgerhartstraaten 25, 1055 KV Amsterdam, The Netherlands , with the kind permission).

Table 2.5.T4: Ethological classification of invertebrate trace fossils (Reproduced from R.W. Frey and S. George Pemberton: Biogenic structures in outcrops and cores. 1. Approaches to ichnology, Table 5, page 83, Bulletin of Canadian Petroleum Geology, vol. 33, no.1, © 1985, Canadian Society of Petroleum Geologists, Alberta, Canada with kind permission)

DEFINITION	CHARACTERISTIC MORPHOLOGY	EXAMPLES
Resting Traces (cubichnia): Shallow depressions made by animals that settle on to or dig in to the substrate surface. Emphasis is on reclusion, traces or escape structures. May include shallow, ephemeral domiciles,	Trough-like relief, recording to some extent the lateroventral morphology of the animal. Ideally, structures are isolated, but may intergrade with crawling	Asteriacites Medousichnus Pelecypodichnus *Rusophycus*
Crawling traces (repichnia): Trackways and epistratal or interstratal trails made by organisms travelling from one place to another. Emphasis is upon locomotion. Secondary activities may be involved.	Linear or sinuous over-all structures, some branched; foot prints or continuous grooves, commonly annulated, complete form may be preserved, or may appear as cleavage relief.	*Aulichnites* *Cruziana* *Diplichnites* *Scolicia*
Grazing traces (pascichnia): Grooved, patterned pits, and furrows, many of them discontinuous, made by mobile deposit feeders or algal grazers at or under the substrate surface. Emphasis is upon feeding behavior analogue to 'strip mining'.	Unbranched, non-overlapping, curved to tightly coiled patterns or delicately constructed spreiten dominate; patterns generally reflect maximum utilization of food resources; complete form may be preserved; over-all structure to be planar.	*Helminthoids* Lophoctenium Nereites *Phycosiphon*
Feeding structures (fodinichnia): More or less temporary burrows constructed by deposit feeders; the structures may also provide shelter for the organisms. Emphasis is on feeding behavior analogus to 'underground mining'. Some tend to be gradational with dwelling structures.	Single branched, unbranched, cylindrical to sinuous shafts or U-shaped burrows, or complex, parallel to concentric burrow repetitions (spreiten structures); walls not commonly lined, unless by mucus. Oriented at various angles with respect to bedding; complete form may be preserved.	Chondrites Gyrophyllites Phycodes *Rosselia*
Dwelling structures (domichnia): Burrows, boring or dwelling tubes providing more or less permanent domiciles, mostly for hemi-sessiles, suspension feeders or in some cases,	Simple, bifurcated, or U-shaped structures perpendicular or inclined at various angles to bedding, or branched burrow or boring systems having vertical or horizontal components;	*Diplocraterion* Ophiomorpha Skolithos *Trypanites*

contd.

carnivores. Emphasis is on habitation. Secondary activities may be discernible.	burrow walls typically lined. Complete form may be preserved.	
Escape structures (Fugichnia): Lebensspuren of various kinds modified or made anew by animals in direct response to substrate degradation or aggradation. Emphasis is upon readjustment, or equilibrium between relative substrate position and the configuration of contained traces. Intergradational with other behavioral categories.	Vertically repetitive resting traces; biogenic laminae either en echelon or as nested funnels or chevrons; U-in-U spreiten burrows; other structures reflecting displacement of animals upward or downward with respect to original substrate surface. Complete form may be preserved, especially in aggraded surface.	Nested funnels, U-in-spreiten, Down-warped laminae

Fig. 2.5.5 F depicts preservation pattern of trace fossil Arenicolites curvatus. Sedimentary surface containing U-tubes (1) has been degraded (2), leaving behind mucus-lined tube fragments as an intraformational conglomerate; sediment has filled the tubes.

Fig. 2.5.5 G depicts movement pattern of trace fossil Diplocraterion yoyo. All have been truncated to common erosion surface. Repeated phases of erosion and sedimentation (1–6) evidently led to development of the various types. (1) show development of burrow (a). With degradation of surface, this tube migrates downward, and at intervals, new tubes (b, c) are constructed (2, 3). Sedimentation follows (4,5), but some tubes are abandoned, and erosion reduces them to a common base (6).

Second example is from Pleistocene beach sediments along the Georgia-Florida coastal plain documenting how relative rates of sedimentation and erosion can be deciphered from burrow structures (Fig. 2.5.6) (Howard 1978).

Fig. 2.5.6 A represents the upper part of a resistant thick-walled burrow in which the near-surface portion of the burrow is constricted, delicate, thin tube terminating at the surface. With renewed sedimentation the response of the organism is to first lengthen its constricted portion of the burrow to the new surface (Fig. 2.5.6 B). Subsequently, the organism extends the larger part (Fig. 2.5.6 C). When erosion occurs, the constricted part of the burrow and to some extent, the permanent burrow may be removed, exposing the permanent portion of the burrow to the substrate (Fig. 2.5.6 D). When the permanent portion of the burrow is exposed to the substrate, the upper part will be closed-up by the occupant and a "bypass branch" will be constructed to the substrate (Fig. 2.5.6 E). With renewed sedimentation, the previous closed part of the burrow will be reactivated while the bypass structure will be closed-off and abandoned.

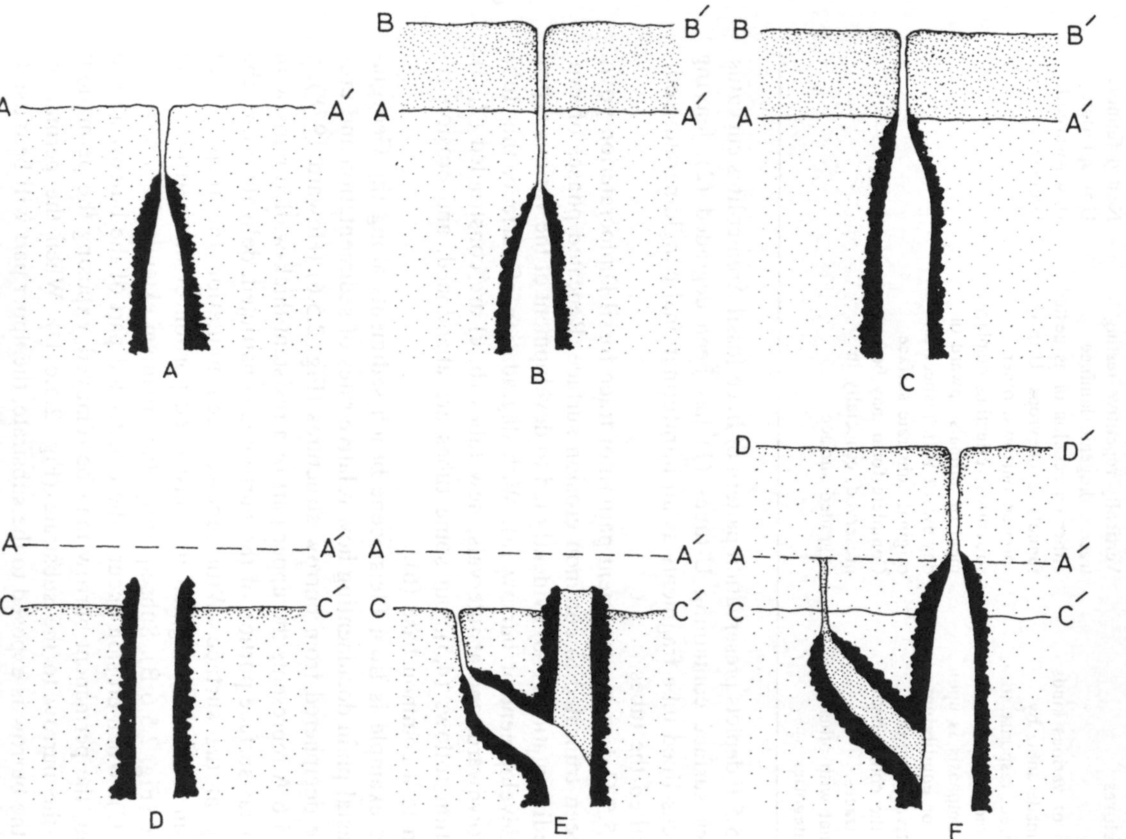

Fig. 2.5.6: Response of Callianasa burrows to erosion and deposition. Details are in the text (Reproduced from J. D. Howard: Sedimentology and trace fossils, *in*, Paul B. Basan, ed., Trace Fossil Concepts, SEPM Short Course No. 5, Fig. 7, © 1978, Society of Sedimentary Geology (SEPM), Tulsa, Oklahoma, U.S.A. with the kind permission).

Frequent occurrence of such 'bypass structures' and their subsequent abandonment indicates rapid progradation of a beach with each termination surface indicating the former position of a beach surface. The example is taken from a Pleistocene outcrop at the Florida-Georgia state-line where several branches of Ophiomorpha burrows are found to terminate at progressively younger bedding planes (Fig. 2.5.7). Furthermore, the presence or absence of the constricted part of the burrow may be interpreted as normal or erosional surface respectively. Preservation of upper constricted part of burrow is rare in rock record. Rather the termination of a thick-walled main burrow against a particular surface is more common. Tracing out of such horizons, where main burrows abut, helps in identification of an otherwise obscure erosional surface (Fig. 2.5.8).

Understanding Nature of Substrate

Construction of deep and strong burrow-walls, fortified by mucus/pellets, usually indicates instability of the substrate caused by the constant shifting and shuffling of particulate material. Again, when the walls grade from thick, peletoidal structures towards the top to thinner, smooth structures further down within a single burrow system are indications of increasing sediment coherence or substrate stability at depth (Frey 1978).

Rate of lithification of the substrate can also be understood from the study of trace fossils. Interpenetrating burrows, abundant deep feeding burrows and burrows that skirt shells and other obstructions in the substrate indicate that the sediments were still soft. But when boundaries of trace fossils are sharp, abrupt and cut through hard individual grains, these are identified as borers and at the same time signal early diagenesis and lithification of host sediments. Furthermore, euxinic or anaerobic conditions of the depositional basins can be gauged by the presence or absence of trace fossils respectively (Seilacher 1964; Rhoads 1975).

In order to withstand water turbulence, water temperature, and salinity, organisms construct deep vertical burrows. Marine organisms, inhabiting areas of high water turbulence, are mostly suspension feeders and many construct deep vertical stable burrows to overcome the mobility of the substrate. Again, shallow marine areas are the places where fluctuations in salinity and temperatures are high and to ward off those effects, organisms construct deep vertical burrows, walls of which are fortified by mucus and fecal pellets.

Trace fossils as Geopetals

Like many physical sedimentary structures, trace fossils can also be used as a way-up indicator. Fig. 2.5.9 shows some trace fossils that can be used as geopetals (Crime, 1975).

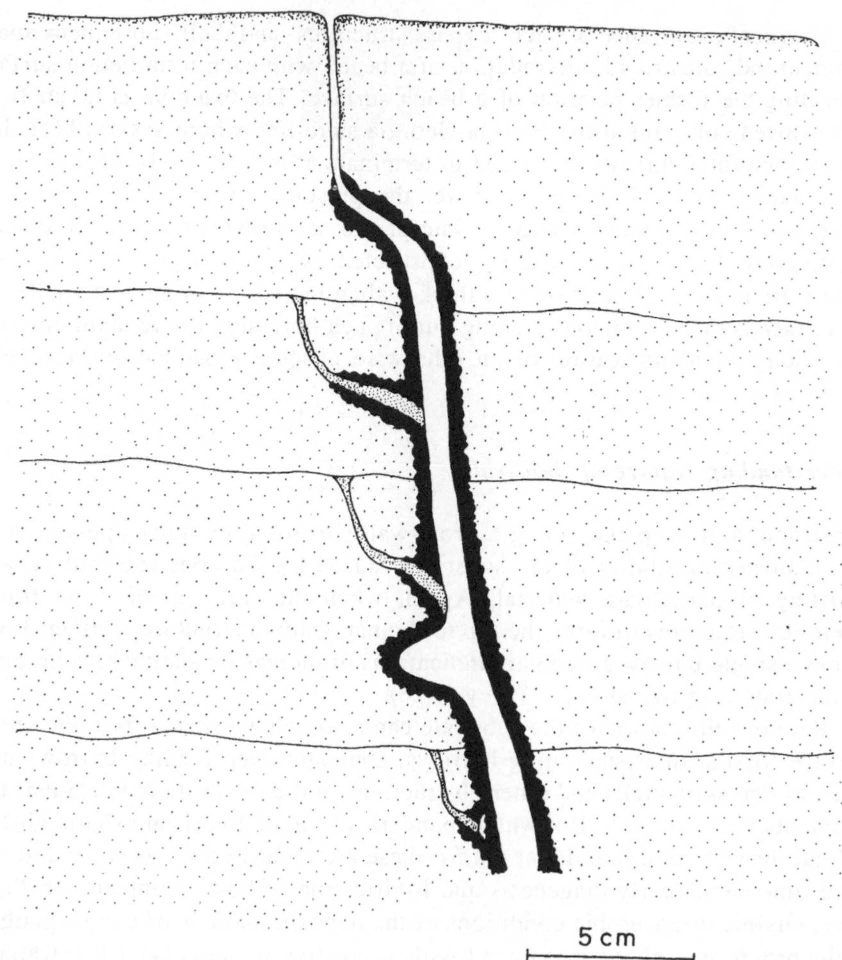

Fig.2.5.7: Development of the burrow Ophiomorpha in response to conditions of rapid deposition (Reproduced from J. D. Howard: Sedimentology of trace fossils, in, Paul B. Basan, ed., Trace Fossil Concepts, SEPM Short Course No. 5, Fig. 8, © 1978, Society for Sedimentary Geology (SEPM), Tulsa, Oklahoma, U.S.A. with the kind permission).

Amongst burrows, the U-burrow, hanging down from the top surface, is the easiest amongst way-up indicators; the surface against which the limbs of the U-shaped burrows terminate indicates the top (Fig. 2.5.9 A). The spreiten (composite traces resulting from multiple activities of the organisms, e.g., feeding forays) that characterize the U-shaped burrows can also be used as way-up indicators, since by origin they are concave-upward backfill-laminae (Fig. 2.5.9 B, C). Some burrowers (Monocraterion burrows) have the habit of deflecting surrounding sediment downward, producing a series of funnel-shaped structures stacked one above another (Fig. 2.5.9 D). The concavity of the laminae points

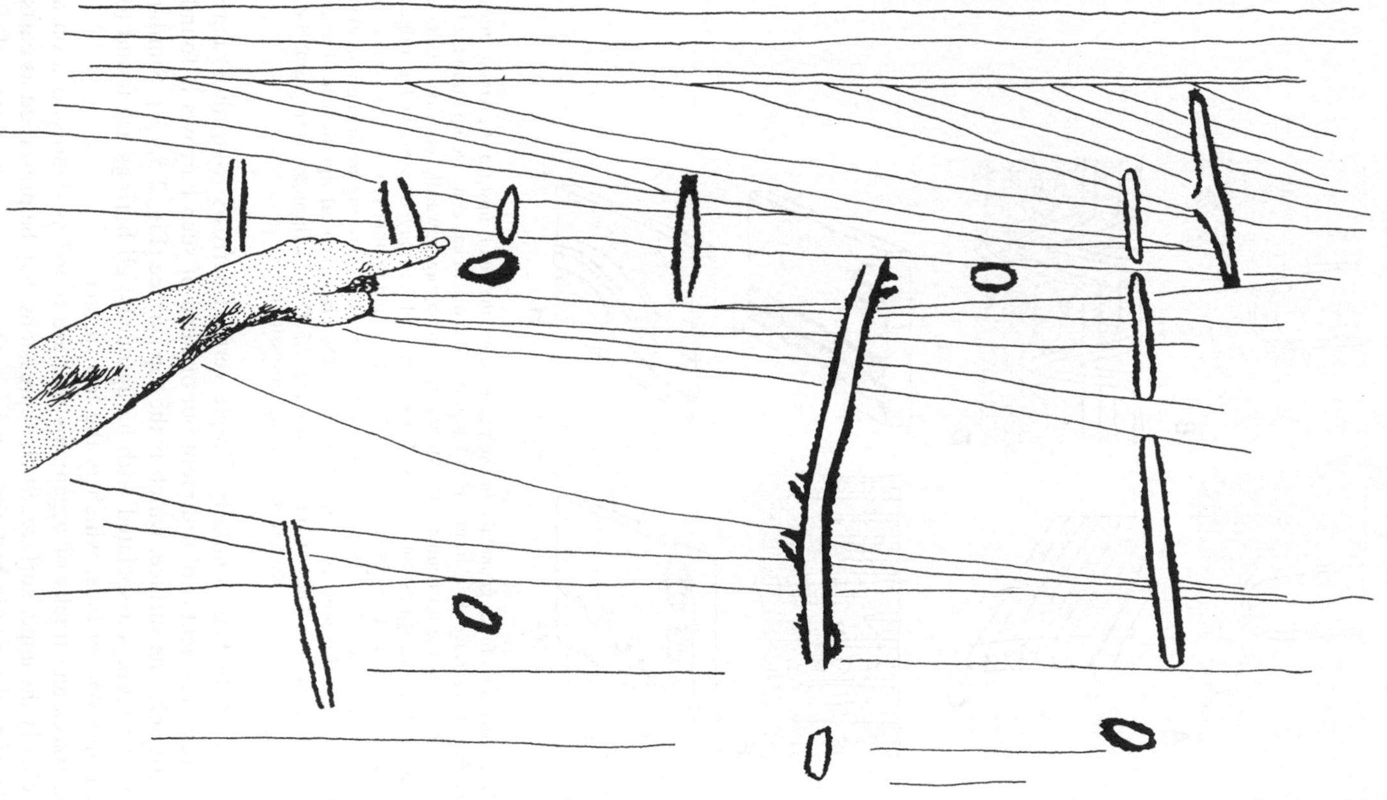

Fig.2.5.8: Truncated Ophiomorpha burrows which indicate position of erosional bedding plane surfaces (Reproduced from J. D. Howard: Sedimentology and trace fossils, in, Paul B. Basan, ed., Trace Fossil Concepts, SEPM Short Course No. 5, Fig. 9, 1978, Society for Sedimentary Geology (SEPM), Tulsa, Oklahoma, U.S.A. with the kind permission).

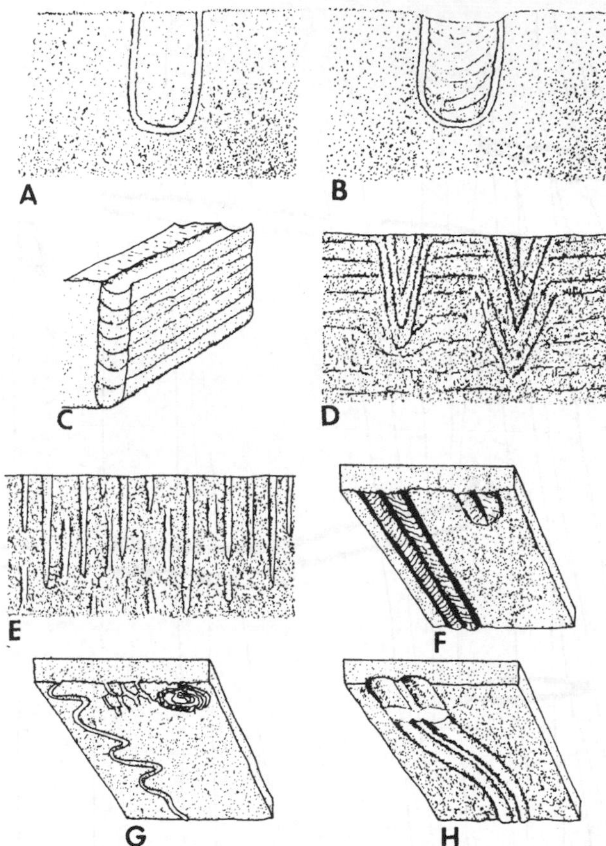

Fig. 2.5.9: Trace fossils useful as geopetals. In each case, top of the bed is upward. A, Arenicolites hanging down from top surface; B, Corophioides hangs down and also has concave-up laminae; C, Teichichnus also has concave-up laminae; D, Monocraterion having upwardly concave stacked funnels; E, Skolithos ending against top surface; F, Cruziana and Rusophycus preserved as ridges on lower surface of sandstone; G, Helminthopsis, Paleodictyon, and Spirophycus also preserved as ridges on lower surface of turbidites; H, Scolicia formed between beds showing confusingly similar upper and lower surfaces (Reproduced from T. P. Crimes: The stratigraphical significance of trace fossils, *in*, R. W. Frey, ed., The Study of Trace Fossils, Fig. 7.3, © 1975, Springer-Verlag GmbH & Co. KG , with the kind permission).

the top surface of the bed. Similarly, simple vertical burrows (Skolithos) may penetrate at different depths of sediment but one end of such burrows is found to end against only one surface, which is the top surface (Fig. 2.5.9 E). Similar is the case with borings associated with hardgrounds; all borings are found to end against only one surface, which is the top surface.

Again, traces and tracks of organisms that are found as depressions on a surface is clearly the upper surface. Such depressions may be preserved as casts on the sole of a siliciclastic bed and can be used as a way-up indicator (Fig.

2.5.9 F and G). Some traces, however, formed between beds may have confusingly similar upper and lower surfaces (gastropod trace Scolicia, Fig. 2.5.9 H), and naturally such traces do not have any geopetal significance.

Ichnofacies and Paleoenvironment

The most significant contribution that has emerged from the studies of ichnofossils for more than last three decades is the concept of recurring ichnofacies. It was Seilacher (1964, 1967) who, after careful observations of behavioural activities and association of numerous ichnofossils from the rock record and their stratigraphic distributions, observed that a relatively small number of distinct communities of trace fossils recur vertically through space and time from the Cambrian through Recent and indicated their implications in paleoenvironmental interpretations. Each is limited to a rather narrow facies range, although they may occur in sandstone as well as shale or limestone (Seilacher, 1967). It is for these distinct 'prototype' assemblages of trace fossils that the term ichnofacies is used. This means that a particular ichnofacies recurs in space and time provided the requisite set of environmental parameters recur.

While the initial ichnofossil model proposed by Seilacher (1967) had an undertone of relationships with water depth (Fig. 2.5.10), ichnologists now realize that the various associations of trace fossils (ichnofacies) are controlled by many other factors besides bathymetry alone, e.g., sediment grain size, substrate character and consistency, energy of the depositional environment, salinity of water, oxygen concentration of bottom water and interstitial water, availability of food. The association between the physical sedimentary structures and biogenic structures form a powerful combination in understanding environmental gradients (Frey et al., 1990). Occasionally one may find anomalous associations of ichnofacies vis-a-vis environment. For example, Cruziana or Skolithos ichnofacies may appear outside the zone of defined original paradigm (Table 2.5.T5). This deviation from idealized ichnofacies association does not indicate the weakness of the ichnofacies study, but most likely indicates departure from 'normal' environmental situation (Frey et al., 1990).

Seven ichnofacies, with a distinct gradation of ecological types, have been identified (though some identify nine ichnofacies), and shown in Table 2.5.T5. It is necessary to point out that i) the occurrence of a particular ichnogenus does not necessarily indicate the presence of ichnofacies of the same name, ii) an ichnofacies may be identified in the rock record without the presence of the namesake. This point is important. Cruziana is a tracefossil created mainly by trilobites, which is extinct since post-Paleozoic. Yet the name of the ichnofacies is Cruziana ichnofacies and has been reported from post-Paleozoic rock record (Bromley and Asgaard, 1979). This is because of the fact that the basic behavioural pattern of the facies still persists to the present day (Frey and Seilacher, 1980).

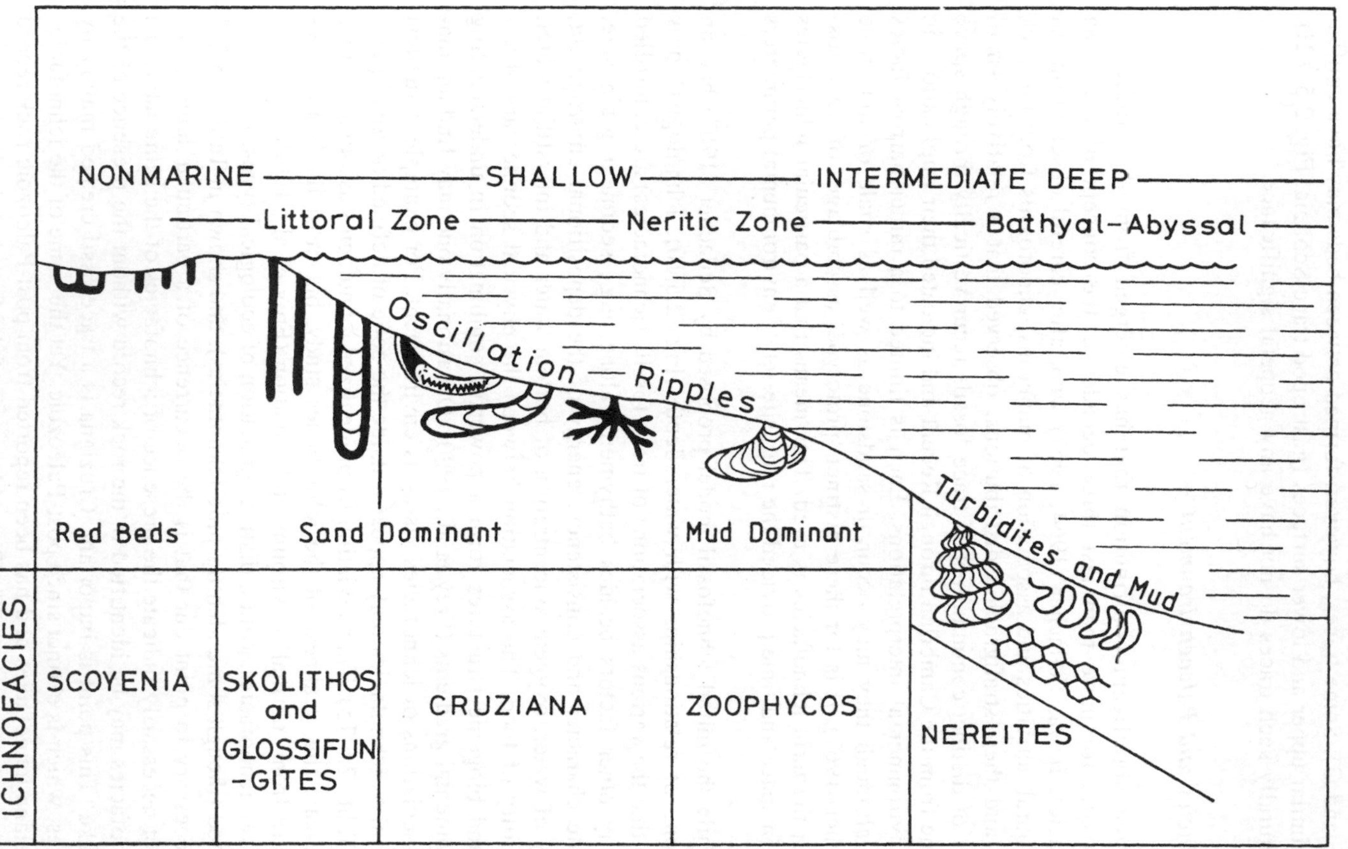

Fig. 2.5.10: Bathymetric sequence of trace fossil communities (Reproduced from Adolf Seilacher: Bathymetry of trace fossils. Marine Geology, Vol. 5, Fig. 2, © 1967, Elsevier Science-NL, Sara Burgerhartstraat 25, 1055 KV Amsterdam, The Netherlands, with the kind permission).

Burrows, Burrow Linings and their Significance

It is commonly found that many burrows are internally lined by 1) mucus impregnation of host sediment, 2) pellets shoved into the wall of the burrows, 3) detrital particles selected and cemented into place by mucus, and 4) chitiniphosphatic secretions by organisms. Why are burrows internally lined? This is an effort by organisms to fortify their dwelling places (dominichnion), particularly in nearshore areas of shifting substrates. This further indicates that the organisms that inhabited such vertically lined burrows were suspension feeders. If these burrows are funnel-shaped at their apertures suggest that the organisms were detritus feeders.

Extensive horizontal burrows without any lining and having a fill that contrasts with the surrounding matrix suggest that the organisms that thrived in the substrate were deposit feeders (pascichnia) in a stress-free environment. Presence of meniscate structure further strengthens this interpretation. Presence of horizontal spreiten also suggests that the organisms are deposit feeders. Unbranched horizontal non-overlapping curved to slightly coiled patterns of tunneling indicates that the animals were not only deposit feeders but also used these open horizontal galleries as dwelling places (fodinichnia).

Occasionally one may find concentrically lined burrow lining. Why? When sediment is washed down into the burrows, the animals, in order to keep the size of the burrow constant, shove the extra sediment into the wall of the burrow and plaster it into place with mucus for reinforcement. Thus periodic influx of sediment into burrows and repeated shoving operation by the animals create multi-layered concentric structures (Frey, 1973).

Burrow casts are also important with respect to depositional processes and animal response. It can be passive, gravity filling of an abandoned burrow, or can be actively filled by the animals (known as back-fill) behind them as they move forward through the substrate. The type of actively filled (back-filled) burrows indicates feeding or foraging behaviour of the animals.

U-shaped, blade-like pattern in burrows, known as spreiten representing the composite traces of multiple feeding forays of the organisms (Figs. 2.5.2 and 2.5.11), is often noticed. Close examination of the structure reveals fine laminae that conform, both in cross- and tangential sections, to the distal bow of the U-tube, and represents a product of the transverse shift of the tube through the sediment. Spreiten is formed 1) by a growing organisms that periodically enlarges its dwelling place, 2) in an attempt to maintain a constant level with respect to substrate surface in response to aggradation or degradation of the substrate (Goldring, 1964), and 3) by a deposit feeding organism that systematically probe the sediment surface in close swaths.

Table 2.5.T5: Recurring tracefossil associations and their general environmental implications (Reproduced from R. W. Frey and S. George Pemberton: Biogenic structures in outcrops and cores. 1. Approaches to ichnology, Bulletin of Canadian Petroleum Geology, vol. 33, Table 7, page 93, © 1985, Canadian Society of Petroleum Geologists with the kind permission).

TYPICAL BENTHIC ENVIRONMENTS	CHARACTERISTIC TRACEFOSILS
Scoyena Ichnofacies: moist to wet, pliable, argillaceous to sandy sediments at low energy sites; either very shallowly submerged lacustrine or fluvial deposits periodically emergent, or water-side sub-aerial deposits periodically becoming submergent; intermediate between aquatic and non-aquatic, non-marine environment.	Small, horizontal, lined, back-filled feeding burrows; curved to tortuous, unlined feeding burrows; sinuous crawling traces; vertical, cylindrical to irregular shafts; tracks and trails. Invertebrates mostly deposit feeders or predators; vertebrates are grovellors, predators or herbivores. Invertebrate diversity very low, yet some traces may be abundant. Vertebrate tracks may be diverse and abundant around water bodies.
Trypanites Ichnofacies (hard substrate): consolidated marine littoral and sub-littoral omission surfaces (rocky coasts, beach rock, hardgrounds, reefs) or organic substrate (beds of shell or bone).Bioerosion is as important as, and indeed accelerates physical erosion of the substrate. Intergradational with *Glossifungites* ichnofacies; somewhat intergradational with the *Teredolites* ichnofacies (woodground borings).	Cylindrical to vase, tear-, or U-shaped to irregular domiciles of endoliths, oriented normal to substrate surfaces, or shallow anastomosing systems of borings (sponges, bryozoans); excavated mainly by suspension feeders or 'passive' carnivores. Raspings and gnawings of algal grazers etc. (chitons, limpets, echinoides). Diversity moderately low, although borings or scrapings of given kinds may be abundant.
Glossifungites Ichnofacies (firm substrate): firm but unlithified marine littoral and sub-littoral omission surfaces, especially semi-consolidated carbonate firm grounds, or stable, coherent, partially dewatered muddy substrates either in protected, moderate energy settings or in areas of somewhat higher energy where clastic, semi-consolidated substrates offer resistance to erosion. Final sedimentary record typically consists of a mixture of relict and palimpsest features.	Vertical, cylindrical, U-shaped or tear-shaped borings or boring-like structures, or sparsely or densely ramified dwelling burrows; [protrusive spreiten in some developed mostly through growth of animals. Fan-shaped *Rhizocorallium* or *Diplocraterion.* Many intertidal species (e.g., corals) leave the burrows to feed; others are mainly suspension feeders. Diversity typically low, but given kinds of structures may be abundant.
Skolithos Ichnofacies (shifting substrate): lower littoral to infra-littoral, moderate to relatively high energy conditions; slightly muddy to clean, well-sorted, shifting sediments; subject to abrupt erosion or deposition. (Higher energy increases physical reworking and obliterates biogenic sedimentary structures, leaving a preserved record of physical stratification).	Vertical, cylindrical or U-shaped dwelling burrows; protrusive or retrusive spreiten in some, developed mainly in response to substrate aggradation or degradation (escape or equilibrium structures); forms of *Ophiomorpha* consisting predominantly vertical or steeping inclined shafts. Animal chiefly suspension feeders. Diversity is low, yet given kind of burrows may be abundant.
Cruziana Ichnofacies: Infralittoral to circalittoral substrates; below daily wave base but not storm wave base, to somewhat quieter off-	Abundant crawling traces, both epi-, and intrastratal; inclined U-shaped burrows having mostly protrusive spreiten (feeding swaths, soft-

shore type conditions; moderate to relatively low energy; well-sorted silts and sands, to inter-bedded muddy and clean sands, moderately to intensely bioturbated; negligible to appreciable, although not necessarily, rapid sedimentation. A very common type of depositional environment, including estuaries, bays, lagoons, and tidal flats, as well as, continental shelves or epeiric slopes.	sediment *Rhizocorallium*); forms of *Ophiomorpha* and *Thalassinoides* consisting of irregularly inclined to horizontal components; scattered vertical cylindrical burrows. Animals may include mobile carnivores and both suspension and deposit feeders. Diversity and abundance generally high.
Zoophycos Ichnofacies: Circalittoral to bathyal, quiet water conditions or protected epeiric sites; muds or muddy sands rich in organic matter but more or less deficient in oxygen. Epeiric sites reflect somewhat stagnant waters. Offshore sites are below storm wave base to fairly deep water, in areas free of turbidity flows, or significant bottom currents. Where relict or palimpsest substrate are present, especially if swept by shelf-edge or deeper-water contour currents, this ichnofacies may be omitted in the transition from infra-littoral to abyssal environments.	Relatively simple to moderately complex, efficiently executed grazing traces and shallow feeding structures; spreiten typically planar to gently inclined, distributed in delicate sheets, ribbons, or spirals ('flattened' form of *Zoophycos* in pelitic sediments, *Phycosiphon*). Animals virtually all deposit feeders. Low diversity; given structures may be abundant.
Nereites Ichnofacies: bathyal to abyssal, mostly quiet but oxygenated waters, in places interrupted by down-canyon bottom currents or turbidity flows (flysch deposits); or highly stable, very slowly accreting substrate. In flysch or flysch-like deposits, pelagic mud typically are bounded above and below by turbidite. In more distal regions, the record is mainly one of continuous deposition and bioturbation. (The stable deep-sea floor is not universally bioturbated, at least not equally intensively at every site).	Complex grazing traces and patterned feeding-dwelling structures, reflecting highly organized and efficient behavior; spreiten structures typically nearly planar, although *Zoophycos* forms are spiraled, multi-lobated, or otherwise very complex. Numerous crawling-grazing traces and sinuous feacl castings (*Neonereites, Helminthoida, Cosmorhaphe*), mostly intrastratal. Animals chiefly deposit feeders or 'scavengers', although many may have 'farmed' microbe cultures with their more or less permanent, open domiciles (*Paleodiictyon*). Diversity and abundance significant in flysch deposits, less so in more distal regions.

A gradation from vertical to incline to patterned horizontal burrows is observed both in rock record and recent sediments. This gradation signifies a shift in energy levels and is, in turn, a response to the availability of food. Near-shore areas are turbulent regions where food particles remain in suspension. Organisms, mostly suspension feeders, produce fortified vertical burrows as suggested earlier. With decreasing energy level that usually coincides with increasing water depth, food particles settle in the substrate and the organism mine through the substrate creating horizontal back-fill burrows. In quiet deep water, organisms thrive in a stress-free environment, and graze through the substrate producing unbranched, non-overlapping curved to slightly coiled

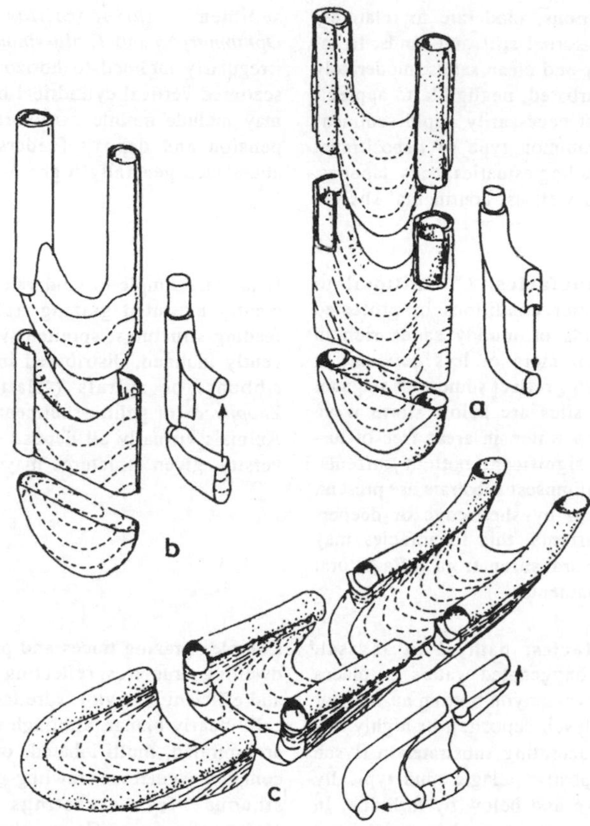

Fig. 2.5.11: Showing structural types of rhizocorallid spreit burrows. **(a)** Protrusive in vertical direction. **(b)** Retrusive in vertical direction. **(c)** Protrusive in horizontal, plus retrusive in vertical direction. (Reproduced from Adolf Seilacher: Bathymetry of trace fossils. Marine Geology, Vol. 5, Fig. 4, © 1967, Elsevier Science - NL, Sara Burgerhartstraat 25, 1055 KV Amsterdam, The Netherlands, with the kind permission).

patterns, as stated earlier. Such 'open strip' mining through the substrate reflects maximum utilization of food resources.

Visual Estimation of Degree of Bioturbation

Biogenic activities cause bioturbation in sediments and consequent disruption of primary depositional features. The intensity of bioturbation is dependent on many variables, e.g., rate of sedimentation, rate of erosion, population density of organisms, degree of physical energy in the form of waves and currents, supply of nutrients, consistency of the substrate, depth of water (Bromley, 1975; Ekdale et al., 1984; Frey, 1973, 1975; Hantzschel and Frey, 1978; Howard, 1975, 1978; Howard and Frey, 1975, 1985; Sarjeant 1975).

Attempts have been made in the past to measure bioturbation based on mathematical modeling of the processes of mixing or in a semi-quantitative assessment of the resultant mixed sediment using pattern-recognition techniques. Shortcomings of these techniques have been discussed by Taylor and Goldring (1993), have proposed a bioturbation index (BI) where "each grade of bioturbation is clearly defined in terms of burrow density, amount of burrow overlap and the sharpness of the original sedimentary fabric" (Table 2.5.T6) and "each grade is allocated a numerical value and descriptive term so that an index can be graphically plotted on a logging proforma and used in a description" (Taylor and Goldring, 1993).

Table 2.5.T6: Bioturbation Index (BI) where each grade is described in terms of the sharpness of the primary sedimentary fabric, burrow abundance and amount of burrow overlap (Reproduced from A.M. Taylor and R. Goldring: Description and analysis of bioturbation and ichnofabric. Jour. Geol. Soc., London, vol. 150, Table 1, © 1993, John Wiley & Sons. Ltd., with permission)

GRADE	PERCENT BIOTURBATED	CLASSIFICATION
0	0	No bioturbation.
1	1 - 4	Sparse bioturbation, bedding distinct, few discrete traces and/or escape structures.
2	5 - 30	Low bioturbation, bedding distinct, low trace density, escape structures often common.
3	31 - 60	Moderate bioturbation, bedding boundaries sharp, traces discrete, overlap rare.
4	61 - 90	High bioturbation, bedding boundaries indistinct, high trace density with overlap common.
5	91 - 99	Intense bioturbation, bedding completely disturbed (just visible), limited reworking, later burrows discrete.
6	100	Complete bioturbation, sediment reworking due to repeated overprinting.

2.6 Paleosol and Paleokarst

It is common experience that rocks exposed on the earth's surface are constantly being subjected to different kinds of weathering processes—physical, chemical and biological (Fig. 2.6.1). Amongst the processes, chemical weathering is by far the most important and all pervasive. The chemical reactions that are brought about between minerals and water produce dissolved constituents, solid residues and formation of new minerals. Physical and biological weathering play supplementary roles in the overall chemically dominated weathering processes. The mixture of solid inorganic and organic products of weathering, commonly known as soil, forms a thin veneer on the earth's surface. Similar processes of soil formation must have been operational on past landscapes, products of which are known as paleosols.

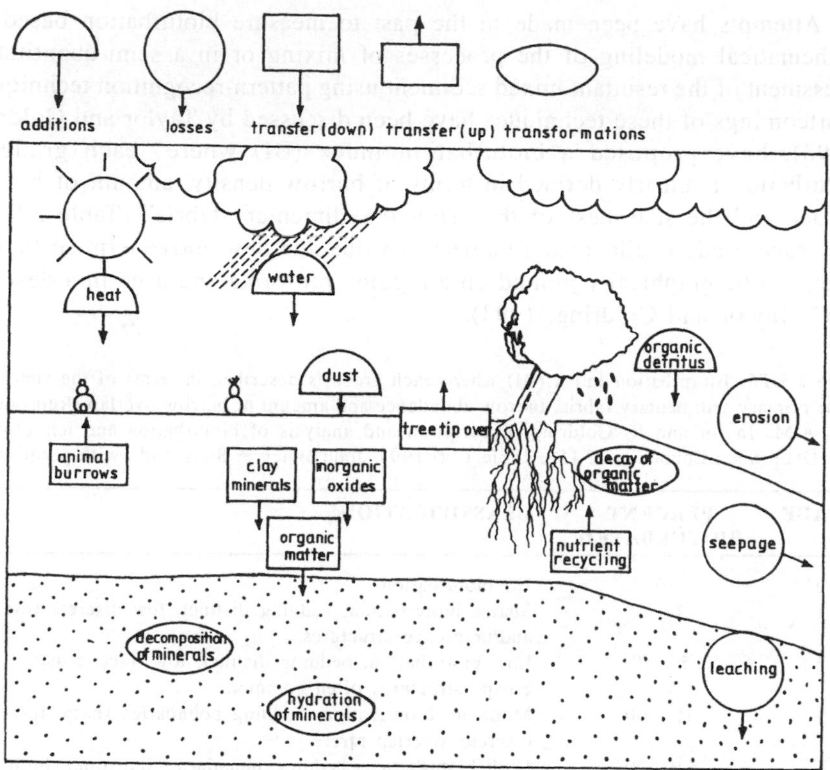

Fig. 2.6.1: Cartoon showing factors that can vary in the formation of soil profile - subdivided into those that involve gains, losses, transfer and transformation of material (Reproduced from Surface Processes; Weathering to Diagenesis. A second level course, Block 5, Fig.22, © 1981, The Open University Educational Enterprises Ltd., England with the kind permission).

Another kind of landform is produced through chemical weathering processes, known as karst; this type of landform is typically associated with carbonate and other easily soluble rocks. There is a net loss of calcium carbonate in karst facies compared to soil, and develops in all types of climates but most common in semi-arid climate. Esteban and Klappa (1983) have defined a karst as a "diagenetic facies, an overprint in sub-aerially exposed carbonate bodies, produced and controlled by dissolution and migration of calcium carbonate in meteoric waters, occurring in a wide variety of climatic and tectonic settings, and generating a recognizable landscape". Paleokarst refers to karst features developed on landscapes of the past.

Paleosols are reported from a variety of continental depositional settings starting from aeolian (Soreghan et al., 1997), palustrine (Tandon et al., 1995; Wright and Platt, 1995; Tandon and Gibbling, 1997), to deltaic (Fastovsky and McSweeney, 1987; Arndorff, 1993). Paleosols have also been reported from marginal marine rock records (Lander et al., 1991; Wright, 1994 b) as also in marine rocks when these are subaerially exposed (Driese et al., 1994; Webb, 1994). However, most commonly recorded paleosol is from alluvial deposits.

Since presence of paleosol and paleokarst in rock record provides much sought-after geologic information, these are discussed separately with emphasis on field criteria to enable one to identify them in the rock record.

2.6.1 Paleosol

Paleosol is a fossil soil formed over a landscape of the past and formed through a multitude of individual reactions in a parent rock (Valentine and Dalrymple, 1976). To understand and interpret a paleosol meaningfully, it is important to understand how a modern soil forms and its characteristics. Hence, a brief review of modern soils and various processes involved in the genesis of soil are briefly presented below with the hope that the record of the past can be better understood in terms of the present (actualistic principle).

Soil is a mixture of mineral matter, organic matter, soil water and air, formed through the surficial modifications of the exposed bedrock or fresh sedimentary deposits. The soil-forming processes are sequential geochemical differentiation of the parent material and pass through a series of development stages (Yaalon, 1960). While the parent material and vegetation provide the necessary solid ingredients to soil development, it is the climate that governs soil development by controlling the temperature and the precipitation. The principal factors in soil formation are CLIMATE and BIOSPHERE. These two factors in association with PARENT MATERIAL, TOPOGRAPHY and TIME mould the constitution of the soil body. Parent material as a factor of soil formation plays a passive role. Different parent materials can give rise to same kind or types of soil provided other soil forming factors remain reasonably constant over a long period of time. As a corollary, similar parent materials give rise to different types of soil where principal factors vary. Since seventy five percent of the land surface is covered by sedimentary rock, these rocks rather than igneous and metamorphic rocks are the more common parent materials of ancient and modern soils. The intensity of weathering of a parent rock is in a large measure controlled by the climate. In arctic and desert regions, physical agents of weathering (diurnal changes in temperature) are potent, while chemical weathering is inhibited to a large extent because of paucity of water. It is in the humid tropics and subtropics that the intensity of chemical decomposition of parent rock reaches its acme because of ample water. Thorough decomposition of parent rock(s) takes place with removal of soluble reaction products. Basic igneous and related rocks leave behind the oxides of Al and Fe and sometime Mn. Most of the SiO_2 leaches into the groundwater. Acid rocks leave behind some quartz silica together with the oxides of Fe and Al.

In humid temperate climate, the intensity of weathering is not as intense as in the tropics with the result that the soluble products of chemical weathering are not removed as rapidly as in the tropics. Some of these products recombine to form new insoluble compounds. Kaolin and similar formations are the principal end products. Because of the high clay content, a good portion of the soluble salts are absorbed and retained.

Biosphere supplies the organic material in the form of leaves, stems, branches, roots and trunks of plants which are ultimately decomposed and assimilated into the soil as various organic acids; this, in turn, increases the dissolving and decomposition capacity of soil water of the parent material.

Topography as a factor of soil development has local rather than regional influence. Steep slopes do not allow soil to develop thick soil cover since these are prone to erosion. Glaciated regions do not encourage soil development, not so much because of freezing temperature that inhibit chemical weathering, but because of removal through scraping by glacial movement.

The pedogenic processes that operate can be broadly classified as (i) physical processes (aggregation, translocation, freezing and thawing, wetting and drying, expansion and contraction), (ii) chemical processes (hydration, hydrolysis, solution, clay mineral formation, oxidation, reduction), and (iii) biological processes (humification of organic matter, microbiological oxidation and reduction of inorganic substances, water and ion uptake, fragmentation, aggregation, nitrification and nitrogen fixation) and these processes are summarized in Fig. 2.6.2 and Table 2.6.T1. Buol et al. (1980) have described these various processes in detail.

Soil Horizons: Characters and Designation

Two overlapping trends are found to operate during soil as pedogenesis proceeds and they are: **horizonation** and **haploidization**. Horizonation includes proanisotropic processes and conditions by which initial materials are differentiated into soil profiles with many horizons. Haploidization, on the other hand, includes proisotropic processes and conditions by which horizonation is inhibited or decelerated or by which horizons are mixed or disturbed (Buol et al., 1980). This layered pattern in soils, known as **soil horizons**, is endowed with certain morphological, chemical, and physical characteristics, e.g., colour, texture, structure, consistency, voids etc. Horizonation is the main result of soil formation and their sequence in soil profile is not random; it is the major feature for the characterization and classification of soils and paleosols and for their field recognition. The succession of horizons, as viewed in the exposed anatomy of the soil/paleosols body, is known as **SOIL PROFILE**. This reorganization process leading to horizons depends on the balance between three processes: (i) gains and losses of material from the system, (ii) transfer of

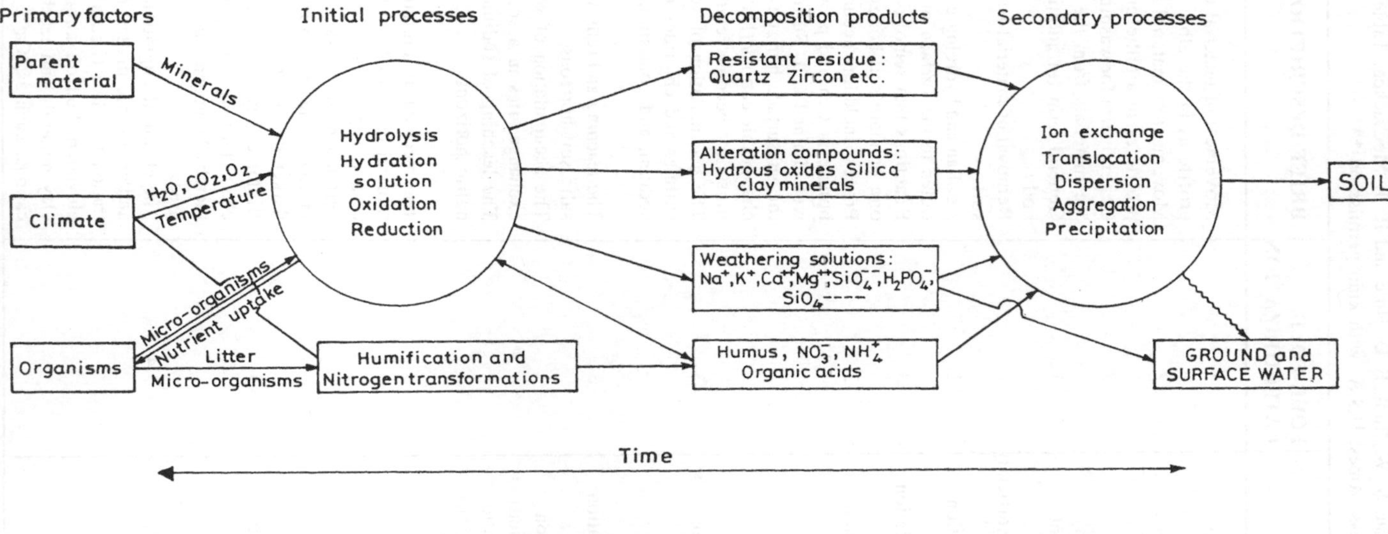

Fig. 2.6.2: Schematic diagram showing reaction and processes of soil profile development from parent materials. Three main groups of processes - chemical weathering, biological activity and profile morphogenesis - are shown inscribed in circles. Usually several of these processes proceed simultaneously. Different combinations, variable rates and accentuation of some of the processes are responsible for differences in the formed soils (Reproduced from D. H. Yaalon; Some implications of fundamental concepts of pedology in soil classification, 7th International Conf. of Soil Sciences, Vol. 16, Fig.1, page 120, © 1960, International Society of Soil Science, Wageningen, The Netherlands with the kind permission).

Table 2.6.T1: Some processes of soil formation (Reproduced from Soil Genesis and Classification, 2nd Edition, S. W. Buol, F. D. Hole and R .J. McCracken, Table 6.1, © 1980, Iowa State University Press, Ames, U.S.A. with kind permission)**

TERM	FOUR-FOLD CATEGORIZATION	BRIEF DESCRIPTION
1a. Eluviation	3	Movement of material out of a portion of a soil profile as in the albic zone
1b. Illuviation	3	Movement of material in to a portion of a soil profile as in argillic or spodic horizon.
2a. Leaching (depletion)	2	General term for washing out or eluviating soluble materials from the solum.
2b. Enrichment	1	General term for addition of material to a soil body
3a. Erosion (surficial)	2	Removal of material from the surface layer of a soil.
3b. Cumulization	1	Aeolian and hydrologic additions of mineral particles to the surface of a soil solum.
4a. Decalcification	3	Reactions that remove calcium carbonate from one or more soil horizons.
4b. Calcification	3	Process including accumulation of calcium carbonate in Cca and possibly other horizons of well-drained soil. Sufficient water present in thc subsurface horizon may remove some exchangeable cations (Na^+, K^+, Mg^+) but not Ca^{++}; thus Ca^{++} become relatively enriched.
5a. Salinization	3	The accumulation of soluble salts, such as, sulfates and chlorides of calcium, magnesium, sodium, and potassium in salty horizons.
5b. Desalinization	3	The leaching and removal of soluble salts from salic soil horizons.
6a. Alkalization (solinization)	3	The accumulation of sodium ions on the exchange sites in a soil.
6b. Dealkalization (solodization)	3	The leaching of sodium ions and salts from natric horizons.
7a. Lessivage	3	The mechanical migration of small clay-size mineral particles from the A to the B horizons of a soil, producing in the B horizons relative enrichment in clay (argillic horizon) that is light in colour and a subsurface clayey zone that is darker in colour.
7b. Pedoturbation	3	Biologic, physical (freeze-thaw and wet-dry cycles) churning and cycling of soil materials, thereby homogenizing the solum in varying degrees.
8a. Podzolization (silication)	3,4	The chemical migration of aluminium and iron and/or organic matter, resulting in the concentration of silica (i.e., silication) in the layer eluviated. Podzolization causes destruction of clay minerals and leaching of exchangeable cations, such as, Ca^{++}, Mg^{++}, K^+, and Na^+.

8b. Laterization (Desilication, feralization, ferritization, allitization)	3,4	Intense deep weathering resulting in thick uniform soil profiles (depleted of exchangeable cations).The chemical migration of silica out of the soil solum and thus the concentration of sesquioxides in the solum (goethite, gibbsite etc.). As in oxic horizons, with or without formation of ironstone (laterite, hardened plinthite) and concretions.
9a. Decomposition	4	The breakdown of mineral and organic materials.
9b. Synthesis	4	The formation of new particles of minerals and organic species.
10a. Melanization	1,3	The darkening of light-coloured minerals of initial unconsolidated materials by admixture of organic matter (as in the dark Al or mollic or umbric horizons).
10b. Leucinization	3	The paling of the soil horizons by disappearence of dark organic materials either through transformation to light-coloured ones or through removal from the horizons.
11a. Littering	1	The accumulation in the soil surface of organic litter and associate humus to a depth of less than 30 cm.
11b. Humification	4	The transformation of raw organic material in to humus and soluble organic acids (humic, fulvic). Transformation may take place through biogenic comminution of large organic structures to fine, amorphous organic matter and chemical oxidation of organic matter to simpler compounds, e.g., CO_2.
11c. Paludization	4	Process regarded by some workers as geogenic rather than pedogenic, including the accumulation of deep (>30 cm) deposits of organic matter as in mucks and peats (histosols).
11d. Ripening	4	Chemical, biological and physical changes in organic soil after air penetrates the biogenic deposit, making it possible for microbial activity to flourish.
12a. Braunification, rubifaction, ferrugination	3,4	Release of iron from primary minerals and dispersion of particles of iron oxide in increasing amount, their progressive oxidation or hydration, giving the soil mass brownish, reddish brown and red colours respectively.
12b. Gleization	3,4	The reduction of iron under anaerobic 'water logged' soil conditions with the production of bluish to greenish-gray matrix colours, with or without yellowish brown, brown and black mottles and ferric and manganiferous concretions. Anaerobic micro-organisms play an important role in reducing oxidized minerals. The process of gleying tends to retard mineral weathering.

** The four categories are: 1, additions to a soil body; 2, losses from a soil body; 3, translocation within a soil body; 4, transformation of material within a soil body.

material upwards and downwards, and (iii) transformations of one type of material into another. The balance achieved is dependent on a variety of factors, e.g., parent material, climate, types of organisms, geomorphology and time (Buol et al. 1980). The process of horizonation obliterates all primary features of the parent rock (Fig. 2.6.3).

In some cases, the contrast between horizons are dramatic and self-evident, while in others, it is very subtle. It is important to note that the horizon sequence in any one soil change from one type of soil horizon to another and is a reflection on basic changes in soil-forming factors as pointed out earlier.

Soil horizons do not develop equally in soil profiles; it occurs at different stages of development depending upon complexity of vegetation and time. Retallack (1988) has suggested a qualitative classification of products of paleosol development (Table 2.6.T2).

The variety of soil horizons, defined on the basis of materials comprising them, are labeled with a shorthand system of letters shown in Table 2.6.T3. Their field nomenclatures are more descriptive than genetic and are very useful in building up a quick mental image of the types of soil horizons present in a vertical section.

Classification of Modern Soils

There are at least 10 different classifications of soil (FitzPatrick, 1980). But there is no international agreement on adopting a single classification because (1) soils are very complex in character, (2) there is a considerable lack of knowledge about many soils, and (3) the aim of classification is not clear. It is, however, necessary to have a classification of modern soils for better understanding of paleosols (Retallack, 1990).

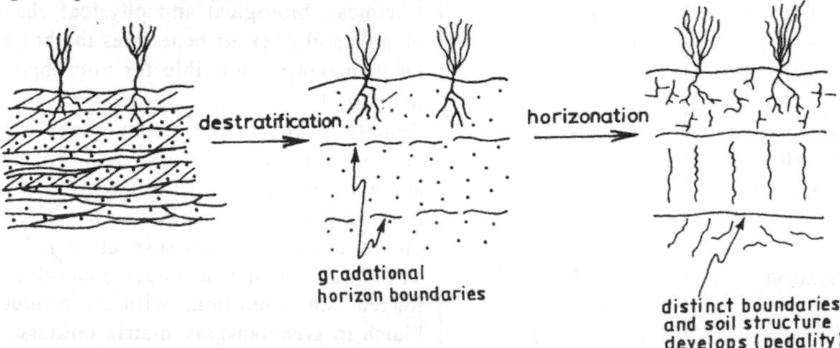

Fig. 2.6.3: Structural changes during pedogenesis. Initially the sediment is destratified (haploidized) but anisotropy later develops as horizonation occurs. With time, horizon distinctness typically increases; distinctive soil structure (pedality) also develops to some soils, such as, vertisols where mixing processes are dominant (Reproduced from V. Paul Wright: A guide to early diagenesis in terrestrial settings, *in*, Diagenesis III: K. H. Wolf and G. V. Chilingarian, eds., Fig. 12-3, page 598, © 1992, Elsevier Science-NL, Sara Burgerhartstraat 25, 1055 KV Amsterdam, The Netherlands, with the kind permission).

Table 2.6.T2: Stages of paleosol development through combined effect of time and climate (Reproduced from Field Recognition of Paleosols: G. J. Retallack, In, Paleosols and weathering Through Geologic Time: Principles and Applications: J. Reinhardt and W. R. Sigleo, eds., Special Paper 216, table 5, page 16, © 1988, The Geological Society of America, Boulder, Colorado, U.S.A. with kind permission)

STAGE	FEATURES
Very weakly developed	Little evidence of soil development apart from root traces; abundant sedimentary, metamorphic or igneous rock textures remaining from parent material
Weakly developed	With a surface rooted zone (A horizon) as well as incipient subsurface clayey, calcareous, sesquioxidic or humic or surface organic horizons, but not developed to the extent that they would qualify as USDA argillic, spodic or calcic horizons or histic epipedon.
Moderately developed	With surface rooted zone and obvious subsurface clayey, sesquioxidic, humic or calcareous or surface organic horizons; qualifying as USDA argillic, spodic, or calcic horizons or histic epipedon, and developed to an extent at least equivalent to stage II of calcic horizons.
Strongly developed	With specially thick, red clayey or humic subsurface (B) horizons or surface organic horizons (coals or lignites) or especially well-developed soil structure or calcic horizons at stage III to IV.
Very strongly developed	Usually thick subsurface (B) horizons or surface organic horizons (coal or lignite) or calcic horizons of stage VI; such a degree of development is mostly found at major geologic unconformities.

The classification of modern soils, briefly given below, is the one proposed by the United States Department of Agriculture (USDA), which has gained wide popularity and acceptance. The U.S. system relies on diagnostic horizons that are identified on the basis of properties such as texture, colour, amount of organic matter, presence of particular minerals, cation exchange capacity, and pH. One drawback of this system is that while this system works well with modern soils it does not work with paleosols because of the system's dependence on soil properties, such as cation exchange capacity or amount of organic matter, which are not preserved in paleosols

The soil taxonomy of Soil Survey Staff (1975) of United State Department of Agriculture is hierarchial and has six levels of categorization: (1) Order, (2) Suborder, (3) Great Group, (4) Subgroup, (5) Family, and (6) Series. The details are given in FitzPatrick (1980), and Retallack (1990). Following are the brief description of 10 orders of soils.

Entisols

Weathering is at an incipient stage with few weathered minerals and surface accumulation of organic litter (Fig. 2.6.4). As such, parent sediments or rocks

Table 2.6T3: Descriptive shorthand for labeling soil horizons (Reproduced from Field Recognition of Paleosols: G. J. Retallack, in, Paleosols and Weathering through Geologic Time: Principles and Application- J. Reinhardt and W. R. Sigleo, eds., Special paper 216, Table 1, page 13, © 1988, The Geological Society of America, Boulder, Colorado, U.S.A. with kind permission).

HORIZON	TERM	DESCRIPTION
MASTER HORIZON	A	Usually has roots and a mixture of organic and mineral matter forms the surface of those paleosols lacking an O horizon.
	E	Underlies an O or A horizon and appears bleached because lighter coloured, less organic, less sesquioxidic or less clayey than underlying material.
	B	Underlies an A or E horizon and appears enriched in some material compared to both underlying and overlying horizons(because darker coloured, more organic, more sesquioxidic or more clayey) or more weathered than other horizons.
	K	Subsurface horizon so impregnated with carbonate that it forms a massive layer.
	C	Subsurface horizon, slightly more weathered than fresh bedrock; lacks properties of other horizons, but shows mild mineral oxidation, limited accumulation of silica, soluble salts or moderate gleying.
	R	Subsurface horizon, slightly more weathered than fresh bedrock; lacks properties of other horizons, but shows mild mineral oxidation, limited accumulation of silica, soluble salts or moderate gleying.
GRADATION BETWEEN MASTER HORIZONS	AB	Horizons with some characteristics of A and of B, but with A characteristic dominant.
	BA	As above, but with B characteristics dominant.
	E / B	An horizon predominantly (more than 50 %) of material like B horizon, but with tongues or other inclusions of material like a E horizon.
SUBORDINATE DESCRIPTORS	a	Highly decomposed organic matter.
	b	Burried soil horizon (used only for pedo-relict horizons within paleosols; otherwise redundant).
	c	Concretions or nodules.
	e	Intermediately (between a and i) decomposed organic matter.
	f	Frozen soil with evidence of ice wedges, dykes or layers.
	g	Evidence of strong gleying, such as pyrite or siderite nodules.
	h	Illuvial accumulation of organic matter.
	i	Slightly decomposed organic matter.
	k	Accumulation of carbonates less than for K horizon.

	m	Evidence of strong original induration or cementation, such as, avoidance by root traces from adjacent horizons.
	n	Evidence of accumulated sodium, such as, domed columnar peds of halite casts.
	o	Residual accumulation of sesquioxides.
	p	Plowing or other comparable human disturbance.
	q	Accumulation of silica.
	r	Weathered or soft bedrock.
	s	Illuvial accumulation of sesquioxides.
	t	Accumulation of clay.
	v	Plinthite (in place of pedogenic laterite).
	w	Coloured or structural B horizon.
	x	Fragipan (a layer originally cemented by silica or clay and avoided by roots.
	y	Accumulation of gypsum crystals or crystal casts.
	z	Accumulation of other salts or salt crystal casts.

retain their original characteristics and occur on young geomorphic surfaces, such as flood plains and steep slopes where erosion removes the soil layer as it is formed. As such, Entisols do not have any horizon at all. Entisols can form under varied climates ranging from warm, humid climates to dry or cold climates. Root traces are characteristic in an otherwise little weathered parent material. Such incipient soil formation may be due to availability of very short time.

Inceptisols

These are the soils of humid regions with well-defined soil horizons (Fig. 2.6.4). A typical inceptisols can be imagined as having a light-coloured surface horizon with subsurface horizon having moderately weathered and form in low rolling parts of the landscape in and around a steep mountain front. Accumulation of clays in the subsurface horizon is not high enough to form an argillic horizon (which is a diagnostic feature of alfisols and ultisols). Surface accumulation of organic litter is not high enough unlike histosols. Primary materials are variable and where formed, retain the primary features of parent material (like entisols). Climate can also be variable as in entisols.

Histosols

These are organic-rich soils forming peaty horizons in low-lying permanently waterlogged parts of the landscape (Fig. 2.6.4). Underlying parent material remains relatively free from weathering, except for some leaching or formation of some gley minerals (pyrite, siderite). As such, most of the weatherable minerals and primary features of underlying parent material are preserved.

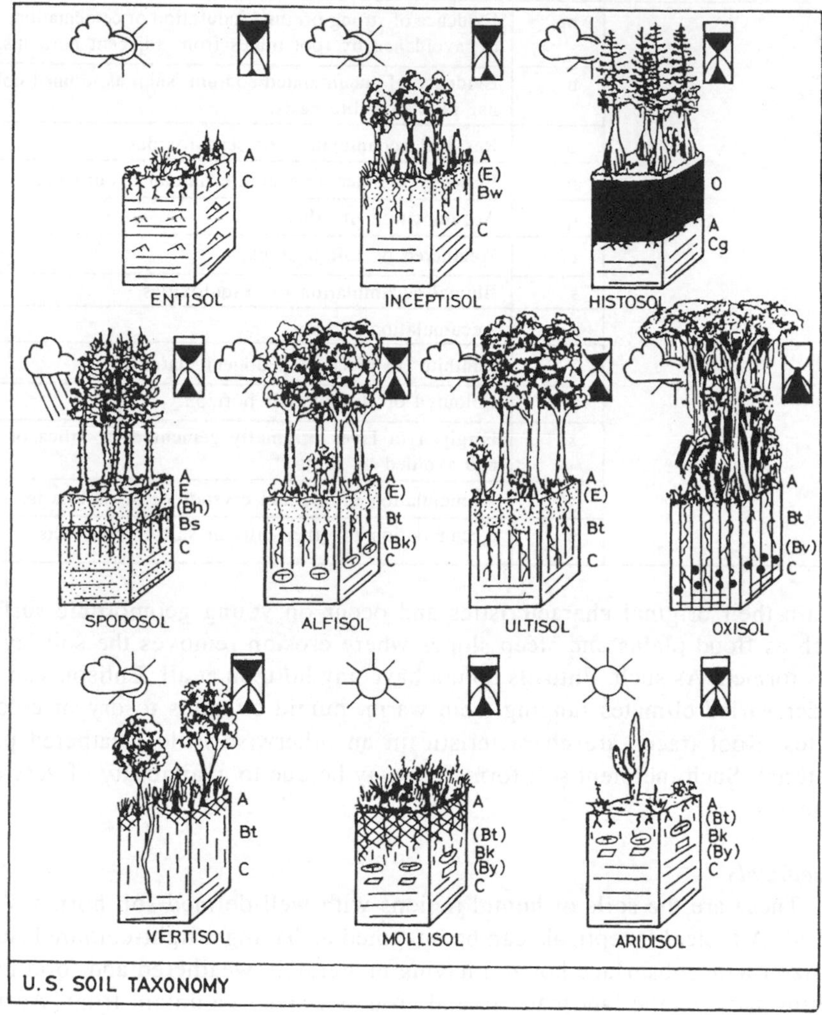

Fig.2.6.4: Cartoons of climate, vegetation and profile form of the various orders of soils defined by the U.S. soil taxonomy (Reproduced from G. J. Retallack; Soils of the Past: An Introduction to Paleopedology, Fig. 5.3, page 102, © 1990, Unwin Hyman, Inc., with the kind permission from Kluwer Academic Publishers, The Netherlands).

Vertisols

Vertisols form a uniform, thick clayey profile having deep wide cracks for a part of the year (Fig. 2.6.4). This cracking produces a hummock-and-swale topography (gilgai micro-relief) and its subsurface expression of a disrupted, festoon-shaped horizon (mukkara structure). Middle horizons usually have a well-defined wedge structure with slickensides. Most vertisols develop on parent materials of intermediate to basaltic composition and may form very rapidly (few hundred years) on shale, claystone or marls having smectitic composition. Usually form in flat terrane at the front of gentle slopes under a dry climate and sparse vegetation. As such, Vertisols have low organic matter and high base saturation. Montmorillonite is the dominant clay mineral, but other clay minerals sometime become dominant.

Mollisols

These soils are dark-coloured, well-developed, well-structured base-rich, surface horizon of intimately admixed clay or organic matter (mollic epipedon) (Fig. 2.6.4). Subsurface horizon may have root traces of grassy vegetation and burrows of different types of soil invertebrates. Mollisols develop under grassland vegetation of low, rolling or flat country under sub-humid to semi-arid climates. Such soils usually develop on base-rich sediments (clay, marl) and rocks (basalt).

Aridisols

This type of soil develops in low-lying areas of arid and semi-arid environments (Fig. 2.6.4). Rainfall being scanty, soils usually have shallow calcareous (calcic), gypsiferous (gypsic) or salty (salic) horizons. These may occur in the form of nodules or as continuous layers. The parent materials of aridisols are varied and may include unconsolidated alluvium, loess and till.

Spodosols

The diagnostic features of this type of soils are a bleached, sandy near-surface horizon, the sub-surface horizon of which is enriched with iron and aluminium oxides or organic matter (Fig. 2.6.4). Formation of such soil is rather rapid (a few hundred years) on quartz-rich sand; it can also form by deep weathering of less felsic sediments or rocks. Spodosols form principally in humid climates when clay and soluble salts are easily washed out from parent material; it has also been reported from the tropics and near the poles.

Alfisols

These are base-rich forested soils with light coloured surface horizon and argillic subsurface horizon rich in exchangeable cations (base saturation greater than 35 %) (Fig. 2.6.4). Presence of carbonate nodules in a horizon deep within the profile indicates base saturation in paleosols. In the absence of carbonate nodules, alfisols can be identified by having (i) base-rich clays (smectite), (ii) the presence of easily weatherable minerals, such as feldspar more than 10 % in 20-200 micron size fraction), and (iii) a molecular weathering ratio of bases to alumina greater than one. Alfisols form in regions with rainfall from sub-humid to semi-arid climates on rocks of intermediate to basaltic composition.

Ultisols

Ultisols represent soil horizons that are deeply weathered, rich in kaolinite and highly weathered aluminous minerals (gibbsite) (Fig. 2.6.4). Unlike alfisols, it is (i) base-poor soils with low reserve of weatherable minerals (less than 10% in 20-200 micron size fraction), (ii) absence of calcareous material and (iii) molecular weathering ratios of bases to alumina are less than one. The characteristic of such soil with deep weathering indicates prolonged process of soil formation (hundreds to thousands of years) and can develop on any rock type under humid and warm climate.

Oxisols

It is a deeply weathered soil having a striking brick-red, yellow or gray colour with trace content of weatherable minerals (Fig. 2.6.4). Oxisols is dominated by kaolinitic clay, oxides of iron and aluminium and small amount of organic matter Sand-sized spherical micropeds of clay are characteristic. Development of such deeply weathered soil profile suggests a humid, tropical to subtropical climate. Formation of such deeply weathered profile takes millions of years, usually develop on a stable continental location, on gentle slopes of plateaus, terraces and plains.

Duricrust

Sub-aerial weathering processes, as briefly discussed above, may create soil/paleosol surfaces that are either (i) soft and friable, or (ii) hard crusts through precipitation of various types of cements. The hard cemented crust, a striking feature of many present-day tropical and subtropical landscape, is known as **duricrust** (also known as pedoderms or pedocrete). Goudie (1973) has defined it as "a product of terrestrial processes within the zone of weathering in which either iron and aluminium sesquioxides (in case of ferricrete or alcrete) or silica (in case of silcrete) or calcium carbonate (in case of calcrete) or other compounds in case of magnesicrete and the like have dominantly accumulated in and/or replaced a pre-existing soil, rock, or weathered material, to give a substance which may ultimately develop into an indurated mass". The accumulation of cements can be (i) relative accumulation i.e., concentration of cementing minerals in the weathering zone associated with simultaneous loss of more labile components of the pre-existing material, and (ii) absolute accumulation i.e., addition of minerals to pre-existing materials. The duricrust may have been generated either by pedogenic (vadose zone) processes, or by the activities of groundwater (phreatic zone). The dominant varieties of duricrust are described as under.

Ferricrete and Alcrete

These are surficial accumulations of the products of rigorous chemical selection where conditions favour greater mobility of alkalies, alkali earths, and silica than iron and aluminium (McFarlane 1976). Aluminium-rich rock, bauxite, is thought to form by long term weathering of silicate minerals at moderate pH in humid, non-seasonal climate on stable land surface (Valeton 1972).

The relative accumulation of Fe and Al due to weathering and extensive desilication results in pedogenic ferricrete. On the other hand, absolute accumulation of Fe and Al in the host material, transported in solution by groundwater, results in non-pedogenic groundwater ferricrete. It is important to discriminate these two types of ferricretes and Table 2.6.T4 shows the differences.

Silcretes

Silcretes are indurated zones of unconsolidated host/parent materials (sediment, saprolite, soil) cemented by various forms of secondary silica, like opal, cryptocrystalline or well-crystallized quartz (Wright 1994a). Cementation and replacement are main operating processes in developing silcrete. Silcrete horizons are usually 1 to 3 meters thick (in some cases 5 meters), and have a well developed vertical joint system with minor sub-horizontal element, producing a columnar appearance.

Chemical weathering of silicate minerals is the primary source of dissolved silica in most of the cases; other sources are a) silica released during replacement of quartz by carbonates, b) quartz dust generated by grain abrasion in aeolian environments, c) atmospheric dust containing quartz particles and d) silica abstracted and concentrated by plants (e.g. grasses) (Summerfield 1983).

Like ferricretes, silcretes can also be pedogenic and groundwater types. Differences between pedogenic and groundwater silcretes are shown in Table 2.6.T5.

Calcretes

It is "a term for terrestrial materials composed dominantly, but not exclusively, of calcium carbonates which occur in states ranging from powdery and nodular to highly indurated, and are the result of displacive and/or replacive introduction of vadose carbonate into greater or lesser quantities of soil, rock or sediment within a weathering profile" (Watts 1977). Calcrete profile has been estimated to cover nearly 13 % of total land surface of the earth (Yaalon, 1988, cited in Wright et al., 1995). Although calcretes are commonly closely associated with calcareous host rocks, there are many occurrences of calcretes where basic volcanic rocks, granite, gneiss, shale, schist, sedimentary rocks etc. serve as host rocks. This is achieved through progressive accumulation of calcium carbonate through the process of cementation, replacement and displace-

Table 2.6.T4: Differences between pedogenic and groundwater ferricretes (Reproduced from Losses and Gains in Weathering Profiles and Duripans - V. Paul Wright, Table 1, Page 112, in, Quantitative Diagenesis: Recent Developments and Application to Reservoir Geology- A. Parker and B W. Sellwood, eds., © 1994, Kluwer Academic Publishers, The Netherlands with kind permission).

PEDOGENIC	GROUNDWATER
Associated with distinct vertical profile with upward increase in Fe.	Little or no profile development
Associated with deeply weathered parent rock	Host may be unaltered other than Fe-oxide cemented
Associated with kaolinitic clay.	May be associated with wide range of clay minerals.
Exhibits wide variety of Fe and Al oxides and oxyhydroxides including goethite, maghemite, gibbsite.	Simple mineralogy with hematite, goethite and rare gibbsite.
Complex fabrics indicating polyphase derivation. Pissoids with Al-goethite cortices, increase in frequency down profile.	Simple typical; single-phased fabrics. Pisoids, if present, apparently decrease in frequency downward.
Profile may be capped by dismantled horizon (brecciated) with altered Al-goethite pisoids.	Top of ferricrete horizon not altered unless exhumed.
Associated with stable land surfaces.	May occur in any substrate including aggrading sediments.
Associated mainly with humid tropical climates with long, dry season but may be relict.	Found under a variety of climates including arid and semi-arid.

ment, as a result of which the host material (siliciclastic) becomes rich in secondary calcium carbonate.

Calcrete profiles show rapid vertical and lateral variations. The idealized profile (Esteban and Klappa, 1983) is shown below:

Hardpan
Sheet calcrete
Glaebular calcrete
Chalky calcrete
Transitional horizon
Host material

Fig. 2.6.5 summarizes the major characteristics of each horizon of the idealized profile, detail discussion of which is available in Esteban and Klappa (1983). One may find many variations of the above-idealized profile. While the uppermost boundary of the profile shows sharp contact (due to erosion), boundaries between horizons vary from abrupt to diffuse. Vegetation usually plays a significant role in the genesis of calcrete. Recently, Wright et al. (1995) have described root-related calcretes from recent sediments and ancient rocks, and have referred to such calcretes as rhizogenic calcretes. Dominant sheet-form

Table 2.6.T5: Differences between pedogenic and groundwater silcretes (Reproduced from Losses and Gains in Weathering Profiles and Duripans - V. Paul Wright, Table 3, page 118, in, Quantitative Diagenesis: Recent Developments and Application to Reservoir Geology - A. Parker and B.W. Sellwood, eds., © 1994. Kluwer Academic Publishers, The Netherlands with kind permission).

PEDOGENIC	GROUNDWATER
Generally 1–2 meters thick.	May be slightly thicker.
May exhibit set of ordered profile features.	Lacks profile.
Exhibits range of macro-features—columnar prismatic, pseudo-brecciated.	Massive, nodular.
Quartz with some opal; quartz typically micro-crystalline, or as overgrowths, euhedral quartz.	Silica chiefly as opal and chalcedony.
Dissolution features present on primary and authigenic silica.	Silica has replaced clays with preservation of earlier host fabric.
Range of vadose features, such as, pendant and stalactitic masses. Range of illuvial features, such as, geopetals, grading, nodule cappings (cutans).	Cements are isopachous.
Typically as simple profile at land surface.	May occur as multiple profiles with gradational margins.

structures in such calcretes are the most striking features where vertical rhizocretions are rare.

There are various sources of calcium carbonate, such as: (i) breakdown of calcium-rich primary minerals, (ii) certain plants which contain calcium within their tissues, (iii) detrital carbonate grains incorporated by colluvial or fluvial processes, (iv) calcareous skeletal grains of soil organisms, (v) calcium carbonate carried in solution by rain water and groundwater, (vi) atmospheric dust, and (vii) sea-water sprays in coastal regions (Klappa, 1980a).

Like other duricrusts, there are two types of calcrete—pedogenic and groundwater types. Table 2.7.T6 shows the differences between the two types.

Field Features of Paleosols to be Studied

Proper field description of paleosols, that are amenable to interpretation, should include the following:

a. Information on grain size is needed for the purpose of classification and interpretation of paleosols. This can be done in the field by using a grain-size comparison chart or samples with proper size designation. Collections of properly oriented samples are required for laboratory checking of grain size.

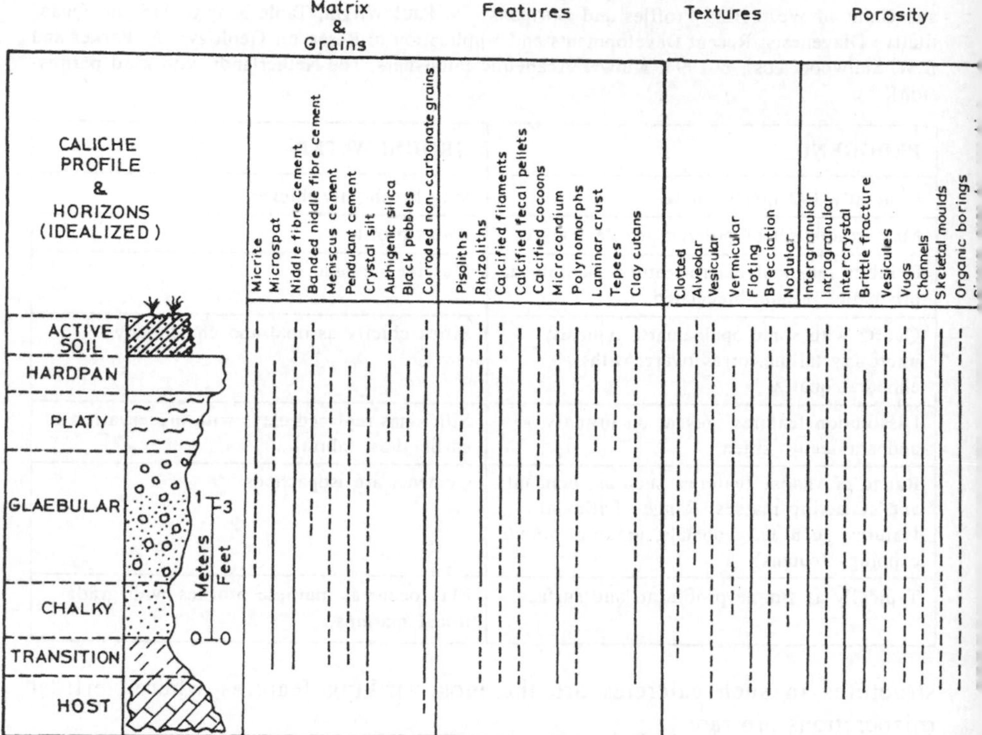

Fig. 2.6.5: Idealized calcrete profile and distribution of major characteristics within the profile (Reproduced from M. Esteban and C. F. Klappa, AAPG Memoir Series No. 33, Fig. 7, © 1983, American Association of Petroleum Geologists, Tulsa, Oklahoma, U.S.A., with the kind permission).

b. Determination of the thickness of different paleosols horizons.
c. Determination of the colour of the paleosols horizons with the help of standardized colour chart, such as, Munsell colour chart.
d. Determination of the nature of reaction of paleosols horizons with acid gives an idea of the degree of calcification of paleosols, if any. Table 2.6.T7 may help to determine the approximate carbonate content of the paleosols. Many paleosols, however, are made up of dolomite where such simple test of reaction with acid cannot be done.
e. Determination of the nature of contact of paleosols horizons with its overlying and underlying beds. This is a very important field feature that helps to understand the genetic relationships between the paleosols horizons. As for example, a sharp horizon suggests erosional contact, while diffuse planar contact suggests a genetically related horizon of single paleosols profile. The following Table 2.6.T8 shows the type of contacts between the paleosols horizons in the field.

Table 2.6.T6: Differences between pedogenic and groundwater calcretes (Reproduced from Losses and Gains in Weathering Profiles and Duripans - V. Paul Wright, table 2, page 116, in, Quantitative Diagenesis: Recent Developments and Application to Reservoir Geology - A. Parker and B. W. Sellwood, eds., © 1994 a, Kluwer Academic Publishers, The Netherland with kind permission.

PEDOGENIC	GROUNDWATER
Generally thin, 1-2 meters. Thicker forms exhibit complex brecciated, pisolitic and peloidal fabric.	May reach thickness of 10 meters or more.
Typically display profile with orderly set of horizons; sharp topped.	Massive, gradational top.
Macro-structures include nodular, massive, laminar, pisolitic, prismatic.	Rarely display laminar horizon; never pisolitic, prismatic.
Rhizocretion may be common.	Rare rhizocretion.
Micro-fabric may exhibit beta-fabrics (biogenic, microbial/root-related carbonates, such as, alveolar, septal structure, needle fibre calcite, *Microcodium*.	Typically alpha-fabrics (densely crystalline).
Vadose fabrics, such as, meniscus and pendant cements.	Absent
Typically finely crystalline.	May show wider ranges of crystal sizes including spheroidal dolomite.
Porosity is low.	Porosity is locally high (>25%), typically horizontally elongated.
Rarely display lateral changes, such as, calcrete to dolocrete to gypcrete. Typically associated with more stable surfaces on floodplains.	May show regular down-dip mineralogical changes. Associated with drainage channels and playas.
Mottling, if present, is minor, reflecting lack of Fe-translocation in generally oxidized setting.	Associated with extensive mottling and Fe-translocation in reducing groundwater.
Most commonly found in finer units in alluvial cyclothem.	More common in permeable coarser units in alluvial cyclothem.

It is very important to prepare a graphic log of a soil profile in the field, showing thickness, grain size, colour, structure, clay content etc. of the different horizons in the profile.

Field Signatures of Paleosols

Amongst many field features that help in the identification of paleosols in the rock record, three most important features are: a) root traces, b) soil horizons, and c) soil structure (Retallack, 1988).

Table 2.6.T7: Scale of acid reaction to approximate carbonate content of paleosols (Reproduced from Field Recognition of Paleosols - G. J. Retallack, Table 2, page 14, in, Paleosols and Weathering Through Geologic Time: Principles and Applications – J. Reinhardt and W. R. Sigleo, eds., Special Paper 216, © 1988, The Geological Society of America, Boulder, Colorado, U.S.A. with kind permission).

CARBONATE CONTENT	REACTION WITH DILUTE ACID
Non-calcareous	Acid un-reactive; often forms an inert bid.
Very weakly calcareous.	Only a little movement within a acid drop which could be flotation of dust particles as much as bubbles.
Calcareous	Numerous bubbles, but not coalescing to form a froth.
Strongly calcareous.	Bubbles forming a white froth, but drop of acid not doming upwards.
Very strongly calcareous.	Drop vigorously frothing and doming upwards.

a) Root traces are strong positive features of paleosols of the post-Devonian age. Klappa (1980b) has defined rhizoliths as "organo-sedimentary structures produced by roots of higher plants and are preserved in rock records through accumulation and/or cementation around, within, or replacement of, by mineral matter". There are five basic types of rhizoliths: i) root moulds, ii) root casts, iii) root tubule, iv) rhizocretion s.s., and v) root petrifactions (Fig. 2.6.6; Klappa, 1980b). Shape, size, branching and orientation of rhizoliths differentiate them from burrow systems (Fig. 2.6.7; Klappa, 1980b). Presence of organic matter in roots and rootlets is a definite criterion to distinguish them from burrows. However, presence of organic matter is dependent upon soil order; organic matter is well preserved in waterlogged lowland environment (gley soil), but not in well-oxygenated environment.

Tubular features of roots and rootlets of higher plants (Figs. 2.6.8 a, b, c and 2.6.9 a, b) are commonly found to be internally filled-in by clay and silt that are washed down into the open hole; in many other cases, holes may be filled-in by calcite, zeolite or chalcedony (Retallack, 1988). Root traces are enhanced by encrustation that forms around them during their growth (Mount and Cohen, 1984; Retallack, 1988; Wright, 1994b). Drab-coloured halos that extend out into the surrounding red-coloured paleosol matrix is another distinctive feature of root traces. In the post-Devonian period, roots and rootlets of higher plants have brought about a major pedo-diagenetic change in the soil fabric, texture and structure.

Roots are responsible for the formation of vertically elongated glaebules (Fig. 2.6.10 a,b) or nodular structures (Fig. 2.6.10 c), platy calcrete layers (Fig. 2.6.11a,b), brecciated texture, formation of some tepee structure, channel and mouldic porosity. *Microcodium* (Fig. 2.6.12) and alveolar septal structure (Fig.

Table 2.6.T8: Sharpness and lateral continuity of boundaries of paleosol horizon (Reproduced from Field Recognition of paleosols - G.J. Retallack, table 3, page 14, in, Paleosols and weathering Through Geologic Time: Principles and Applications - J. Reinhardt and W. R. Sigleo, eds., Special paper 216, © 1988, The Geological Society of America, Boulder, Colorado, U.S.A. with kind permission).

CATEGORY	CLASS	FEATURES
Sharpness	Abrupt	Transition from one horizon to another completed within one inch (2 cm).
	Clear	Transition completed within 1-25 inches (2-5 cm).
	Gradual	Transition spread over 2.5-5 inches (5-15 cm).
	Diffuse	One horizon grading to another over more than 5 inches (15 cm).
Lateral continuity	Smooth	Horizon boundary forms an even plane.
	Wavy	Horizon boundary undulates with pockets wider than deep.
	Irregular	Horizon boundary undulates with pockets deeper than width.
	Broken	Parts of adjacent horizon are disconnected, e.g., by deep and laterally persistent clastic dykes in vertisols.

2.6.13 a,b) are considered to be products of coalesced millimeter-sized rhyzoliths (Klappa, 1980 a,b).

Densely intertwined roots create another distinctive soil horizon – laminar calcrete above a hardpan surface or impermeable rock (Wright, Platt and Wimbledon, 1988; Wright, 1994b; Retallack, 1988). Wright (1994b) has termed such laminar calcrete as 'rhizogenic calcrete', whereas Klappa (1980b) has termed it as 'rhizolite'. A brief description of Microcodium will not be out of place here. *Microcodium* [their radiating microstructure, resembling the present marine chlorophyceae Codium, led Gluck (1912, cited in Freytet and Plaziat, 1982) to coin the term *Microcodium*] is a microcrystalline prism of calcite, having polygonal cross-section, measuring 100–800 μm in length and 20–70 μm in width. It frequently exhibits dark peripheral inclusions delimiting a more limpid central canal or vacuole and recently interpreted as dissolution feature. The calcite prisms are grouped into two types: 1) cylindrical, 'corn-cob' or 'rosette' with prisms disposed in a radiating manner around a hollow axis (axial canal) filled with sparite, and 2) planar or undulatory layers of prisms perpendicular to substrate. This fabric, constituting crystalline aggregate, is sometime called 'colony' or 'aggregate' (Freytet and Plaziat, 1982). Controversies exist concerning (i) its origin—mineral versus organic and (ii) the environment of its formation – marine versus continental (Freytet and Plaziat, 1982).

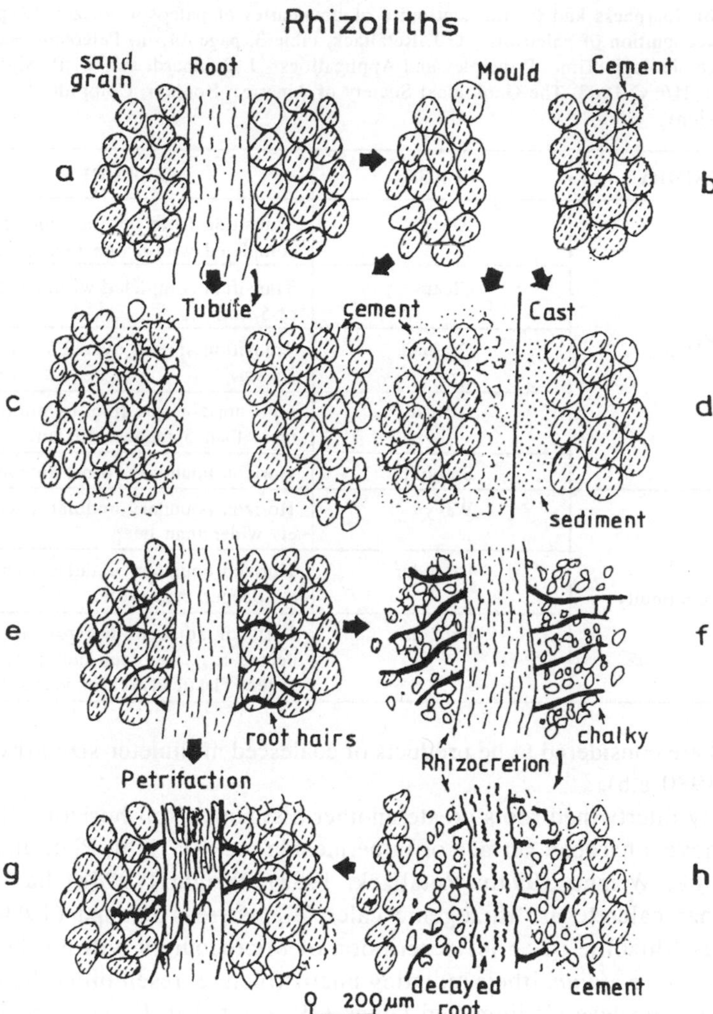

Fig. 2.6.6: Schematic representation of types of rhizoliths and their relationships. **(a)** hairless part of root in lime sand, **(b)** root mould in lime sand, **(c)** Root tubule. Intergranular pore space in proximity of root mould has been filled with low magnesian calcite cement. Remainder of lime sand is uncemented or weakly cemented. Cementation took place either during life span of root mould because of greater porosity and permeability and hence greater availability of percolating $CaCO_3$-bearing solutions through the mould (b &c). **(d)** Root cast formed filling root mould or tubule with sediment and/or calcite cement. **(e)** Root with root hairs in lime sand, **(f)** Rhizocretion. Etching of root hairs of lime sand and localized re-precipitation of dissolved calcite as isolated micrite grains produces a chalky rhizocretion (compare with e). **(g)** Root petrifaction. Partly decayed root and root hairs impregnated with calcite. **(h)** Rhizocretion. Partly decayed root and sloughed off root hairs encased in partly etched lime sand. Root materials are not calcified (contrast with g). (Reproduced from C. F. Klappa: Rhizoliths in terrestrial carbonates: Classification, recognition, genesis and significance. Sedimentology, Vol. 27, Fig. 1, page 614, © 1980b, Blackwell Science, U.K with the kind permission).

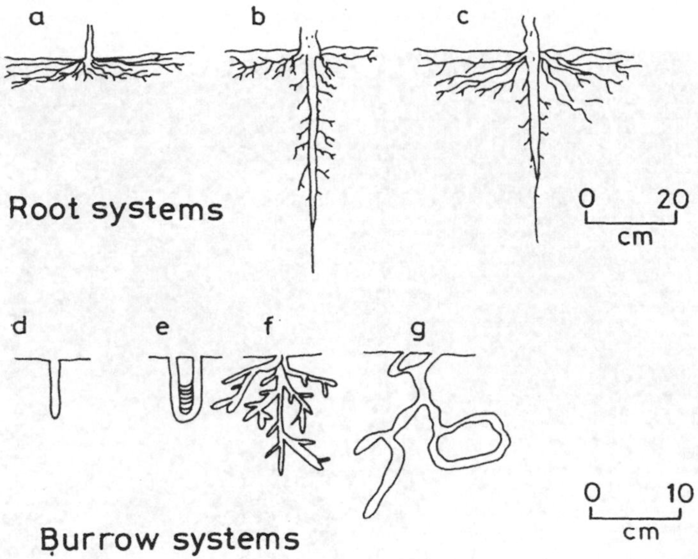

Fig. 2.6.7: Root systems versus burrow systems. **(a)** - **(c)** root systems ; note downward tapering diameters. **(a)** Lateral roots; **(b)** tap root; **(c)** generalized root system with well-developed tap root and laterals; **(d)-(g)** common burrows; **(d)** skolithos; **(e)** U-shaped burrow with spreiten Diplocraterion; (f) branching burrow network Chondrites; **(g)** part of boxwork burrow system Ophiomorpha. (Reproduced from C. F. Klappa: Rhizoliths in terrestrial carbonates: Classification, recognition, genesis and significance. Sedimentology, Vol. 27, Fig. 2, page 614, 1980b, with the kind permission of Blackwell Science, U.K.).

b) The second most important feature for identification of paleosols is the 'soil horizon'. While soil horizons vary both laterally and vertically, there are some consistent features that persist and help in their identification and classification. These have been discussed earlier.

c) The third most important features are soil structures. The individual grain of soil does not, as a rule, exist in its ultimate unit size. Other than sand sized particles, generally the grains are grouped into compound particles by cohesion, adhesion and other such processes. This segregation of particles of the soil mass into variously shaped and sized soil particles forms the unit constituting soil structure. The agents responsible for the formation of structures are clay and organic matter. The adhesive and cohesive properties of clay bind together not only the clay grains but also silts and sands. Clay is thus the basis for the formation of aggregates. Dissolved organic matter forms coatings over the clay-bound aggregates. Upon drying, these glue-like organic coatings become less soluble and attain considerable strength and pliability (Joffe, 1965). The stable aggregate of the soil material is commonly known as **peds**. Peds are commonly surrounded by cutans (clay skin) formed through illuviation or ferrugination. The cutans, surrounding the peds, help in the identification of peds and determine their geometry. When the clay skins are the common type

Fig. 2.6.8 a: Rhizoliths. Vertically oriented taproot rhizoliths and sub-horizontal second-order rhizoliths (Reproduced from M. Esteban and C. F. Klappa, AAPG Memoir Series No. 33, Fig.53, page 30, © 1983, American Association of Petroleum Geologists, Tulsa, Oklahoma, U.S.A, with kind permission).

b: A well-developed rhizocretion in the centre of the photograph. Upper part shows profuse development of the pedogenic glaebules, Lameta Formation, Jabalpur, India. Measuring tape is 5 cm across (Courtesy Dr. Partha Ghosh, Indian Statistical Institute, Calcutta).

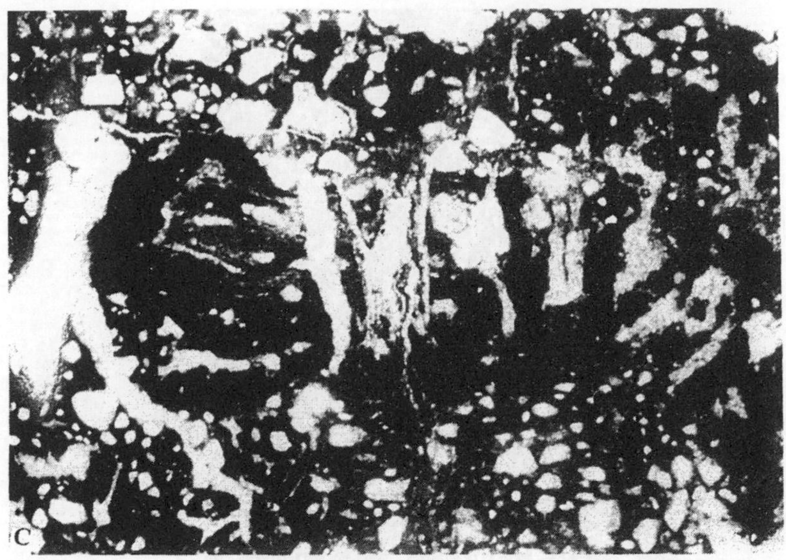

Fig. 2.6.8 c: Photomicrograph showing several vertical to subvertical, downward-tapering spar-filled root moulds and V-shaped cracks cutting across a laminated micrite and microspary host, Lameta Formation, Jabalpur, India. Plane polarized 13X (Courtesy Dr. Partha Ghosh, Indian Statistical Institute, Calcutta, India).

of cutan, it is known as illuviation argillans; however, ferruginized or manganiferous encrusted surfaces of peds are also common as cutan. When peds are not surrounded by cutan, open spaces between peds are commonly filled-in by crystalline calcite, barite, or gypsum (crystallaria in soil terminology). Retallack (1988) has proposed a simplified classification of peds as shown in Fig. 2.6.14.

Glaebule, is another distinctive soil structure formed by the localized concentrations of minerals, rendering it hard and distinct with respect to the surrounding matrix. Glaebule may be calcareous, ferruginous or sideritic in composition and helps in understanding the pH and Eh of the paleosols. For example, (i) ferric nodules and concretions indicate well-drained oxidized soils, (ii) sideritic nodules and concretions indicate water-logged soils, especially those which are marine influenced, and (iii) calcareous nodules and concretions indicate well-drained alkaline soils.

d) Colour mottling, reflecting oxidation-reduction processes operating within a soil, is a very common and characteristic feature found in modern and ancient soil profiles. This is best exemplified in soils where groundwater table is close to the surface, sub-soils remain permanently saturated and top soils can be periodically submerged, depending on the extent of the seasonal fluctuations. Soils that develop this way are known as **gley soils.** Sediments above the groundwater table also develop soils under the influence of percolating

Fig.2.6.9 a: Network of horizontal root moulds on a bedding plane parallel surface, Lameta Formation, Jabalpur, India (Courtesy Dr. Partha Ghosh, Indian Statistical Institute, Calcutta, India).

rainwater. Water may stagnate during a part of the year whenever such soils develop impervious layers. This stagnant water may have a distinct effect on the development of soil; such soils are known as **pseudo-gley soils**. Alternating reduction and oxidation of such pseudo-gley soils cause the segregation of iron and manganese compounds that are found as mottles and concretions; extent of segregation depends on the amount of percolation of water into the soil (Buurman, 1980). This situation creates a reducing condition in the soil atmosphere where ferric iron is reduced to ferrous state. The ferrous iron, being more soluble in water, is readily removed from the soil. The removal of the ferrous iron causes a shift in soil colour from orange-brown to gray-blue and gray-green colours. Mottling is controlled primarily by soil porosity; higher porosity causes higher aeration and naturally higher oxidation and no mottling. Two situations may arise while mottling develop: (i) when the lower part of the soil, below water table, is more gleyed than the upper part, and (ii) when the upper part of the soil, above water table, is more gleyed than the lower part. The former is known as **groundwater gley** and the latter as **surface water gley**

Fig.2.6.9 b: Well-developed rhizocretions, parallel to the bedding plane, mostly elongated carbonate-cemented cylindrical forms; several coalesced masses of carbonate occur on the same surface. Late Quaternary succession in Mahi River section, Rayka, Gujrat, India (courtesy Prof. S. K. Tandon, University of Delhi, Delhi, India).

(Fig. 2.6.15). The difference between the two is more apparent around burrows and root traces, which are more reduced than the soil matrix in surface water gley, but more oxidized in groundwater gley (Retallack, 1990).

Colour mottling is described using three characteristics: contrast, abundance and size of area of each colour. Retallack (1988) has proposed a method for description of mottling (Table 2.6.T9). Yaalon (1966) has suggested a comparison chart that enables one for rapid visual estimate of percentage of mottling or of nodules in the soil profile (Fig. 2.6.16).

Recently, Pimentel et al., (1996) have drawn attention to Paleogene alluvial deposits in Portugal, in which colour mottling, secondary carbonate deposits and prominent colour variations occur both at the bases of the fining-upward sequences and as thick units capping the alluvial sequences. These have been interpreted as the products of shallow, saline, reducing groundwater unrelated to pedogenesis; such non-pedogenic products are prone to be misinterpreted as paleosols. Criteria for their differentiation suggested are: i) thickness – most pedogenic calcretes are 1-2 meters in thickness while groundwater calcretes can be as thick as 20 meters as in the present case, ii) pedogenic calcretes have sharp tops, while it is gradational in case of groundwater calcretes, iii) while pedogenic soil horizons are characterized by features as calcrete texture, irregular fracture and embayed grains, meniscus and pendant cements, these features are conspicuously absent in groundwater calcretes, iv) while soils develop

Fig.2.6.10 a: White chalky, vertically elongated glaebules (nodules) separated by red silt, grading upward into smaller, sub-spherical glaebules, Pleistocene caliche, Tarragona, Spain (Reproduced from M. Estaban and C. F. Klappa, AAPG Memoir Series No. 33, Fig. 43, page 25, © 1983 American Association of Petroleum Geologists, Tulsa, Oklahoma, U.S.A., with kind permission).

b. Plan view of paleosol horizon showing several calcareous glaebules of varying shapes and sizes floating in a calcareous groundmass. Some glaebules have thick brown coating. Lameta Formation, Jabalpur, India (Courtesy Dr. Partha Ghosh, Indian Statistical Institute, Calcutta, India).

c. Stage 3 nodular calcrete with well-developed vertical fabric, Late Carboniferous Sydney Mines Formation, Sydney Basin, Nova Scotia, Canada. Scale 1 meter (courtesy Prof. S. K. Tandon, University of Delhi, Delhi, India).

in fine-grained, less permeable deposits of fining-upward sequence of fluvial deposits, groundwater calcretes occur in coarse-grained alluvium, v) laminar and pisolitic fabrics, rhizocretions, vadose fabrics are features of pedogenic soils that are clearly absent in groundwater calcretes, vi) mottling in groundwater calcretes occurs as pseudogley mottling in reduced zones along fractures and roots within oxidized horizons. During periods of raised water levels, reducing waters migrate up into the fractures and other pores in the oxidized finer-grained sediments. Phases of oxidation and reduction in a zone of fluctuating redox conditions led to the segregation of Fe to give striking pseudogley mottling, representing a type of groundwater gley, vii) presence of intraformational clasts in overlying units, characteristics of soil profiles, are clearly absent in groundwater calcretes.

e) Soil micro-relief in modern and ancient soils is a common and characteristic feature and has been reported by many workers (Allen, 1973, 1974a, b; Goldbery, 1982a, b; Jennings and Sweeting, 1961; McPherson, 1979; Watts, 1977). Various forms of micro-relief structures are **pseudo-anticline** (Fig. 2.6.17) and **gilgai** (Fig. 2.6.18). In cross-section, pseudo-anticline has an overall cuspate synclinal and anticlinal fold which, when reconstructed, are expressed in plan view by a series of parallel surface undulations (Goldbery, 1982a). Wave length and amplitude of these curvi-planar deformed surfaces may vary; Goldbery (1982b), from his study on Lower Jurassic 'Laterite Derivative Facies' of Mishhor and Ardon Formations, Israel, reports a variation in amplitude from 0.30 to 1.25 m, and that of the wave length from 0.40 to 3.40 m.

Fig.2.6.11 a: A well-developed hardpan (platy calcrete) horizon in a paleosol profile, Lameta Formation near Sivni, Jabalpur, India, underlain by a glaebule-rich zone. White patchy zone in the lower part is a gleyed zone. Stick length 1.5 meter. (Courtesy Dr. Partha Ghosh, Indian Statistical Institute, Calcutta, India).

b. Two well-developed platy calcrete horizon (projecting outward) separated by a rhizocretion-rich zone. Note a number of vertically stacked paleosol profiles in the Lameta Formation near Sivni, Jabalpur, India. Stick length 1.5 cm. (Courtesy Dr. Partha Ghosh, Indian Statistical Institute, Calcutta, India).

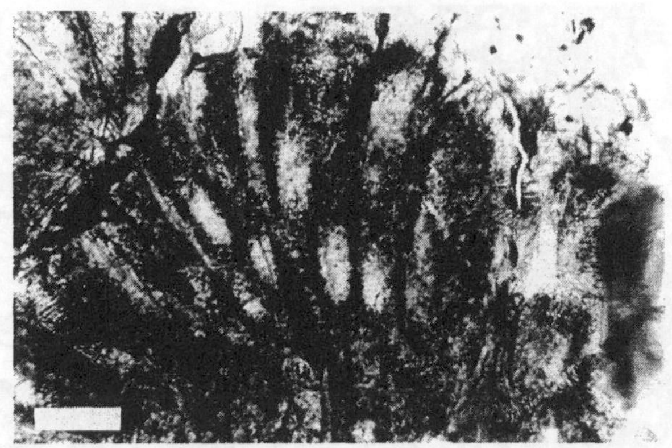

Fig. 2.6.12: Photomicrograph of tangential section of Microcodium with prisms disposed in radial fashion. Note dark peripheral inclusions with more limpid central vacuoles. Scale bar 100 μm (courtesy Dr. P. Freytet, France).

Swelling, particularly smectite-rich clay, and upward buckling of the soil along deeply cracked hummocks forms structures with natural undulations of hummock-and-swale micro-topography, called **gilgai micro-relief** (Fig. 2.6.18) (Paton, 1974) and has been reported from 2200 million years old paleosol in the lowest Dwaal Heuvel Formation near Waterval Onder, Natal, South Africa (Fig. 2.6.19) (Retallack, 1986). Material from hummocks may be eroded and re-deposited in swales, producing highly differentiated undulating layers of swale-fill between the more massive ridge materials (Retallack, 1990). Its sub-surface expression—in the form of disrupted festoon-shaped structure—is called **mukkara structure** (Fig. 2.6.18). Gilgai micro-relief is classified into (i) **nuram gilgai** when depressions are equant in plan, and (ii) **linear gilgai** when depressions are elongate (Paton, 1974). Gilgai micro-relief and mukkara sub-surface structure are indicative of a climate with pronounced dry and wet period and are characteristic feature of vertisols (Paton, 1974). These types of structures not only help in identifying paleosols but also help in reconstructing past climate.

Micro-relief structures are commonly associated with black earth, chernozenic soils of strongly seasonal, sub-humid climatic zones. These have also been reported from semi-arid saline alkaline soils with short period of seasonal rainfall interspersed with long dry season.

Similar to micro-relief structure, tepee structure (tent structure) is reported from peritidal carbonate sediment. It is important, therefore, to differentiate these two similar looking structures. Goldbery (1982a) has suggested following criteria: (i) the striking feature is in the nature of the control fracture systems which are usually closed vertical features defined by closely interlocking slickensided blocks; this is in contrast to the irregular axial voids and injected

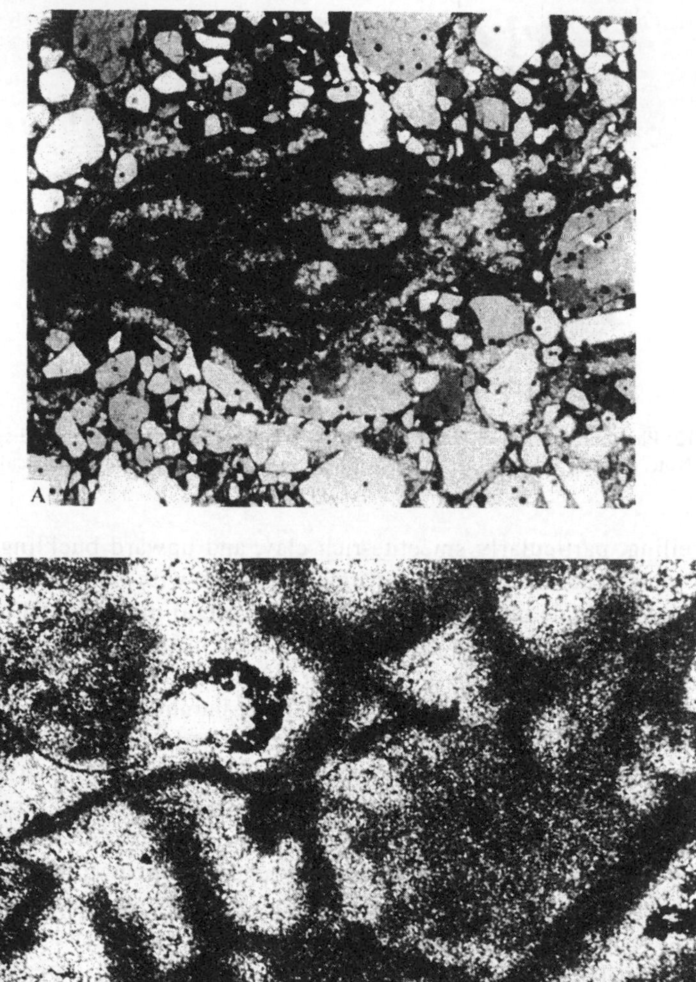

Fig.2.6.13a. Photomicrograph showing alveolar septal structure (in the middle). Alveoli filled with light coloured spars and microspars. Lameta Formation, Jabalpur, India. Oblique polars 26X (courtesy Dr. Partha Ghosh, Indian Statistical Institute, Calcutta, India).

b. A close view of alveolar septal structure. Pleistocene, Mollorca, Spain. Thin section plane polarized light. Scale bar 0.1 mm (Reproduced from M. Esteban and C. F. Klappa, AAPG Memoir Series No. 33, Fig. 66, © 1983, American Association of Petroleum Geologists, Tulsa, Oklahoma, U.S.A., with kind permission).

TYPE	PLATY	PRISMATIC	COLUMNAR	ANGULAR BLOCKY	SUBANGULAR BLOCKY	GRANULAR	CRUMB
SKETCH		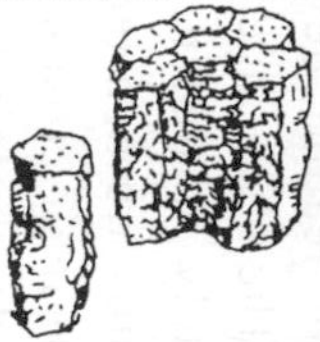	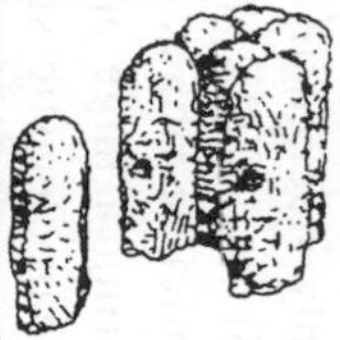		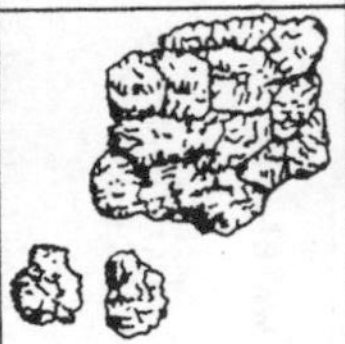	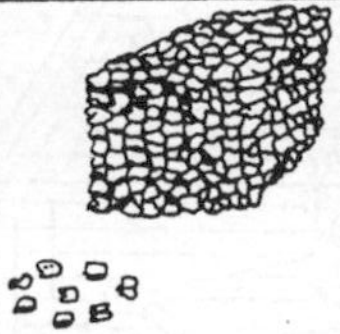	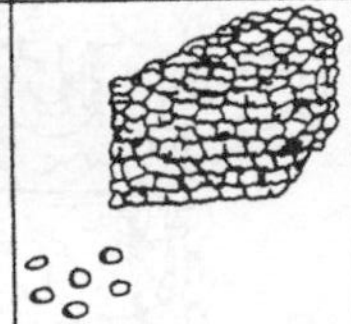
DESCRIPTION	tabular and horizontal to land surface	elongate with flat top and vertical to land surface	elongate with domed top and vertical to surface	equant with sharp interlocking edges	equant with dull intelocking edges	spheroidal with slightly interlock-ing edges	rounded and spheroidal but not interlocking
USUAL HORIZON	E, Bs, K, C	Bt	Bn	Bt	Bt	A	A
MAIN LIKELY CAUSES	initial disruption of relict bedding accretion of ce-menting material	swelling and shrinking on wetting and drying	as for prismatic, but with greater erosion by percolating water, and greater swell-ing of clay	cracking around roots and burrows swelling and shrinking on wetting and dry-ing	as for angular blocky, but with more erosion and deposition of material in cracks	active bioturbation and coating of soil with films of clay, sesquioxides and organic matter	as for granular; including fecal pellets and relict soil clasts
SIZE CLASS	very thin <1mm	very fine <1cm	very fine <1cm	very fine <0.5cm	very fine <0.5cm	very fine < 1mm	very fine < 1mm
	thin 1 to 2mm	fine 1 to 2cm	fine 1 to 2cm	fine 0.5 to 1cm	fine 0.5 to 1cm	fine 1 to 2mm	fine 1 to 2mm
	medium 2 to 5mm	medium 2 to 5cm	medium 2 to 5 cm	medium to 2cm	medium 1 to 2cm	medium 2 to 5mm	medium 2 to 5mm
	thick 5 to 10mm	coarse 5 to 10cm	coarse 5 to 10cm	coarse 2 to 5cm	coarse 2 to 5cm	coarse 5 to 10mm	not found
	very thick > 10mm	very coarse > 10 cm	very coarse > 10 cm	very coarse > 5cm	very coarse > 5cm	very coarse 10mm	not found

Fig. 2.6.14: Classification of soil peds (Reproduced from G. J. Retallack: Field recognition of paleosols, Fig. 9, page 12, *in*, Paleosols and Weathering Through Geologic Time: Principles and Applications: J. Reinhardt and W. R. Sigleo, eds., Special paper 216, © 1988 ,Geological Society of America, Boulder, Colorado, U.S.A., with kind permission).

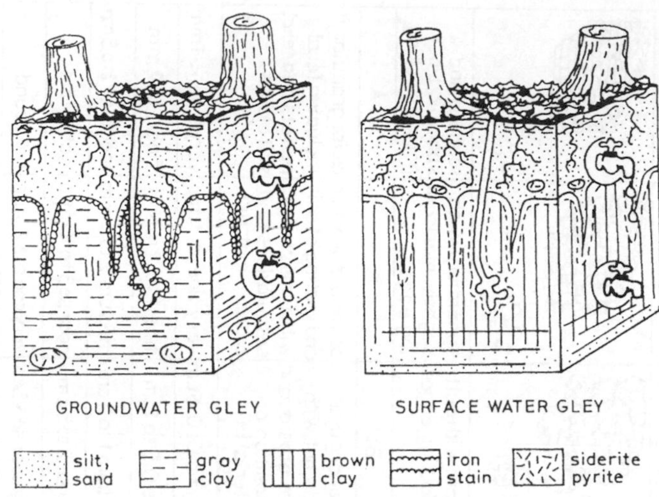

Fig. 2.6.15: Schematic models for the formation of groundwater gley (left) and surface water gley (right) (Reproduced from G. J. Retallack: Soils of the Past: An Introduction to paleopedology, Fig. 4.7, page 72, © 1990, Unwin Hyman, Inc., with the kind permission of Kluwer Academic Publishers, The Netherlands).

filler material of tepees of peritidal carbonates (Burri et al., 1973; Smith, 1974); (ii) soil micro-relief structures do not appear as thick multiple sets, in contrast to tepee structures in carbonates that are commonly stacked in superimposed layers with aggregate thickness of upto 10 meters (Smith, 1974), and (iii) marked differences are seen in the surface plan of the anticline ridges. Whereas soil micro-relief structures are linear and planar covering a large area, polygonal patterns in tepee of peritidal carbonates are the characteristic surface plan.

Slickenside on a fold surface is a characteristic feature, developed due to vertical movement (thrusting) (Fig. 2.6.20 a, b). The striae due to slickenside have a fan-like arrangement oriented approximately normal to the fold axis (Goldbery, 1982b) That these micro-relief structures are non-tectonic and develop penecontemporaneously are evident from (i) their erosional termination surfaces, (ii) incorporation of reworked slickenside clasts into the overlying sediment, and (iii) the conformity of the deformed primary bedding with the folded anticlinal surfaces (Goldbery, 1982b). Various interpretations for the genesis of these micro-relief structures have been offered and summarized in Goldbery (1982b).

f) The soil profile, as discussed earlier, can be used as diagnostic feature of paleosols in rock record, characterized by the illuviation of soluble constituents within the B-horizon. The B-horizon can be found to be highly differentiated into zones, and the C-horizon can be found to imperceptibly grade into the parent rock.

Table 2.6.T9: Method for description of colour mottling (Reproduced from Field Recognition of Paleosols: G. J. Retallack, Table 7, page 18, *in*, Paleosols and Weathering through Geologic Time: Principles and Applications : J. Reinhardt and W. R. Sigleo, eds., Special Paper 216, © 1988, The Geological Society of America, Colorado, U.S.A. with the kind permission)

ABUNDANCE	Few	Mottles occupy less than 2 per cent of the exposed surface.
	Common	Mottles occupy about 2 to 20 per cent of the exposed surface.
	Many	Mottles occupy more than 20 per cent of the exposed surface.
SIZE	Fine	Mottles less than 5 mm diameter in greatest visible dimension.
	Medium	Mottles between 5mm and 15mm in greatest visible dimension.
	Coarse	Mottles greater than 15mm in greatest visible dimension.
CONTRAST	Faint	Indistinct mottles or glaebules visible only upon close examination. Both matrix and mottles have closely related hues and chromas.
	Distinct	Mottles are readily seen with hue, value and chroma different from that of surrounding matrix.
	Prominent	Mottles are obvious and form one of the outstanding features of the horizon; their hue, chroma and value differ from that of the matrix by as much as several Munsell colour units.

g) **Sand wedge** and **ice wedge structures** are sharply tapering, V-shaped in vertical section and are upto several meters deep. These structures result from the winter contraction cracking and expansion during summer; infilling in sand wedge consists of vertically laminated sand and are formed by the repeated filling of frost contraction cracks by the wind blown material. Ice wedges, on the other hand, are filled-in with horizontally laminated fine-grained sediment as well as slumped material (Williams, 1986). These two structures are most useful in identifying periglacial environment. It is generally agreed that sand wedges are products of extreme aridity, while ice wedges are indicative of relatively more humid environment. Such wedge structures, which develop in non-clayey materials that do not expand and crack under wet-dry cycles, are sure signatures of paleosols formed under periglacial environment (Williams, 1986).

h) Hardpan horizons of calcretes are prone to be colonized by lichens, producing spongy, micritic layers of organic-rich and organic-poor millimeter-thick laminations that have been termed as **lichen stromatolite** (Klappa, 1979a) and is a good signature of paleosol. Since these lichen stromatolite closely

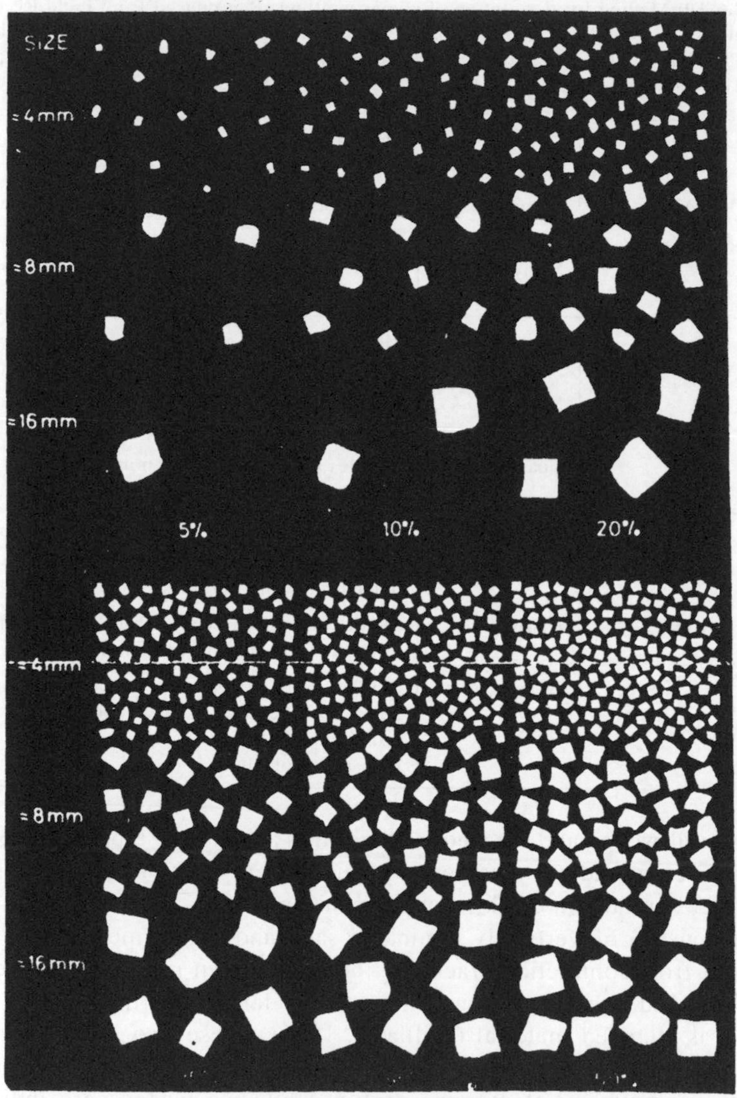

Fig.2.6.16: Chart for the quantitative estimation of mottling and of nodules (Reproduced from D. H. Yaalon: Note. Soil Science, Vol. 102, No.3, Fig. 1, © 1966, Williams & Wilkins, Baltimore, U.S.A. with the kind permission).

resemble cryptalgal laminite, and since these are produced under two distinct environmental settings (sub-aerial vadose versus shallow marine tidal flat), it is necessary to develop criteria to distinguish these in the field. Read (1976) has suggested the following criteria:

Fig.2.6.17: Pseudoanticline in paleosols developed on complex fluvial "Wealdon' facies, Lower Cretaceous, Isle of Wight, southern England (Courtesy Dr. V. Paul Wright, University of Cardiff, U.K).

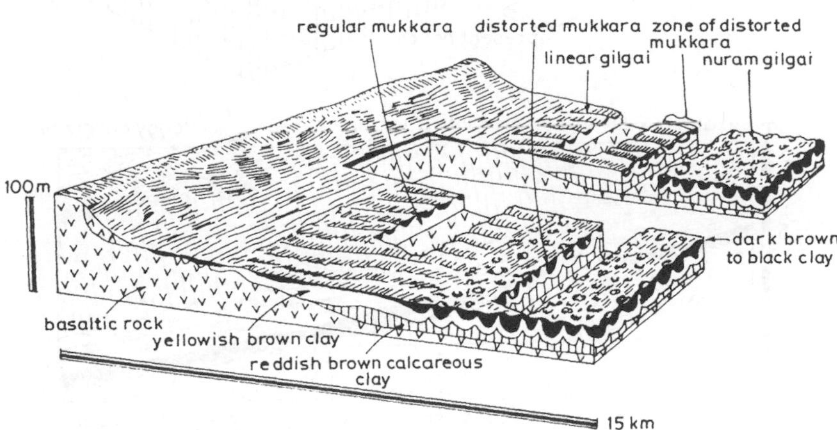

Fig. 2.6.18: Mukkara structure in soils of the Darling Downs, southeastern Queensland, Australia (Reproduced from T. R. Paton: Origin and terminology for gilgai in Australia. Geoderma, Vol. 11, figure 7, © 1974 Elsevier Science-NL, Sara Burgerhartstraat 25, 1055 KV Amsterdam, The Netherlands, with the kind permission).

- Cryptmicrobial stromatolite tends to thicken over highs in contrast to lichen stromatolite that tend to fill the depressions. Moreover, lichen stromatolite tends to follow underlying micro-topography, bending into collapse structures, sinkholes, lining vertical walls of fissures and flooring voids.

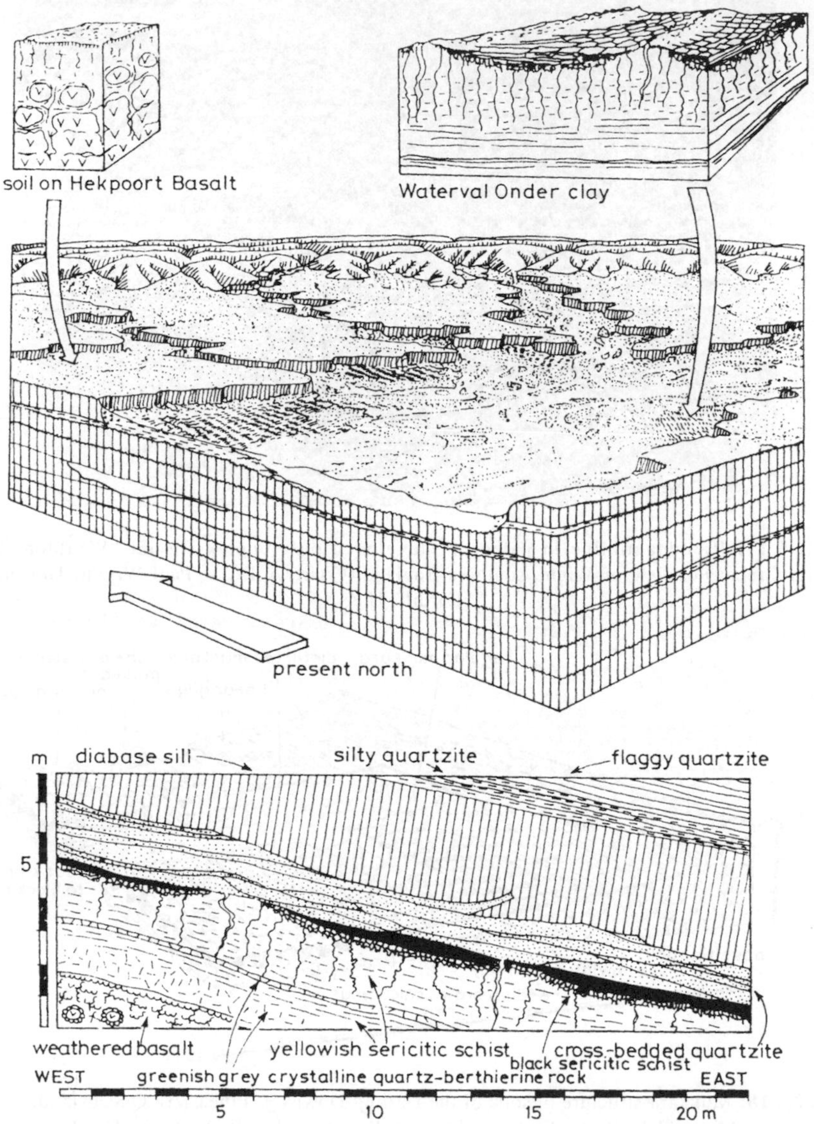

Fig. 2.6.19: Reconstructed paleosol of early Proterozoic age (2200 m.y. old) in the lowest Dwaal Heuval Formation near Waterval Onder, Natal, South Africa, with an annoted field sketch showing the interpretation of its mukkara structure. (Reproduced from G. J. Retallack; Reappraisal of a 2200 m.y. old paleosol from near Waterval Onder, South Africa, Figs. 3 and 14, © 1986a Elsevier Science-NL, Sara Burgerhartstraat 25, 1055 KV Amsterdam, The Netherlands, with the kind permission).

Fig. 2.6.20 a: Note striations on pedogenic slickenside (paleo-pedogenic) surface in paleosols developed on complex fluvial 'Wealden' facies, Lower Cretaceous, Isle of Wight, southern England (Courtesy Dr. V. Paul Wright, University of Cardiff, U.K.).

b: Red calcic vertisols showing pedogenic slickenside surfaces (marked as broken white lines), Late Carboniferous Sydney Mines Formation, Sydney basin, Nova Scotia, Canada. One meter stick for scale (Courtesy Prof. S. K. Tandon, University of Delhi, Delhi, India).

- Laminations in lichen stromatolite are largely due to colour differences than material, whereas cryptalgal laminite show upward-fining, rippled and/or cross-laminated layers and often seen to alternate with calcite and dolomite layers and thin bituminous films.
- Whereas lichen stromatolite is likely to be affected by borings (attesting their indurated nature), cryptmicrobial laminite is commonly extensively burrowed, destroying the primary features in the process.
- The lichen stromatolite is commonly associated with other calcrete features, whereas cryptalgal laminite is found to be associated with different forms of stromatolite with attendant shallow marine facies.
- Terrestrial organisms (land snails, plant roots, insect cases) are associated with calcretes, while these are totally absent in cryptmicrobial laminites.
- Thickness of the calcrete profiles is much thinner (few centimeters to a meter) compared to cryptmicrobial laminites that run considerable thickness.

i) Occurrence of pisolite is a good signature of calcrete profiles.Pisolite in calcrete profiles may apparently look alike to oncolites. Distinguishing features are:

- Calcrete profile containing pisolites lack mechanical sedimentary structures except for reverse grading; nuclei of pisolites are usually lithoclast fragments or broken pisolite fragments. On the other hand, rocks containing oncolites display mechanically formed structures characteristic of current-laid deposits. Broken skeletal fragments form the nuclei of oncolites.
- Pisolite of calcrete profiles show in-place downward growth (meniscus cement), fitted polygonal structures and perched inclusions; these features are absent in beds containing oncolites.

j) Multiple phases of fracturing and their highly irregular orientation are characteristics of recent and ancient soil horizons, particularly the hardpan facies. Such fracturing of the hardpan facies may be in response to (i) periodic wetting and drying, (ii) root activities, and (iii) displacive growth of secondary calcite.

k) Elongate infilling after former soil cracks is characteristic of some paleosols. Joeckel (1994) has described such 1-4 cm wide, slightly undulating, vertical infilling of former soil cracks (Joeckel, 1994, Fig. 3).

l) Sulfate mineralization (gypsum, anhydrite and celestite) is a common feature of paleosol in arid climate (Goldbery, 1982b). Solution activities may remove pedogenic gypsum leaving cavities within paleosol. Microscopic and X-ray studies may help in such situations.

Paleosols in Precambrian Rock Record

The features that help in the identification of paleosols, as discussed above, in the Phanerozoic rock record, may not be that easy in Precambrian rock records.

Plant roots and burrow systems that are commonly associated with Phanerozoic paleosols are absent in the Precambrian. Secondly, metamorphism and tectonic disturbances and hydrothermal alteration have played havoc in obliterating the features of paleosols in the Precambrian rock record. Despite these, paleosols in the Precambrian rock records have been discovered in unprecedented numbers. Following are the points, as highlighted by Retallack (1992), which should be taken into account for the identification of paleosols in the Precambrian.

1. It is important, as a first step, to read in between lines of earlier publications on the Precambrian studies and note the positions of regional unconformities and map these surfaces in detail. As in the Phanerozoic sequences, paleosols are likely to occur here, particularly if the rocks overlying the unconformity surface represent continental environment. Irregularity of a mapped geological contact occasionally records paleo-topography rather than tectonism. Highly irregular contacts in carbonate sequences identify ancient karst surfaces (James and Choquette, 1988).
2. Careful facies analysis of Precambrian sedimentary record is necessary. It is in the rocks of continental environment where paleosols are likely to occur. In the absence of first-order information (root traces and burrow systems), search for other macro-structural features may help in the identification of paleosols in the Precambrian. For example, (i) ice-wedges, sand-wedges (soil features of the permafrost region as discussed earlier), (ii) presence of salt crusts (characteristic of desert playas and coastal sabkhas), (iii) micritization, boring, nodule formation, and (iv) irregular topography (dolines, grikes, cavernous porosity and collapse breccias in carbonate-dominated terrane) may help to identify paleosols in the Precambrian rock record (Williams, 1986; Lowe, 1983; Bertrand-Sarfati and Moussine-Pouchkine, 1983).
3. Terrestrial lava flows are sites where paleosols are likely to occur. Thick clayey horizons in between lava flows may possibly be paleosols. Recently, stoichiometric and well-ordered dolomite of pedogenic origin has been reported from a Quaternary basaltic weathering profile, Kohala peninsula, Hawaii (Capo et al., 2000). Geologic, geochemical, and isotopic data indicate that this dolomite is neither marine influenced nor the result of aeolian input. This, being a physical process, is also likely to occur in Precambrian terrestrial lava flows.
4. While no indisputable microfossils have yet been reported from the Precambrian paleosols, isotope signature may suggest pedogenic microbial colonization. Beeunas and Knauth (1985) and Retallack and Mindszenty (1994) have reported an upward increase in higher stable isotopes for both oxygen and carbon in carbonate rocks of 1.2 Ga age. Such enhancement in lighter isotopes is attributed to photosynthetic organisms. This should be substantiated by concurrent study of illite crystallinity, kerogen maturation

and metamorphic petrography to ensure that such gradient in lighter isotopes are not due to burial and/or metamorphic effect.

5. Soil peds with cutans are good indicators of paleosols and have been reported, amongst others, (i) from Watervaal Onder paleosols, Transvaal, South Africa (Retallack, 1986a), (ii) from ca. 1.74 Ga Thelon paleosols, Canada (Gall and Donaldson, 1990), and (iii) from sub-Huronian Supergroup paleosol near Elliot Lake, Canada (Gay and Grandstaff, 1980). In all these places, blocky angular peds with dark cutans of washed-in clay with opaque oxide grains of amorphous iron and manganese form the characteristic identifying features. Many others have reported presence of blocky soil ped (Duchac and Hanor, 1987; Lowe, 1983; Lowe et al., 1985).
6. Gilgai and Mukkra structures (Fig.2.6.18), common in clay-rich vertisols in sub-humid to semi-arid climates with strong seasonal distribution of rainfall, is a diagnostic feature of paleosols and has been reported by Retallack (1986a) from the Precambrian Watervaal Onder paleosol, Transvaal, South Africa.
7. Nodules and crystals may also form features of paleosols. Since these features may also develop through diagenesis, metamorphism and cooling of lava, extra criteria are needed. These are:

- Voids surrounding the nodules are commonly cement-filled,
- Paleosol nodules preserve easily weatherable minerals that usually do not survive outside,
- Paleosol nodules commonly congregate in horizons at a fixed depth below an erosional surface,
- When paleosols develops on sedimentary rocks, nodules may include fossils, stromatolites, or sedimentary lamination inherited from parent rock.

8. Soil stratification (horizonation) is another physical feature that may help in the identification of Precambrian paleosols. Surface A-horizons, saprolitic C-horizons and incipiently weathered protolith R-horizons are typically the only soil divisions that are usually recognized in Precambrian paleosols (Retallack, 1986b). Such horizonation has been described from paleosols associated with 1.74 Ga Matonabee unconformity where it shows an upper hematitic zone followed downward by less oxidized saprolitic and clay-rich zone overlying an incipiently weathered protolith with no clear zone boundaries (Gall, 1992). **Corestone** is another field feature that helps in the identification of Precambrian paleosols, especially those developed on crystalline bedrocks as opposed to sedimentary parent materials (Retallack and Mindszenty, 1994).
9. Vertical colour gradation is another useful criterion. Sub-aerial exposure of protolith causes oxidation and reddening (rubifaction) of the paleosol in its upper part. A vertical colour gradation from a reddened hematitic top to gray-green illitic and chloritic base has been reported from the paleosols

associated with the Matonabee unconformity (Gall and Donaldson, 1990). Colour mottling due to local redox changes may also develop in paleosols as in Pitz Formation below the Thelon Formation (Gall and Donaldson, 1990).

10. In case of metamorphic changes of paleosols, one has to rely on metamorphic mineral assemblages to identify paleosols in the Precambrian terrane. Ideally, the occurrence of lithologic stratification with aluminosilicate-rich phases (e.g., Al_2SiO_5 polymorphs, corundum, diaspore, pyrophyllite, kaolinite), which increases in abundance towards a lithologic contact (unconformity), may suggest the presence of paleosols. (Gall, 1992). If the aluminosilicate-rich phase goes beyond 20%, it becomes a suspect, since crustal igneous rocks average about 15.6 % Al_2O_3. Table 2.6.T10 lists mineral assemblages for some metamorphosed Precambrian paleosols from India. Moreover, enrichment of various economically exploitable minerals (iron, manganese, gold, uranium, lead, zinc etc.) takes place along paleo-weathering surfaces (Button and Tyler, 1981). One should, therefore, map these mineralized zones carefully to identify Precambrian paleosols.
11. Irregularly oriented slickensided planes of red clay have been reported from Sheigra clay (Retallack and Mindszenty, 1994). Under a petrographic microscope, such clay shows internal lamination and cut other clay skins in a complex fashion (their Fig. 5c). Such features, in soil science, are called **illuviation argillans** (Brewer, 1964) and form by the washing down of clay into cracks. Such low temperature opening and filling of cracks by clay are diagnostic features of paleosols. The contorted irregular form of such filling by clay distinguishes them from hydrothermal and igneous veining.
12. Presence of irregular iron stains around opaque grains and trellis-like grains formed by progressive oxidation of biotite are signatures of ancient weathering and has been reported from the Precambrian paleosols (Retallack and Mindszenty, 1994) and Williams, 1968).
13. Presence of weather-resistant minerals (i.e., quartz, microcline) in a situation where all other minerals have altered to clay, along with shattered and

Table 2.6.T10: Metamorphic mineral assemblages of three Precambrian paleosols from India (modified from Gall, 1992).

Metamorphic Minerals	**Comments** (Paleosols protolith in parenthesis)	**Reference**
Pyrophyllite and/or diaspore ± muscovite ± kaolinite	Metalaterite (granite anatexites)	Sharma, 1979
Sillimanite ± Corundum, Sillimanite - quartz	Metabauxite	Golani, 1989
Quartz-sillimanite-garnet ± K-feldspar	Khondalites (granodiorite-tonalite)	Dash et al., 1987

embayed edges of quartz grains are signs of paleo-weathering (Retallack and Mindszenty, 1994). Absence of hydrothermal minerals negates this as a product of hydrothermal alteration.

2.6.2 Karst Facies

Karst represents a specific landform with characteristic surface morphology, subsurface drainage and collapse features commonly found in carbonate rocks, and a geographic region characterized by this landform. Esteban and Klappa (1983) have defined it as a "a diagenetic facies, an overprint in sub-aerially exposed carbonate bodies, produced and controlled by dissolution and migration of calcium carbonate in meteoric waters, occurring in a wide variety of climatic and tectonic settings, and generating a recognizable landscape".

Karst and calcrete facies are two end members of a facies, both involving mobility of calcium carbonate. While in karst facies there is net loss of calcium carbonate, in calcrete facies there is zero balance of calcium carbonate or a net gain (addition of $CaCO_3$). Calcrete facies forms under an arid or semi-arid climate, while karst facies develops in all climates (preferably in humid climate) with availability of water. Again, while calcrete facies develops in all rock types (igneous, sedimentary and metamorphic), karst facies specially develops in carbonate and evaporite rocks. The main control on karst facies development are i) climate, ii) composition of bedrock, iii) relief (in relation to sea level changes), and iv) time.

Paleokarst refers to karst features developed on past landscape. Features, inferred to represent paleokarst, are moulds, vugs, *in situ* breccia, collapse strata, solution-widened joint (either open or filled-in by clayey mud), red-orange discolouration of rocks, internal sediment, breccia-filled channel pores, boxwork, caves, scalloped and planar erosion surface, breccia interpreted to represent terra rosa, closed depressions interpreted to be sinks and dolomitic shale (considered to be paleosols) (Kahle, 1988).

It is important to ascertain whether the karst features observed on the bedding plane surface were formed soon after deposition or an effect of the recent processes, or were formed at an unknown intervening time. The surest way of confirming that features are 'fossil' features is to find skeletons or calcretes cemented onto, or borings penetrating in to the paleosurface (James and Choquette, 1984).

The process of karst formation can take place both (a) on the exposed bedding plane surfaces and (b) beneath soil cover with different styles of karstic surfaces. Corrosion of rock surfaces is more active and intense under soil cover in the humid tropics (compared to high latitude or desert environment) because of higher temperature, increased availability of organic litter and rainfall. Climate is the major controlling factor in karst development. Porosity and permeability of carbonate rock also control karst formation; well-lithified carbonate rocks show better development of surface features (James and Choquette, 1984).. As a general rule, the various solution sculptures formed on the exposed bedding plane surface are sharp crested compared to those under soil cover that are more rounded (James and Choquette, 1984).

Field Signatures of Paleokarsts

a) Features of Surface Karst

(i) Dolines are small-scale depressions on the surface of carbonate rocks (Fig. 2.6.21); depressions can be bowl-shaped or cone-shaped in vertical sections and are either circular or elliptical in plan view. The diameter (10–10,000 meters) is usually greater than the depth (2–100 meters). Dolines usually occur in clusters in close proximity to each other (Sweeting, 1972). In common geologic parlance, dolines are usually known as sink- holes. Dolines can be grouped as (a) normal solution dolines, (b) alluvial dolines, (c) solution-subsidence dolines, and (d) collapse dolines (Sweeting, 1972). As the dolines grow in dimension, they often coalesce and such coalescent dolines are known as uvalas. Since uvalas form through the growth and ultimate coalescence of neighbouring

Fig. 2.6.21: Bedding plane view of a doline forming a bowl-shaped basin, Romaine Formation (Middle Ordovician), Mingan Island, Quebec. The basin dimensions are 3 meter depth and 20 meter in diameter (Reproduced from A. Desrochers and N. P. James: Early Paleozoic surface and subsurface paleokarst : Middle Ordovician carbonates, Mingan Island, Quebec, *in*, Paleokarst : N. P. James and Phil W. Choquette, eds., Fig. 9.5A, page 188, © 1988, Springer-Verlag GmbH & Co.KG., with kind permission).

dolines, floors of uvalas are undulating and irregular in shape (usually 500–1,000 meters in diameter and as much as 100-200 meters deep). Dolines owe their origin to dissolution beneath a soil cover or subsurface dissolution, formed mainly through activities of percolating meteoric water into joints and fissures and aided by (a) structural weak planes of the country rocks, (b) vegetation, and (c) porosity and permeability of limestone.

(ii) Rundkarren (rounded solution runnels) are dendritic to sub-parallel grooves on carbonate rocks (Fig. 2.6.22) having 10-50 cm width and 12–50 cm in depth and length is variable from a few centimeters to over 10 meters and occur as sub-parallel grooves. These are round to flat in cross-section, closely spaced and separated by rounded and smooth crest ridges and troughs. These are basically drainage features formed under a cover of peat or vegetation, and are normally related to the slope of the rock. On horizontal surfaces, these are short and dendritic, and long and sub-parallel on inclined surfaces. Rundkarren are fewer in well-jointed limestone and occur prominently where joints are fewer.

(iii) Rillenkarren are finely chiselled tunnels or grooves having rounded troughs and sharp fine ridges (Fig. 2.6.23); these are usually about 1–2 cm deep, 1–2 cm wide and usually less than 50 cm long. Formation of this type of karren is rather rapid and appears on rocks that are exposed to rain/melt water for only a few years or even months, in their most perfect form they look like flutes.

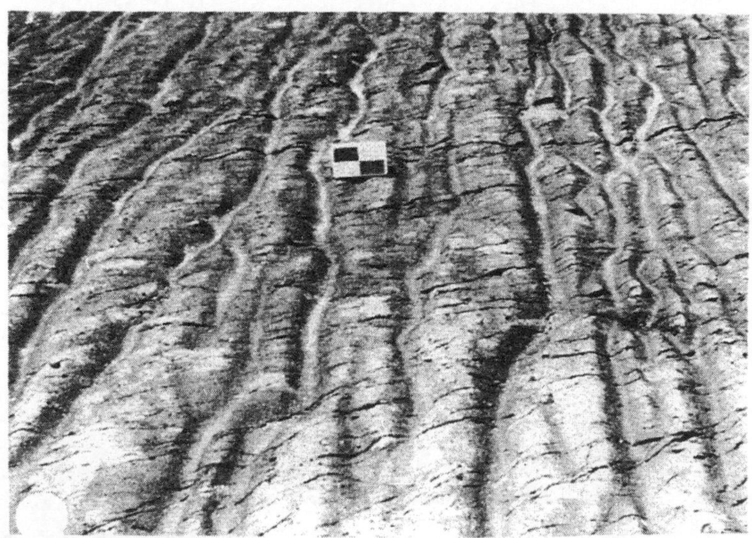

Fig. 2.6.22: Closer view of rundkarren consisting of rounded grooves that are separated by rounded crest ridges, Post-Romaine paleokarst, Mingan Island, Quebec. Field note book is 20 cm (Reproduced from A. Desrochers and N. P. James: Early Paleozoic surface and subsurface paleokarst: Middle Ordovician carbonates, Mingan Island, Quebec, *in*, Paleokarst: N. P. James and Phil W. Choquette, eds., Fig. 9.5C, page 189, © 1988, Springer-Verlag GmbH & Co. KG., with the kind permission).

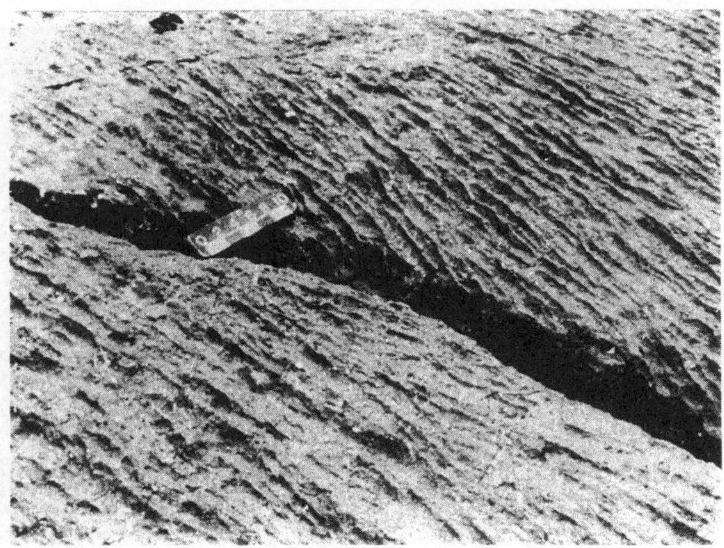

Fig. 2.6.23: Rillenkarren developed on Ordovician limestone, western Newfoundland. Scale in centimeter (Reproduced from N. P. James and Phil W. Choquette: Diagenesis 9 - Limestone - The meteoric diagenetic environment, Geoscience Canada, Vol. 11, No. 4, Fig. 16, © 1984, The Geological Association of Canada with the kind permission).

(iv) Rinnenkarren is another solution-sculptured feature on carbonate rock surfaces that are larger in size and shape than rillenkarren (being up to 40 cm deep and about 40–50 cm wide and may be meters in length). Their sides are steep to vertical with sharp edges and are randomly oriented unlike rundkarren. The troughs are usually rounded with sharp crest.

(v) Trittenkarren is another karstic feature that resembles a heel (Fig. 2.6.24). It is asymmetrical in vertical section with a flat tread (20-40 cm long) and a vertical riser of 5–15 cm. It is a bedding plane feature occurring as semi-circular and laterally coalescent form; it is commonly organized in steps and oriented perpendicular relative to other karren forms. It is thought to result from solution activities on steeply inclined surface exposed to rain water. It often occurs in conjunction with rillenkarren; while rillenkarren forms on gentler, steeper slopes, trittenkarren forms on Kamenitzas are circular to oval in outline with flat bottom and rounded edges and are distinctly solution basins (Fig. 2.6.25). Occasionally, kamenitzas may have steep sides with irregular and sharp edges that in places may overhang (Fig. 2.6.26). Such structures rarely occur alone and are commonly associated with dendritic rundkarren. In dimensions, these are small basins having diameters of a few centimeters to over 3 meters.

(vi) Scalloped and erosional surfaces, geometrically, show the form of a series of upwardly concave smooth curved depressions (10 cm to over 1 meter

Fig. 2.6.24: Bedding plane view of trittenkarren on top of a calcarenite cycle, Middle Ordovician carbonates, Mingan Island, Quebec. Hammer is 30 cm long. (Reproduced from A. Desrochers and N. P. James: Early Paleozoic surface and subsurface paleokarst: Middle Ordovician carbonates, Mingan Island, Quebec, *in*, Paleokarst: N. P. James and Phil W. Choquette, eds., Fig. 9.18A,page 207, © 1988, Springer-Verlag GmbH & Co, KG., with the kind permission).

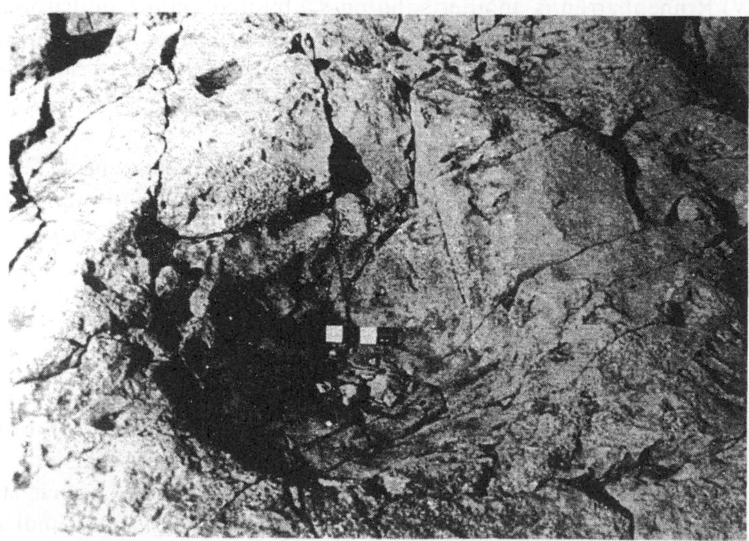

Fig. 2.6.25: Bedding plane view of kamenitza on top of the Romaine Formation, Middle Ordovician. Mingan Island, Quebec. Scale in centimeter (Reproduced from A. Desrochers and N. P. James: Early Paleozoic surface and subsurface paleokarst: Middle Ordovician carbonates, Mingan Island, Quebec, *in*, Paleokarst: N. P. James and Phil W. Choquette, eds., Fig. 9.5D, page 189, ©1988, Springer-Verlag GmbH & Co. KG., with the kind permission).

Fig. 2.6.26: Cross-section view of kamenitza, overlain by skeletal limestone. Note the steep to vertical sides of the sculptured kamenitza characterized by sharp edges. Hammer is 30 cm long (Reproduced from A. Desrochers and N. P. James: Early Paleozoic surface and subsurface paleokarst: Middle Ordovician carbonates, Mingan Island, Quebec, *in*, Paleokarst: N. P. James and Phil W. Choquette, eds., Fig. 9.11A, page 197, © 1988, Springer-Verlag GmbH & Co. KG., with the kind permission).

wide and a relief of about 30 cm), with intervening steep, locally overhanging walls and sharp-crested ridges (Fig. 2.6.27). Scalloped surfaces have extremely sharp boundaries with underlying rocks (Cherns, 1982; Read and Grover, 1977). Many scalloped/planar surfaces are encrusted by calcareous algae and bryozoans. Rare microscopic and larger macroscopic borings extending to a few millimeters below the contact are reported (Read and Grover, 1977).

Commonly, the scalloped irregular surfaces in carbonate rocks are seen to pass laterally into planar erosional surfaces (Fig. 2.6.28). Planar surfaces are commonly featureless surfaces except for centimetre deep pits and represent unconformity surfaces. Such surfaces indicate that (i) the carbonate on which the planar surfaces developed is fine-grained, and (ii) the surfaces were initially covered with sediments at the time of formation of planar surfaces (Pluhar and Ford, 1970). Similar planar surfaces have also been reported from the Middle Ordovician limestones, Mingan Island, Quebec, by Desrochers and James (1988), from Middle Ordovician limestone, Virginia, by Read and Grover (1977), and from the Silurian Eke Formation, Gotland, Sweden, by Cherns (1982).

Scalloped/planar surfaces apparently look alike to hard-grounds and stylolitic surfaces, and hence need criteria to differentiate them. Read and Grover (1977) and Cherns (1982) have suggested the following criteria:

Fig. 2.6.27 a: Scalloped stack-like mass (outlined with felt marking pen) between New Market Fenestral Limestone and overlying Lincolnshire skeletal limestone, Penn Laird, Virginia. Scale in centimeters (Reproduced from J. F. Read and G. A. Grover: Scalloped and planar erosion surfaces, Middle Ordovician limestones, Virginia: Analogues of Holocene exposed karst or tidal rock platforms. J. Sediment. Petrology, Vol. 47, No. 3, Fig. 2B, page 958, © 1977, Society for Sedimentary Geology (SEPM), Tulsa, Oklahoma, U.S.A. with the kind permission).

- The scalloped/planar surfaces are free from mineral impregnation and micritization that are commonly associated with hard-ground surfaces.
- The scalloped/planar surfaces show rapid lithologic and biotic changes across the boundary in contrast to hard-ground surfaces where these remain broadly similar.
- Scoured sidewalls and undercut scalloped surfaces are neither encrusted nor bored and the erosion surfaces show truncation of grains; sediments beneath scalloped/planar surface generally lack bioturbation texture, even though the overlying sediments may be intensely bioturbated. This is in sharp contrast to hard-ground surfaces where the overlying and the underlying beds are commonly intensely bioturbated.
- The smooth basinal forms and overlying pinnacles that characterize the scalloped contacts are generally absent from hard-grounds.

(vii) Fissures (kluftkarran) are sometimes present below paleokarstic surfaces that are commonly sediment-filled. These fissures form a complex network and are the most likely product of fracturing, jointing and subsequent solution widening (Fig. 2.6.29). Vertical fissures are found to be planar to slightly curved and are occasionally cut by minor secondary fissures. In

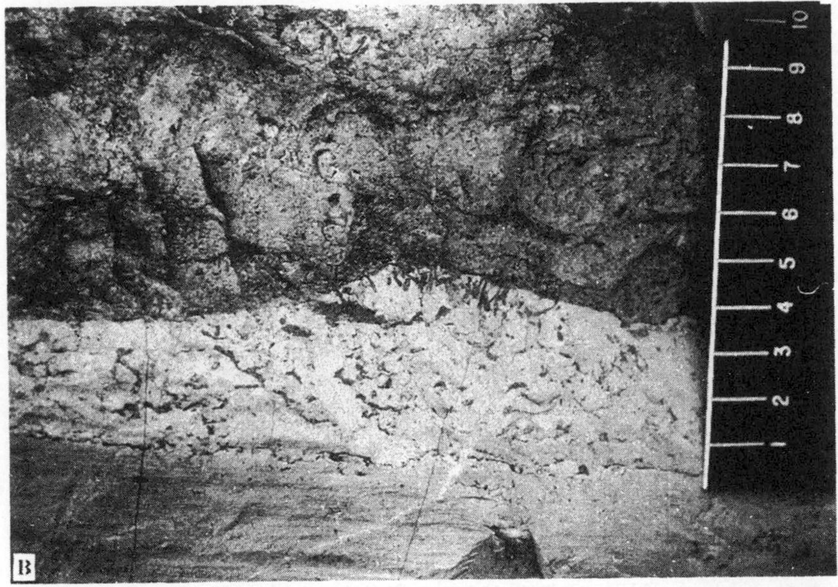

Fig. 2.6.27 b: Scalloped surface being affected by borings extending down a few millimeters beneath the contact, near Eggleston, Virginia (Reproduced from J. F. Read and G. A. Grover: Scalloped and planar erosion surfaces, Middle Ordovician limestones, Virginia: Analogues of Holocene exposed karst or tidal rock platforms. J. Sediment. Petrology, Vol. 47, No. 3, Fig. 3C, page 959, © 1977, Society for Sedimentary Geology [SEPM], Tulsa, Oklahoma, U.S.A. with the kind permission).

England, such solution-widened joints are known as **grikes,** and are filled-in by carbonate cements, commonly known as speleothem flowstone (Fig. 2.6.30 a, b), with occasional filling by banded fibrous dolomite. Roofs are encrusted in a fairly even manner, while the floors of the cavities are covered by pisolites (**cave pearls**) of 2–5 mm in diameter (Fig. 2.6.31). The grikes remain mainly restricted to the upper part of the karst profile that, on plan view, display diamond pattern aligned parallel to the joint system. Horizontal cavities are usually associated with grike system.

(viii) Mammilated surface is a broad irregularly spaced hollow and swells upto 1 meter across, developing below a soil cover (Fig. 2.6.32). Hollows are funnel-shaped or cylindrical pits upto 1.5 meters diameter. Described first by Walkden (1974), mammilated surfaces are occasionally stained red with iron oxide. The outer regions of such surfaces may show presence of pyrite/marcasite crystals or may have an irregular layer of brown laminated carbonate replacing the limestone fabric. That these mammilated surface are not products of recent origin can be discounted by the fact that

- The base of the clay horizon, below which mammilated surface develops, should be dry in freshly dug quarries,

Fig.2.6.28: Vertical cliff exposure showing planar erosion surface capping a calcarenite cycle that separates overlying skeletal muddy limestones (L), Mingan Island, Quebec (Reproduced form A. Desrochers and N. P. James: Early Paleozoic surface and subsurface paleokarst: Middle Ordovician carbonates, Mingan Island, Quebec, *in,* Paleokarst: N. P. James and Phil W. Choquette, eds., Fig. 9.18D, page 208, © 1988, Springer-Verlag GmbH & Co. KG., with the kind permission).

Fig. 2.6.29: Field photograph showing sand-filled grikes(s) below unconformity surface, Dismal Lakes Group, Canada. The white notebook is 16 cm long (Reproduced from C. Kerans and J. A. Donaldson: Proterozoic paleokarst profile, Dismal Lakes Group, N.W.T., Canada, *in*, Paleokarst: N. P. James and Phil W. Choquette, eds., Fig. 8.10, page 176, © 1988, Springer-Verlag GmbH & Co. KG., with kind permission).

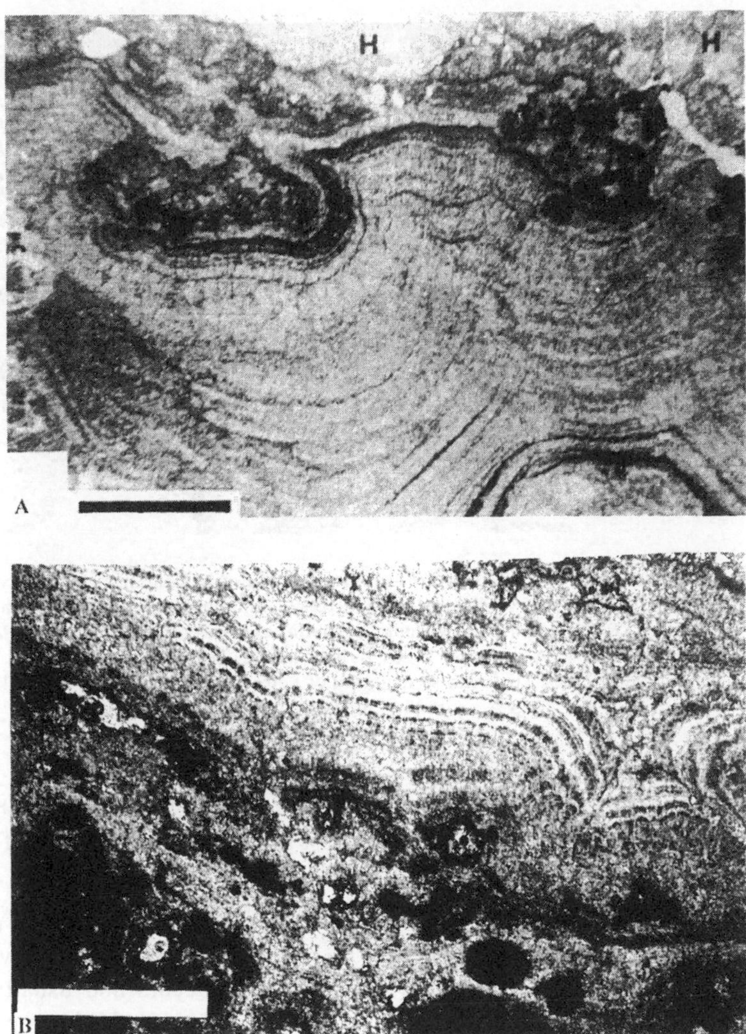

Fig.2.6.30 a: Photomicrograph showing laminated flowstone (speleothem) from grike. Laminar fibrous dolostone crust grows downward from the contact with microsparite host dolostone (H). Scale bar is 2 mm. (Reproduced from C. Kerans and J. A. Donaldson: Proterozoic paleokarst profile, Dismal Lakes Group, N.W.T., Canada, *in*, Paleokarst: N. P. James and Phil W. Choquette, eds., Fig. 8.9A, page 175, © 1988, Springer-Verlag GmbH & Co. KG., with the kind permission).

b: Photomicrograph of laminar flowstone (speleothem) of Pleistocene age. Thin section, plane polarized light. Scale bar is 1 mm. (Reproduced from M. Esteban and C. F. Klappa: Subaerial exposure environment, *in*, Carbonate Depositional Environment: P. A. Scholle, D. G. Bebout and C. H. Moore, eds., Fig. 26, page 16, AAPG Memoir Series No. 33, © 1983, American Association of Petroleum Geologists, Tulsa, Oklahoma, U.S.A., with the kind permission).

Fig.2.6.31: Cave pearls and laminated cavern sediment, southern Appennines, Italy. Cave pearls are micrite-rich and occur in layers with characteristic coarsening upward grain size arrangements (Reproduced from M. Esteban and C. F. Klappa: Subaerial exposure environment, *in*, Carbonate Depositional Environment: P. A. Scholle, D. G. Bebout and C. H. Moore, eds., AAPG Memoir Series No. 33, Fig. 28, © 1983, American Association of Petroleum Geologists, Tulsa, Oklahoma, U.S.A., with the kind permission).

- The mammilated surface is not related with joint system, and
- Axes of the solution pits in dipping limestone or dolomite should be perpendicular to the bedding plane in contrast to vertical axes in case of recent origin.

(ix) Dissolution pipes are cigar-shaped tubes having a depth of upto 4 meters and width upto 1 meter (Fig. 2.6.33 a, b). Pipes are generally vertical irrespective of the dip of the host strata in which it occurs and usually have circular cross-section. Density of these vertical pipes is very high; in extreme cases, pipe hollows become dominant over solid rock. The most remarkable features are their regularity and smoothness of their walls. Pipes are commonly found filled and covered by coarse-grained sediment or sandy terrestrial red clay. Occurrences of these dissolution pipes have been reported both from recent sediments (Lundberg and Taggart, 1995) and ancient rock records (Brunsden et al., 1976; Walkden and Davies, 1983; Webb, 1994).

b) *Features of sub-surface Karst*

Under-saturated meteoric water creates extensive sub-surface dissolution features that vary in dimension from small pores to large caves (Choquette and

Fig. 2.6.32: Gently rolling mammillated surface. Upper Visean (Carboniferous) limestones, Derbyshire, England. Hammer (centre foreground) is 30 cm. (Reproduced from G. M. Walkden: Paleokarstic surfaces in Upper Visean (Carboniferous) limestones of the Derbyshire block, England. J Sediment. Petrology, Vol. 44, Fig. 3, © 1974, the Society for Sedimentary Geology (SEPM), Tulsa, Oklahoma, U.S.A. with the kind permission).

Pray, 1970). The characteristic precipitates that adorn these dissolution voids are flowstone (Fig. 2.6.30 a, b), dripstone, cave popcorn, cave pearls (Fig. 2.6.31) etc., discussed earlier. Besides, these voids contain mechanically emplaced sediments and large voids by breccia (for detail discussion on paleokarst, its genesis and products see Craig, 1988; Desrochers and James, 1988; Esteban and Klappa, 1983; Ford, 1988 James and Choquette, 1984; Kerans and Donaldson, 1988; Lohmann, 1988; Meyers, 1988;Wright and Smart, 1994).

It is important to discriminate between **interstratal** and **subjacent karst**. Interstratal karst refers to wholesale dissolution of the soluble unit, such as, evaporites in the sub-surface and subsequent collapse forming breccias, known as collapse breccia (Fig. 2.6.34). Such interstratal karst breccias are characterized by the absence of speleothms or mechanically emplaced sediment (Wright, 1994a). Subjacent karst, on the other hand, refers to dissolution features developed at the contact between the overlying permeable bed (usually siliciclastic) and the less permeable underlying carbonate. The overlying permeable bed allows water to percolate downward and corrode the top of the less permeable carbonate bed forming solution features of various dimensions (Wright, 1994b).

Subjacent karst produces features closely resembling a paleokarstic surface and hence, needs discrimination. Wright (1982) has suggested some simple criteria as under:

Fig. 2.6.33 a: Cross-section of dissolution pipe in cross-bedded coarse sandstone exposed by cliff erosion, Punta Higuero, Puerto Rico. The pipe is filled with coarse clasts of marine origin including coral fragments. Notebook is 20 cm (Courtesy Dr. Joyce Lundberg, Carleton University, Canada).

- The paleokarstic surface should show features indicating very near-surface exposure, or overlain by paleosols, or other terrestrial deposits.
- The truncation of the paleokarstic surface by the overlying bed indicates that solution and erosion developed before burial.
- The overlying bed should not display any feature indicative of solution activities, such as, pipes that connect to the underlying karstic surface. This negates post-burial solution processes that develop subjacent karst.
- The beds that overlie paleokarstic surface should not show any sign of collapse into the underlying paleokarstic surface suggesting solution after burial.

Fig. 2.6.33 b: Looking down on dissolution pipes in coarse sandstones, East Island, Puerto Rico. The pipe is being removed by sea spray erosion. Notebook is 20 cm (Courtesy Dr. Joyce Lundberg, Carleton University, Canada).

Breccia forms the most spectacular feature of the paleokarstic facies and is one of the best field criteria. Breccia associated with paleokarst facies has a wide variety of external form: (i) irregular sheet and patches in case of mantling breccia, (ii) cylindrical form associated with sinkholes and dolines, (iii) more or less tabular form with irregular roofs and regular floors associated with cave-roof collapse (Fig. 2.6.34).

c) Other associated features of paleokarst

The paleokarst surfaces of Phanerozoic rock record are characterized by the presence of borings with circular aperture oriented perpendicular to the paleokarst surfaces. Beside, pebble lags are occasionally found to have accumulated in local depressions on paleokarst surfaces. These are poorly sorted clasts commonly having a size range of 1–20 cm. The accumulation of clasts may possibly

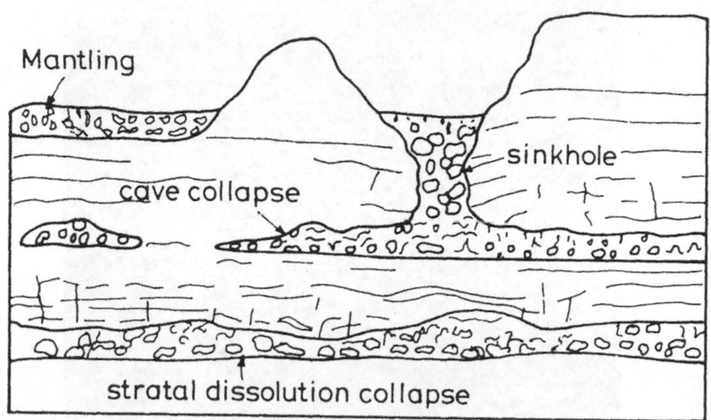

Fig. 2.6.34: A sketch outlining common types of surface and subsurface breccia in karst terrain (Reproduced from Phil W. Choquette and N. P. James: Introduction, *in*, Paleokarst: N. P. James and Phil W. Choquette, eds., Fig. 5, page 10, © 1988, Springer-Verlag GmbH & Co. KG., with kind permission).

represent regolith (residual soil material) that may have collected in local depressions.

While karstic landforms are commonly reported form recent and ancient carbonate rock sequences, identical morphologies have recently been reported from terrigenous quartz-rich rocks and quartzites (Wray, 1997, and references cited therein). Solution landforms in sandstone and quartzite include large bedrock pinnacles and towers (Figs. 2.6.35 and 2.6.36), corridors (Fig. 2.6.37), grikes (Fig. 2.6.38), solution basins (Fig. 2.6.39), runnels (Fig. 2.6.40), and even stalactites (Fig. 2.6.41) and stalagmites (Fig. 2.6.42). The occurrences of these various solutional features in siliciclastic rocks explode the myth that the quartzose rocks are physically and chemically resistant rocks. Secondly, these solutional landforms in quartzose rocks are not only restricted in tropical climatic zones but do also occur in temperate and sub-tropical climatic settings. This implies that the structural attitude of bedrocks is more important in the development of karstic features than the climatic control (Wray, 1997). While such karstic features in quartzose rocks are reported from the present-day earth surface, their ancient equivalents are yet to be reported; the lack of record of such solution features in ancient rock records may possibly be apparent than real.

Utility of Paleosol Study

1. It helps in paleo-environmental study. Comparative data from modern and fossil root studies suggest that in arid climate gross root morphologies have a strong correlation with availability of water. Modern plants, which have

Fig. 2.6.35: Residual sandstone towers of Proterozoic Kambolgie Sandstone on the Arnhem Land Plateau, Northern Territory, Australia. Towers are about 15 meters high in highly weathered quartz sandstone (Courtesy Dr. R. A. L. Wray, University Wollongong, Australia).

Fig. 2.6.36: Joint-bounded towers in very highly weathered Permian Nowra Sandstone, Monolith Valley, Budawang Range, south of Wollongong, NSW., Australia. Towers are about 50 meters high. Note the solution flutes on tower walls (Courtesy Dr. R. A. L. Wray, University of Wollongong, Australia).

Fig. 2.6.37: 10 meter deep corridor (grike) in quartzose sandstone. Rock is highly weathered, base of grike is solid rock and there has been no tectonic movement. Grampians Range, Victoria, Australia (Courtesy Dr. R. A. L. Wray, University of Wollongong, Australia).

continual access to surface water, develop shallow and often matted root systems. In contrast, plants growing over sub-surface water sources develop substantial vertical root systems in order to tap the sub-surface water. Hence, fossil root cast morphologies accurately reflect these tendencies, reflecting paleo-groundwater conditions and paleo-climate (Cohen, 1982).

2. Clay mineralogy has been used to detect and interpret a climatic change (Robert and Kennett, 1994; Singer, 1980, 1984; and Curtis, 1990). Recently, Wright et al., (1991b) have interpreted shift in the "monsoonal system from predominantly dry-type monsoonal regime to one with either a greater net precipitation or more concentrated distribution of rainfall" from the presence of unusual clay minerals assemblages of kaolinite/(chlorite-like material)/smectite mixed-layer clay in Arundian alluvial paleosol.

Fig. 2.6.38: Solution-widened joints (grike, in highly weathered Permian Snapper Point Formation Jervis Bay, NSW, Australia. No tectonic movement (Courtesy Dr. R. A. L. Wray, University of Wollongong, Australia).

3. Paleosols are most widely used in fluvial/alluvial sequences for estimating sedimentation rate, paleo-geomorphology etc. (Friend et al., 1989; Marriott and Wright, 1993). Paleosols may be especially useful for recognizing ancient episodes of incision and terracing. Bestland (1997) has been able to identify several episodes of terrace formation, from the presence of strongly developed paleosol separating each episode, indicating long pauses in sedimentation. Estimation of sediment accumulation rate from paleosol, however, suffers from limitations: i) variability in the rate of formation of calcrete because of variation in local rainfall and the availability of airborne carbonate dust, ii) amount of erosion subsequent to soil formation. With erosion, accumulation rate can be underestimated. Again, if the paleosols are composite (commonly found in many stratigraphic sections) accumulation rates are liable to be overestimated. Hence caution is necessary in estimating sediment accumulation rate.

Fig.2.6.39: Solution basins in weathered highly quartzose sandstone, Permian Snapper Point Formation, at Jervis Bay, NSW., Australia. Hammer for scale (Courtesy Dr. R. A. L. Wray, University of Wollongong, Australia).

4. It provides evidence of former vegetation (post-Silurian time) against which the degree of adaptation of limbs and teeth of associated fossil vertebrates can be assessed (Retallack, 1990).

5. It helps in making an intelligent guess about water flow and their retention in paleosols. The more copiously rooted and burrowed a paleosol, the more porous and permeable it may have been (Retallack, 1990).

6. It helps in predicting paleo-water table in paleosols. Most of the creatures that thrive within soil are air breathers excepting few (e.g., crabs) that live in water-filled burrows. As such these creatures are not expected to burrow below the water table. Hence, the lowest termination in a densely burrowed paleosol can be taken as the usual level of the paleo-water-table (Retallack, 1976).

7. It helps in understanding whether groundwater in paleosols was free flowing or stagnant. The presence of ferruginous concretion indicates oxygenated, free-flowing water, while siderite nodules indicate a stagnant, poorly oxygenated water (FitzPatrick, 1980; Retallack, 1990).

8. Repeated occurrence of calcrete layers in stratigraphic succession, developed on shallow marine carbonates, may indicate repeated sea-level lowstands. Caution is, however, necessary. Recently, Rossinsky et al., (1992) have reported occurrence of calcrete layers that have formed simultaneously in the surface and subsurface during a single sea-level lowstands. They have termed such subsurface calcrete layers as penetrative calcrete layers, formed in close association with downward penetrating and laterally spreading root systems. Since

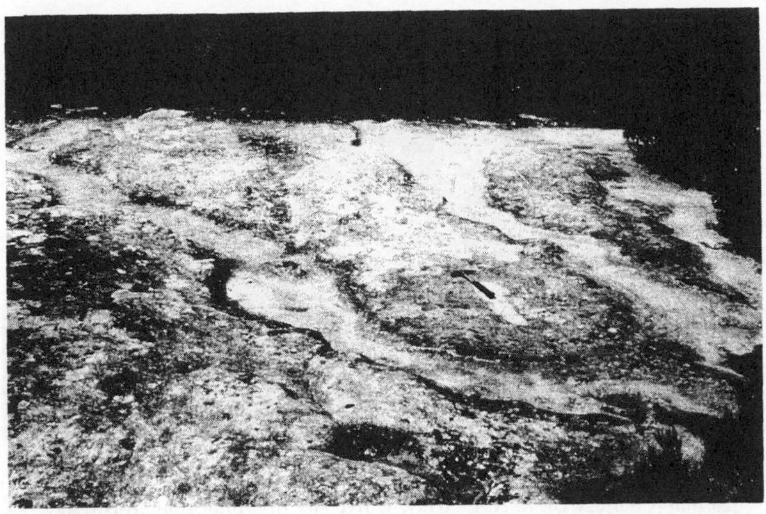

Fig. 2.6.40: Solution runnels, 1–7 meters deep, in highly quartzose Nowra Sandstone at Pointers Gap near Milton, south of Wollongong, NSW., Australia (Courtesy Dr. R. A. L. Wray, University of Wollongong, Australia).

such surface and subsurface calcrete layers look physically alike (except for some geochemical differences), they are likely to be misinterpreted to have formed due to repeated sea level lowstands (Beach and Ginsburg, 1980; McNeil et al., 1988). Rossinsky et al., (1992) have suggested criteria for their discrimination:

- Sub-aerial calcretes signal facies change, while subsurface calcretes do not.
- Sub-aerial calcretes show upward truncation of various structures, while subsurface calcretes show none.
- Sub-aerial calcretes usually occur at the top of the shallowing-upward sequence, while the subsurface calcretes do not.
- While sub-aerial calcretes form regionally wide layers, subsurface calcretes show lateral variability controlled by topography.
- Subsurface calcretes show presence of rhizoliths between two or more calcrete layers, but not in sub-aerial calcretes.

9. The evolution of Precambrian atmosphere can be addressed by studying paleosols (Zbinden et al., 1988; Retallack, 1990). The isotopic composition (C^{13}, O^{18}) of the soil carbonate can be used as a potential indicator of the atmospheric pCO_2 levels, paleo-temperature and paleoecology (Alam et al., 1997; Cerling et al., 1989; Emiliani, 1955, 1972; Emiliani et al., 1974). This is based on the idea that the carbon isotopic composition of soil carbonates reflects, in large part, the nature of the associated vegetation type (C_3, C_4, CAM) (Cerling, 1984, 1991). C_4 plants (xerophytic plants, shrubs, grasses etc.,)

Fig. 2.6.41: Opal A and chalcedony stalactites, Blue Mountains, west of Sydney, Australia. Coin 23 mm (Courtesy Dr. R. A. L. Wray, University of Wollongong, Australia).

Fig. 2.6.42: Top view of opal-A and chalcedony stalagmites, Blue Mountains, west of Sydney, Australia (Courtesy Dr. R. A. L. Wray, University of Wollongong, Australia).

are known to be better adapted to hot and arid condition of high water stress than C_3 plants (Cerling et al., 1988, 1989). Low temperatures favour C_3 grasses and high temperature C_4 grasses. Oxygen isotopic composition is also temperature dependent; the heavier O^{18} represent higher evaporation rate and low humidity, reflecting high water stress condition, meaning high temperature. Conversely, the higher isotopic composition reflects a lower evaporation and relatively wetter climatic conditions. Caution is necessary in interpretation, since (i) re-crystallization of carbonates, (ii) carbonates that are not pedogenic, *sensu stricto,* and (iii) the assumption that plants utilizing C_3 or crassulacean acid metabolism (CAM) pathways did not become dominant until the Miocene, may reflect on interpretation (Wright and Vanstone, 1991a).

10. A comparison of Precambrian paleosols with surficial alteration of equivalent or greater age of Moon, Mars, Venus and other planetary bodies may elucidate the early history of our solar system (Retallack, 1990).

11. Study of paleosols also provides evidence for the origin and early evolution of life on land (Retallack, 1986b, 1990).

12. Paleoweathering surfaces are hunting grounds for exploration geologists, since a host of minerals (e.g., gold, uranium, lead, zinc, iron, manganese, fluorite etc.) are commonly found to be closely associated with such paleoweathering surfaces. (Button and Tyler, 1981).

13. Succession of paleosols bounded by erosion surfaces in fluvial sediments can be interpreted as terrestrial equivalents of the unconformity bounded units of sequence stratigraphy. Interpretation of these paleosol packages as non-marine sequences is not only based on correlation with sea-level changes but on correlation with global climatic change (Bestland et al., 1997; Tandon and Gibling, 1997).

14. Paleosols can be used to quantitatively estimate ancient mean annual precipitation (MAP) and mean annual temperature (MAT). Ancient climates are usually interpreted from their modern soil analogs. For example, Oxisols and Argillisols indicate wet equatorial climates in which MAP and MAT are high and have little seasonal variation. Similarly, Argillisols, Spodosols, and Gleysols indicate humid mid-latitude climates, where MAP>1000 mm. Calcisols indicates dry subtropical climate where MAP<1000 mm. (Mack et al., 1993)

APPENDIX

Microscopic Features of Paleosols

While discussion on microscopic features of paleosols do not fall under the present structure of the book, nevertheless enumeration of microscopic features, with accompanying references, bring the discussion on paleosols to its logical end. Association of one or more of the following microscopic features, together with the macroscopic features discussed earlier, establish paleosols on firm ground.

a. Microscopic features of rhizoliths (Brewer, 1964; Cohen, 1982; Esteban and Klappa, 1983; FitzPatrick, 1980; Goldstein, 1988; Klappa, 1980b; Mount and Cohen, 1984).
b. Tangential needle fibres (Esteban and Klappa, 1983; Goldstein, 1988; James, 1972; Wright, 1984).
c. Alveolar septal structure (Esteban, 1974; Goldstein, 1988; Harrison, 1977; Klappa, 1979b).
d. Ribbon spar (Goldstein, 1988; James, 1972).
e. Isopachous spar (Goldstein, 1988).
f. Early, clear coarse spar (Goldstein, 1988).
g. Blackened grains (Goldstein, 1988; Strasser, 1984; Ward et al., 1970).

h. Random needle fibre cement (Esteban and Klappa, 1983; Goldstein, 1988; James, 1972).
i. Coated grains (Calvet and Julia, 1983; Esteban and Klappa, 1983; Goldstein, 1988).
j. Glaebules (Brewer, 1964; Calvet and Julia, 1983; Goldstein, 1988).
k. Micritization (Bathurst, 1966; Goldstein, 1988; James, 1972; Kahle, 1977).
l. Early cracking (Freytet and Plaziat, 1982; Goldstein, 1988; Klappa, 1980a).
m. Microcodium (Freytet and Plaziat, 1982; Goldstein, 1988; Klappa, 1978).
n. Microscopic soil structure (Brewer, 1964).

2.7 Event Beds

Sedimentary basins are commonly filled-in in two ways: one, more common, by gradual filling by terrigenous and non-terrigenous sediments from different sources obeying the Principle of Uniformitarianism; second, by infrequent events of rapid deposition by processes that are sudden, unpredictable, and convulsive (Clifton (1988) has preferred the term 'convulsive' over other terms, such as, 'catastrophic', 'cataclysmic'). Sedimentary records of such non-regular and rare convulsive episodes are known as event beds. Careful stratigraphic analysis reveals that the sedimentary record is mostly a record of episodic geologic events (Dott, 1983). The various convulsive processes are: severe river flood, turbidity current, seismic shocks, giant tsunami, violent volcanic eruption, large bolide impact. Convulsive events are essentially instantaneous in geologic time-scale; hence, its products are essentially time-parallel and can be used for chronostratigraphic correlation.

Besides these various physical events of episodic sedimentation, purely chemical events may also generate important episodic deposits. An oxygen crisis, or a brief period of salinity crisis is commonly reflected by a widespread mass mortality of organisms in the sedimentary record (Dott, 1996).

Amongst these episodic processes, the turbidity currents and their resultant products are well studied in the field and in laboratories; turbidites (a sedimentary record of turbidity current) account for a great portion of the total sedimentary body from the Proterozoic to the Recent. But the products of other convulsive events are not well understood, and mostly go unrecognized in the stratigraphic record. There could be two reasons for this: one, geologists do not look for it in the field with the care that it deserves; second, later geological processes might have erased out the features of event beds by bioturbation, or erosion subsequent to deposition. The best places where such event deposits are likely to be preserved are lake basins, anoxic basins with higher salinity, and deep-sea basins.

What are the geological implications of investigating the products of such rare episodic events? Their identifications in the stratigraphic record help in regional and local stratigraphic correlations, in drawing correct paleogeographic pictures and in basin analysis. It is only the last two decades that record an upsurge in the study of event beds (see, Clifton, 1988; Einsele and Seilacher,

1982; Einsele, Ricken and Seilacher, 1991,) and various criteria are now emerging for their identification in the rock record. Despite the fact that the turbidity current, its hydrodynamics, and the resultant products are well understood, a brief description will not be out of place here.

Turbidity current, a variety of gravity mass flow, is generated as a result of the density contrast created by sediments in suspension. It differs from other types of current in that without the suspended sediment, no turbidity current would flow. Turbulence is a must for turbidity current generation - no turbulence, no turbidity current. Suspended sediments make the water in the turbidity current denser than ambient water and consequently flows down the slope of the basin. The resultant product of turbidity current is known as turbidite. Turbidite forms a great portion of the total sedimentary record in both modern and ancient deep-sea basins (Kelts and Arthur, 1981).

Collectively, a group of turbidites in the outcrop will appear as regular bedded sandstone and mudstone. This rhythmic sequence may be all sandstone with thin shale partings on one end of the spectrum and dominantly shale with thin sandstone beds on the other. The sandstone/shale ratio may be 5:1 in the proximal near-source area to 1:1 or less in distal area (away from the source). Each sandstone bed is laterally persistent and consistent in thickness and has a sharp base but a gradational top. On the basis of consistent internal and external structures, Bouma (1962) proposed a structural subdivision of turbidite facies that is widely known as "Bouma sequence" (Fig. 2.7.1). The ideal Bouma sequence is structured into five divisions (**A** → **B** → **C** → **D** → **E**) with each division being characterized by a distinct set of structures and textures. Division A of the Bouma Sequence represents graded division, division B represents lower parallel division, division C represents ripple and convolute division, division D represents upper parallel division, and division E represents pelagic and hemipelagic division. In rock record, however, one would rarely come across a complete "Bouma sequence"; bottom cut out sequences are common. Whatever may be the truncated sequence, the important fact is that they rarely occur in reverse order.

Sandstone beds are internally graded with either distribution grading or coarse-tail grading. In distribution grading, the whole particle size population shows a change from bottom upward; in coarse-tail grading only the coarsest fraction shows a change from bottom upward. Amalgamated beds of sandstone are not uncommon, but in each case are separated by erosional surfaces. Other diagnostic criteria of turbidites, beside graded beds with sharp bases and gradational tops, are sole marks (where flute marks are common), tool marks (grooves, striations, prod marks, bounce marks are most common) and fossil and ichnofossil associations (shallow water forams are found associated with microfaunas that are of deep-water origin; Nereites ichnofacies is diagnostic of deep quiet water). The paleocurrent pattern, determined from various sole structures, is also complex; in proximal region it shows large variations but broadly

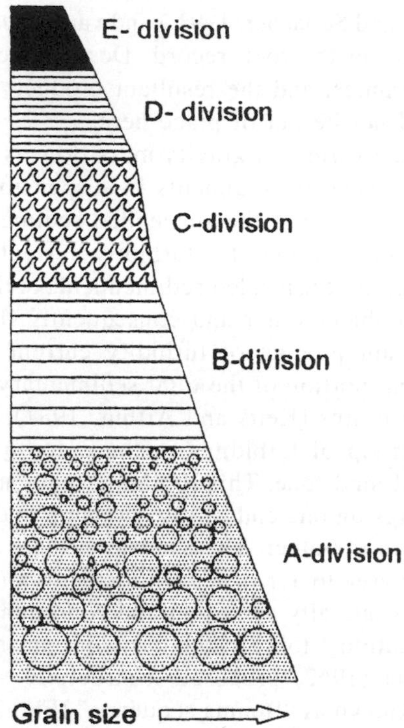

Fig. 2.7.1: Ideal sequence of structures in a turbidite bed (the Bouma Sequence).

a radial pattern. This radial pattern of paleocurrent is helpful in delineating several submarine fans, identifying points of sediment input. In the distal basinal region, flow pattern is more uniform and may be longitudinal.

The characteristics, outlined above, help to identify a turbidite bed that can be reasonably described by the Bouma sequence. But one may come across other resedimented facies, in association with "normal" turbidites. Walker and Mutti (1973) have classified them as under:

a) massive sandstone, commonly containing "dish" structures, are believed to be a product of grain flow/fluidized sediment flows,
b) coarse grained, massive and pebbly sandstones that are commonly ungraded, irregularly bedded with scoured bottoms. Such beds are termed as "fluxoturbidite",
c) conglomerate, which can either be massive, lacking visible structures, or stratified having distinct clast orientation and imbrication, indicating deposition from turbulent flow.

Resedimented conglomerate and other features of turbidite facies have been discussed in detail by Allen (1971); Bouma (1962); Bouma and Hollister (1973);

Dott (1963); Dzulynski and Walton (1965); Hampton (1972); Kuenen and Migliorini (1950); Middleton (1966a, b, c and 1967); Middleton and Hampton (1973); Sanders (1965); Walker (1965, 1967,1973), amongst many others.

Recently, Shanmugam (1997) has challenged the idea of all deep-sea sands as being products of turbidity current. He has laid more emphasis on the rheology of flows in order to distinguish turbidity current from other types of flows, such as, debris flow, bottom-reworking current and geostrophic current. Re-examination of many published data, based on detailed description of 6402 meters of conventional cores and 365 meters of outcrop sections, has led Shanmugam (1997) to reinterpret the deposits as products of sandy debris flow and bottom-current reworking with only a minor percentage of true turbidites. What are then the signatures of products of turbidity current? These are (i) presence of sharp basal contact, (ii) normal size grading, and (iii) gradational upper contact; the sands may then be reasonably explained as products of turbidity current. Shanmugam (1997) has also challenged the "common practice of interpreting deep-water sands that contain flutes and scour surfaces as turbidites"(page 216). This is because (i) small-scale scour surfaces and flutes and sand layers covering them may represent products of two distinct processes, and (ii) large scale erosional surfaces may have been produced by mass movements (e.g., slump scars), and not just by turbulent flows. However, the task of distinguishing these two in outcrops is almost impossible. Shanmugam (1997) has stressed that interpretations of origin of deep-water sands should be based on their internal depositional features, and not on their erosional basal contact or sole marks.

As Seilacher and Aigner (1991) have indicated 'flow regimes along the bottom (because of temporal turbulence) do change vertically in a regular pattern. The result is an event bed with an erosive base followed by graded succession of grain sizes and depositional structures (Bouma sequence). Since this basic principle is applicable to all kinds of turbulent events, individual deposits of river floods (inundites), of storms (tempestites) and of turbidity current (turbidites) share similar characteristics'. It is thus important to recognize features which help to identify these event beds in the stratigraphic record. Recently, Myrow and Southard (1996) have proposed a predictive model after considering a spectrum of possibilities and shown on a triangular diagram where the end members are i) geostrophic currents, ii) wave oscillations and iii) density-induced flow (Fig. 2.7.2). The model attempts to predict the nature and association of internal stratifications and sole marks, corresponding to various flow conditions. Emphasis has been laid on various associations of sole marks, since these give an insight in to the mechanisms of storm transport of sediment on to the shelf. In an earlier publication, Myrow and Southard (1991) emphasized on internal stratification styles for a full range of combined flows between purely unidirectional and oscillatory flows (Fig. 2.7.3). When these two models are combined together and applied in the field form a powerful tool in the identification of storm-dominated sequence. Since criteria for identifying turbidites are well known, and briefly outlined above, no further elaboration on this will be made.

Characteristics of Event Beds

1. Hunter and Clifton (1982) have suggested three kinds of evidences that indicate a bed of storm origin: a) evidence of strong flow before and during the first stages of deposition characterized by erosional surfaces overlain by hummocky cross-stratification, b) evidence of rapid deposition followed by deposition at a declining rate characterized by the formation of planar bed, and c) evidence of oscillatory flow characterized by the development of wave-ripple bedded facies at the top. Presence of wave-ripple is the strongest evidence of oscillatory flow and helps to distinguish storm deposit from deep-water turbidites. What is hummocky cross-stratification? It was Gilbert (1899) who first recognized, illustrated, and interpreted the structure from the Silurian Medina Formation, U.S.A that is now known as hummocky cross-stratification; but this structure went unnoticed amongst geologists for more than half century until it was resurrected by Campbell (1966) who named this structure as 'truncated wave ripple' and described it as 'sets of parallel laminae that conform to the surface form of wave (oscillation) ripples. The laminae peak upward or curve over the crest of the ripple, continue downward into the adjacent trough, and again curve upward to the next crest'. Harms et al. (1975, p. 87-88) first coined the term 'hummocky cross-stratification' for those 'truncated wave ripple' structures described by Campbell (1966) and listed the characteristics as 'i) lower bounding surfaces of sets are erosional and commonly slopes at angles less than 10^0, though dips can reach 15^0, ii) the laminae above these erosional set boundaries are parallel to that surface, or nearly so, iii) laminae can systematically thicken laterally in a set, so that their traces on a vertical surface are fan-like and dip diminishes regularly, and iv) the dip direction of erosional boundaries and of the overlying laminae are scattered' (Fig. 2.7.4 a, b). Basal scouring produces unoriented hummocks and swales that are subsequently 'mantled by laminae of fine particles swept over these irregularities' (Dott and Bourgeois, 1982). The height of the hummocks may vary between 10 and 15 cm and the spacing (measured from hummock to hummock or swale to swale) is usually between one to few meters. In plan view, hummocks and swales generally are three dimensional and radially symmetrical. Ancient examples of hummocky cross-stratification generally occur in coarse siltstone to fine sandstone and are rarely observed in medium to coarse sandstone (Duke, 1984). The stratification within the fill of scoured depressions commonly is asymmetric, inclined more steeply at one side of the scour than at the other. Also, the locus of points defining the deepest part of the depression at successive stages of fill tends to

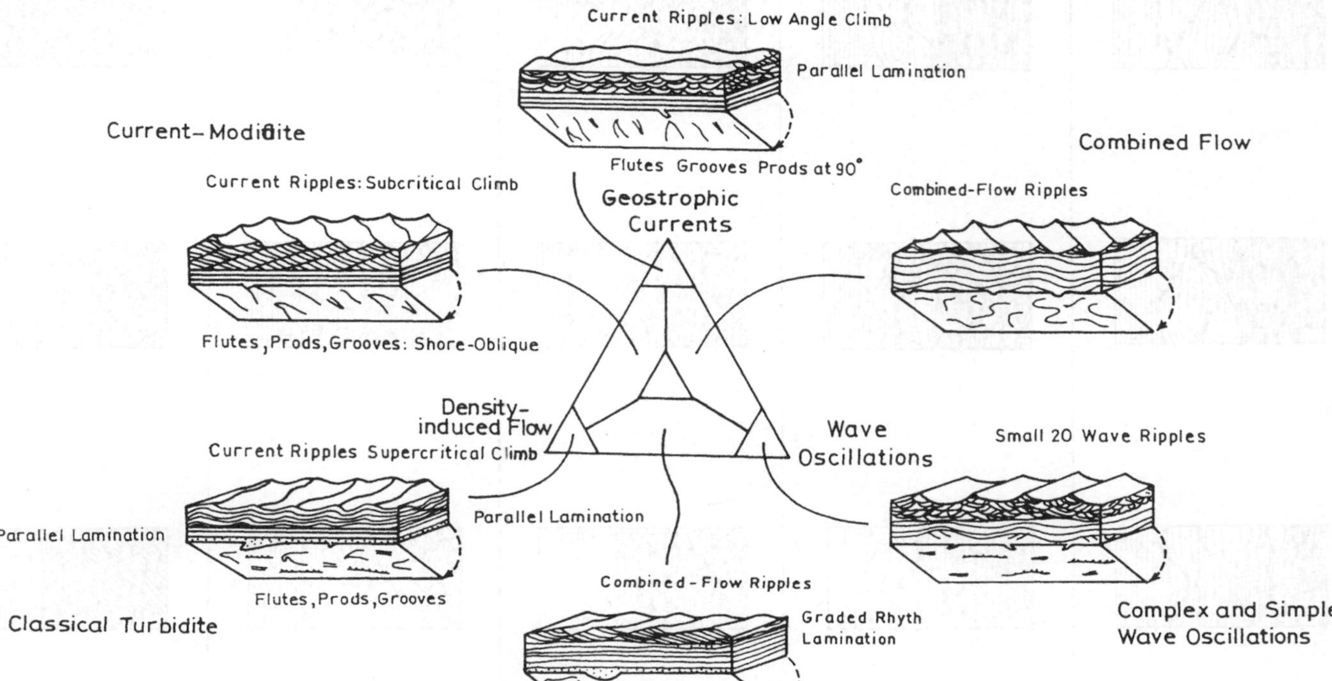

Fig. 2.7.2: Predictive model of the origin of stratification and sole markings for various possible combinations of storm processes. The triangular diagram has three corners representing density-induced flow due to excess weight forces, geostrophic currents and wave oscillations (Reproduced from P. M. Myrow and John B. Southard: Tempestite deposition. J. Sediment. Research, Vol.A/66: 875-887, Fig. 7, © 1996, the Society for Sedimentary Geology (SEPM), Tulsa, Oklahoma, U.S.A. with the kind permission).

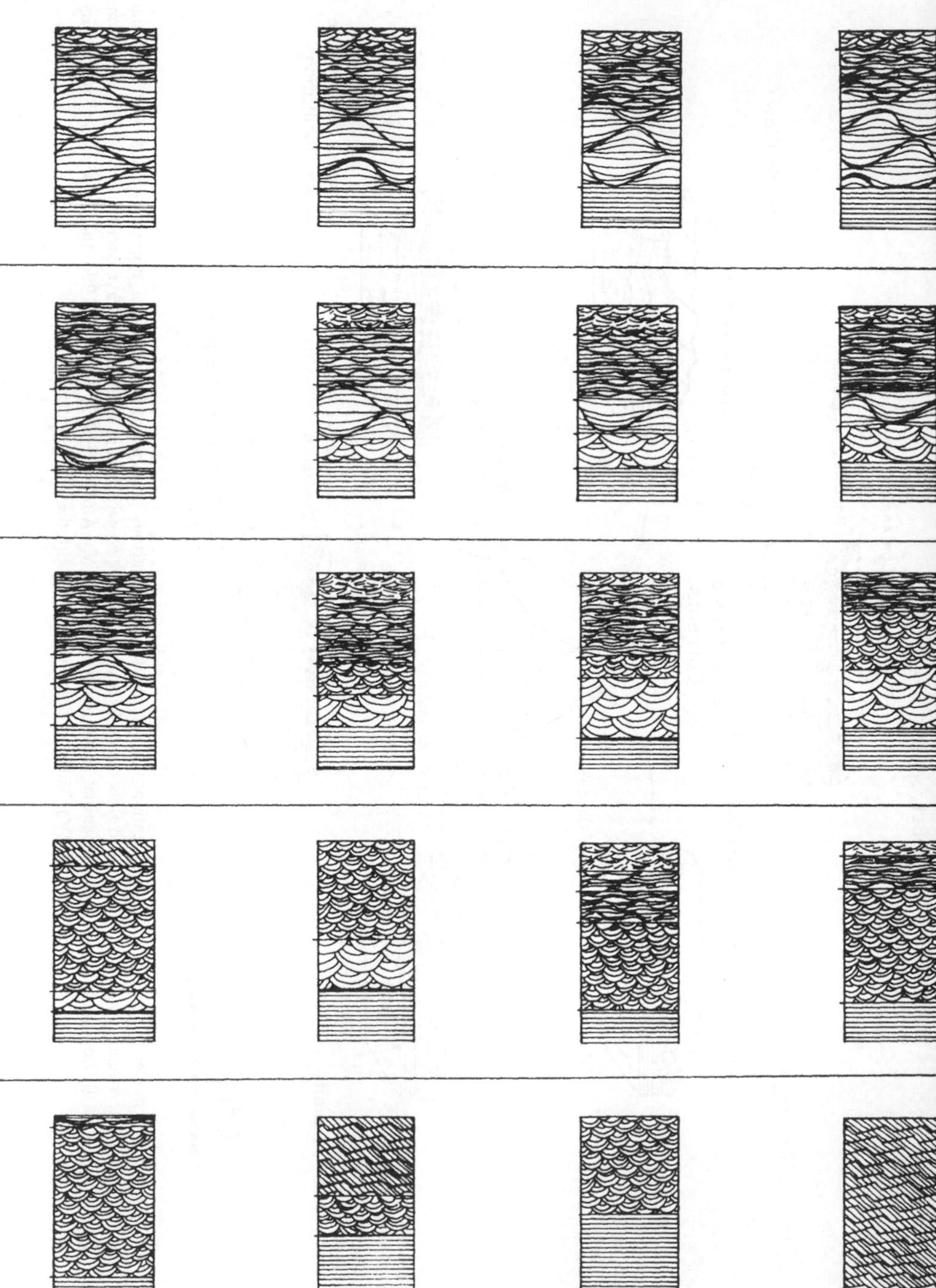

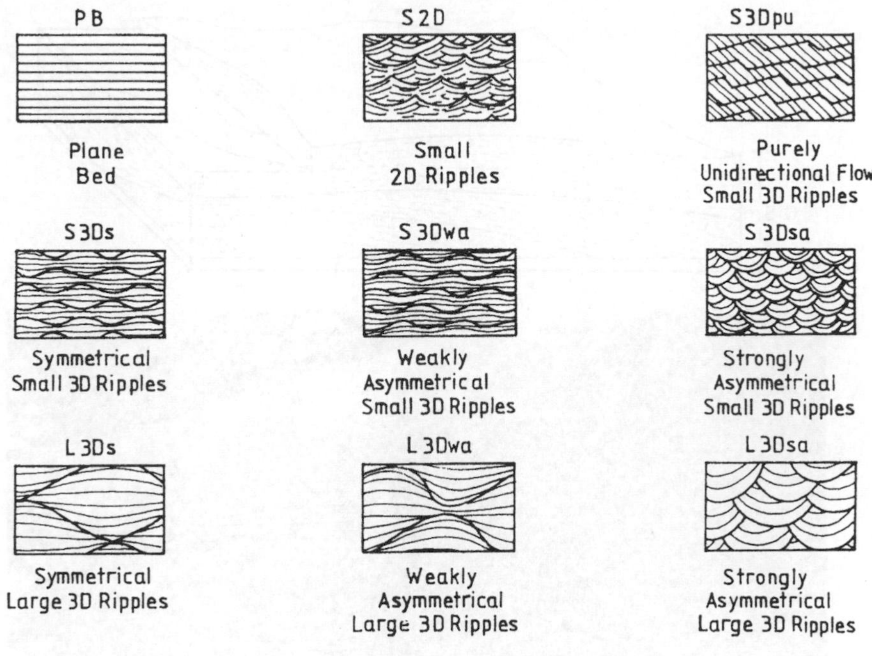

Legends for Fig. 2.7.3

Fig. 2.7.3: The twenty qualitatively different kinds of vertical stratification sequence produced by the model. (Reproduced from P. M. Myrow and John B. Southard: Combined-flow model for vertical stratification sequence in shallow-marine storm-deposited beds. J. Sediment. Petrology, Figs. 3 and 5, Vol. 61/2, © 1991, the Society for Sedimentary Geology (SEPM), Tulsa, Oklahoma, U.S.A. with the kind permission).

be inclined rather than vertical (Hunter and Clifton, 1982). While the angle of inclination of foresets of hummocky cross-stratification usually is low, occasionally this may be as high as 50^0 (Hunter and Clifton, 1982); this high inclination of foresets is the best evidence of rapid deposition. Analogus to the idealized Bouma sequence, an idealized hummocky cross-stratification sequence, in descending order, is as follows (alphabets in brackets are notations) (Dott and Bourgeois, 1982):

Zone of burrowed mudstone/siltstone (M),
Zone of symmetrical ripple forms and ripple cross-laminae (X),
Zone of flat laminae (F),
Zone of hummocky cross-stratification with several second-order truncation surfaces within (H),
First-order scoured surface with/without sole structures.

There are many variations from the ideal hummocky unit as depicted by

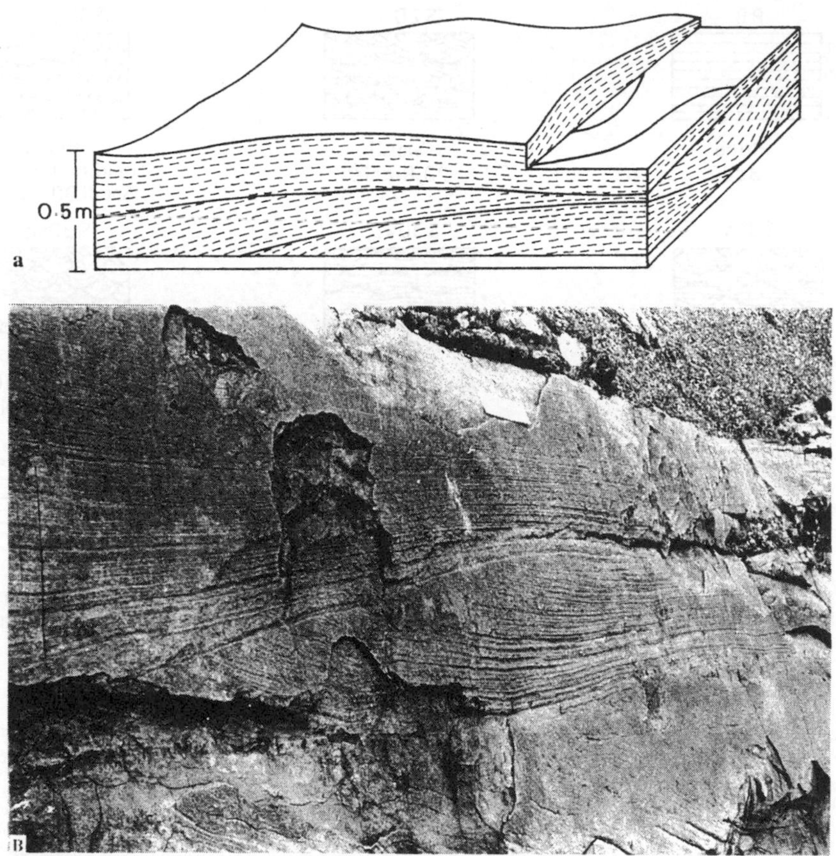

Fig. 2.7.4 a: Hummocky cross-stratification. This form is common in coarse siltstone or very fine sandstone within the lower shoreface and offshore facies. (Reproduced from J. C. Harms, J. B. Southard, D. A. Spearing and R. G. Walker: Depositional Environments as Interpreted from primary Sedimentary Structures and Stratification Sequence, Fig. 5.5, SEPM Short Course No. 2, © 1975, Society of Sedimentary Geology [SEPM], Tulsa, Oklahoma, U.S.A. with the kind permission).

b. Photograph of hummocky cross-stratification, Proterozoic Chaibasa Formation, Ghatsila, India. Diagonal scale in the top center is for scale.

Dott and Bourgeois (1982, see their Fig. 5). If storms follow in quick succession, it may produce amalgamated hummocky cross-stratified sandstones, each separated by an erosional surface with occasional presence of remnant mudstone lenses that survived erosion, and often display abundant bioturbation as shown by Dott and Bourgeois (1982, see their Fig. 8). The presence of scattered pebbles within hummocky cross-stratification, with erosional surfaces at the base and absence of ripple bedding in the hummocky cross-stratified sandstone, attest to the fact that hummocky cross-stratification is a product of high velocity flow (but not necessarily unidirectional flow). That the storm surge gradually decreased after the formation of the hummocky cross-stratified bed is evident from the upward decrease in grain size, and vertical sequence of sedimentary structures (Hunter and Clifton, 1982). Hummocky stratified fine sandstone and coarse siltstone typically contain abundant mica and carbonaceous detritus (Dott and Bourgeois, 1982).

While Dott and Bourgeois (1982) have visualized hummocky cross-stratification solely as a scour-and-drape structure, Walker et al. (1983) visualize it to be a growing bedform. Following publications are useful for complete comprehension of this structure besides those already cited (Bourgeois, 1980; Brenchley, 1985; Brenchley, et al., 1979; Dott and Bourgeois, 1982; Duke, 1985; Hamblin and Walker, 1979; Hunter and Clifton, 1982; Hobday and Reading, 1972; Kumar and Sanders, 1976; Leckie and Walker, 1982; Swift et al., 1983; Tillmann et al., 1985; Walker et al., 1983; Wright and Walker, 1981).

While the majority of hummocky cross-stratification are reported from shallow marine sedimentary rocks, the structure has also been reported from lake sediments (Late Pleistocene Lake Bonneville in North Utah, U.S.A.; Triassic Narrabeen Formation, Australia; Permian Tomago Formation, N.S.W., Australia). These occurrences of hummocky cross-stratification in lake deposits are interpreted to have formed from less violent terrestrial wave cyclones (Duke, 1985). Hence, the presence of this structure does not necessarily indicate hurricane-affected shallow marine environments. It is important to look into the regional geologic setting in the process of reconstructing a correct paleogeographic picture.

Spatial variations in sedimentary structures have also been reported from storm sequences. Handford (1986) has described a case where parallel laminae spatially changes ⇒ climbing-ripple cross-laminae ⇒ low relief hummocky cross-strata overlain by parallel laminae and climbing-ripple cross-lamina ⇒ parallel laminae with some bioturbation structures at the top.

Leckie (1988) has described another motif of storm facies where coarse-grained wave ripples occur in close lateral and vertical association with hummocky cross-stratification. Their close associations have been interpreted to form under hydrodynamically similar—oscillatory or combined—flows during storm surges.

While the presence of hummocky cross-stratification is a positive signal of storm sequence, its absence does not indicate otherwise. Reineck and Singh (1972) and Aigner (1982) have described a vertical sequence of graded beds starting with an erosional base followed by parallel laminated and wave-ripple laminated sets as a product of storm surges. Parallel lamination passing gradationally to wave-ripple at the top amply demonstrates that the wave ripple is not a product of later reworking.

While graded beds intercalated with mud are usually taken as signature of storm deposits, this is also produced on the shallow-marine shelves in the absence of any large storm waves, e.g., graded beds on the Brazilian shelf (Figueredo et al., 1982). Caution is therefore necessary in this respect. Morton (1988) has argued that strong decelerating currents, produced by the interactions of semi-permanent current with tidal, wind-driven, density-driven currents, are also capable of reworking the shelf sediments and selective sorting, producing thin graded beds.

Sheet sandstone has been interpreted as the product of storm surges, possibly aided by ebb tidal currents. Goldring and Bridges (1973) have described thin, fine to very fine sheet sandstone of 5–70 cm thick from sandy shelf environment. Each bed displays a suite of sedimentary structures: i) undulose to flat sole, ii) plane or low angle lamination and iii) wave-rippled tops. While some sandstone tops may be gradational, the majority shows penecontemporaneous erosion and it is this feature that tells it apart from turbidites sequence. Load casts are conspicuously absent. Laterally, thin sandstone can persist from a meter to several kilometers. Organisms in them are mostly suspension feeders and ichnofossils are mostly escape burrows.

Handford (1986) has described a sequence of thin to medium, even-bedded and continuous black lime mudstone interbedded with dark coloured shale from lower Fayettville Shale and Pitkin Limestone, Arkansas. Black shale has been interpreted to be the background sediment where black lime mudstone is allocthonous carbonate mud of shallow water origin. It has been interpreted that this shallow water carbonate mud was stirred up into suspension by periodic storms, carried offshore and deposited as the particles settled out from a deeper and quieter water column (Fig. 2.7.5). Such mixing of sediments has been termed "punctuated mixing" by Mount (1984).

Similar mixing may arise when siliciclastic sediments from near-shore regions are stirred up by storm-generated currents, carried offshore and deposited interbedded with carbonate sediments (background sediment) of sub-tidal environment, as has been documented by Brenchley et al. (1979), Fairchild (1980), Kelling et al. (1975), Kreisa (1981) and Tucker (1982). Interbedded sandstone is graded, parallel laminated and capped by cross-laminations. Sandstone may often contain limestone intraclasts indicating deposition from decelerating currents. It is important, however, to note that mixed sediments are not always of storm origin, since there are other processes which may produce such couplets, e.g., facies mixing, in-situ mixing and source mixing (Mount, 1984).

Fairchild et al., (1997) have described a 65 metre storm-deposited Neoproterozoic carbonate ramp deposits from the Xinmincum Formation, north China. The features, suggestive of storm processes, are "abundance of quasi-parallel lamination, and the variable occurrence of coarse basal lenses, erosion surfaces and gutter casts". Coarse basal lenses are intraclastic with variable clast orientation. Dominance of quasi-parallel lamination rather than hummocky cross-stratification implies "strong unidirectional flow component (consistent

Fig. 2.7.5: Interbeds of pelmicrite and shale. The pelmicrite was derived from shallow water environment and emplaced in deep-water. Upper Castletown Formation, upper part of Lower Ordovician, New York, U. S. A.

with the presence of gutter casts, Myrow, 1992a, b) superimposed on a long period oscillatory component" (Arnott, 1993; Fairchild et al., 1997).

Molina et al. (1997) have also described carbonate tempestite from the pelagic facies of middle and late Jurassic of Beltic Cordilleras, southern Spain. Association of tempestites and pelagic facies are rarely described from the rock record. Emplacement of tempestites in the pelagic facies has been commonly explained as relative sea level falls (lowstand phases). The carbonate storm facies, described here, comprising of calcarenite and calcisiltite beds, start with normal internal organization of horizontal or slightly wavy parallel lamination, hummocky cross-stratification and low angle cross-lamination related to symmetrical ripples at the top.

Davies et al. (1989) have described from the Holocene of the Gulf Coast of Florida a storm sequence characterized by graded storm facies, homogenous storm facies and fluvial storm facies. Graded facies, representing higher energy condition, is formed by landward transport of Gulf-derived shelly sediments. Homogenous storm facies, representing lower energy condition, is formed by reworking of bay sediments that is much thinner sediment units than graded beds. Fluvial storm facies represents deposition of land-derived terrigenous sediments delivered by streams in flood conditions due to extreme rainfall associated with storms.

Aigner (1985) has summarized the work of Reineck (1977) and Wunderlich (1979) on storm deposits in the inner part of German Bay between Busum and

Helgoland. Most commonly storm deposits start with a sharp irregularly scoured base (resembling gutter cast) with impact marks. The irregular erosional surface is covered by reworked shell debris (internally graded), followed by parallel laminated sands (internally weakly graded) that in turn are covered by wave-rippled top (see also Hobday and Morton, 1984). In the proximal shallow-water region, storm layers tend to be highly amalgamated and even "cannibalistic"(succeeding storm layers receive their sediments from the preceding ones by erosion).

Nelson (1982) has described a sequence of shallow-water graded sand layers interbedded with mud from the Yukon delta shoreline, northern Bering, Alaska. Graded sand layers individually show from, base to top, (i) plane parallel lamination of upper flow regime, (ii) cross-lamination, (iii) plane parallel lamination of lower flow regime, and (iv) mud, analogus to Bouma Tac turbidite sequence. On closer examination, graded sand layers show proximality trend (Aigner, 1982, 1985) from coarser and thicker beds possessing trough cross-lamination and plane parallel lamination in the proximal region to the thinner silt beds containing flat and ripple-drift lamination, commonly bioturbated and associated with shell and pebble lags in the distal region. These graded sand layers with the above association, have been interpreted as products of waning storm-surge-associated currents. These storm sequences, though mimicking many of the features of thin-bedded turbidites, can be distinguished by (i) the predominance of trough cross-lamination in the proximal part, and (ii) gradation to common shallow marine fossils, storm lag layers, and profuse starved ripples and bioturbation in the distal areas.

2. Multidirectional paleocurrent patterns are considered as an important criterion for the recognition of tempestites, in contrast to turbidites that display unidirectional paleocurrent pattern (Gray and Benton, 1982).

3. Amongst sole marks, flute and groove casts are more commonly associated with turbidite and inundite facies because of unidirectional flow dynamics. Tempestite, on the other hand, is characterized by the presence of pot and gutter casts. The pot and gutter casts form near the wave base and their long axes are oriented parallel to the coastline and to the storm tracks (Aigner, 1985). That gutter cast is a product of oscillatory flow is indicated by: a) the bi-directionality of the prod marks which commonly occur at the base of the gutter casts, and b) the trend of the wave ripples, occurring at the top of the gutter casts, is commonly perpendicular or at high angle to the axis of the gutter (Aigner, 1985; Seilacher, 1982; Seilacher and Aigner, 1991).

It should, however, be noted that flutes are not exclusively associated with turbidites and despite claim to the contrary, abundant occurrence of flutes have been reported from ancient storm-influenced deposits (Hamblin and Walker, 1979; Leckie and Krystinik, 1989; Myrow, 1992a).

Similarly, while pot and gutter casts have been shown to represent products of oscillatory or combined flows, in many others, analysis of pot casts and sinuous gutter casts with helical grooves and associated with flute marks indicate formation by storm-generated flows in which early stages were dominated by unidirectional flow. Experimental studies have never shown that flute marks, sinuous gutter casts or pot casts can form under pure oscillatory or combined flows (Myrow and Southard, 1996).

4. Load casts are more commonly associated with turbidite and inundite facies than in tempestites. Seilacher and Aigner (1991) are of the opinion that "erosional and depositional phase of a storm are locally symmetrical. In current events, however, erosional and depositional peaks shift so that in distal zones sand may be deposited on unproportionally soft mud that had escaped equivalent erosion ".

5. Convolute lamination, possibly an expression of thixotropic dewatering of saturated sediment, along with unidirectional ripple or climbing-ripple lamination are commonly associated with turbidites. This is in sharp contrast to the predominantly wave-influenced structures in tempestites (Kreisa, 1981).

6. Climbing-current ripples are characteristically associated with turbidites, but are a rarity in tempestites. This reflects basic differences in flow dynamics between the two-current dominated versus wave dominated situation in turbidite currents and storm surges respectively.

7. Spill-over ripples (Fig. 2.7.6), which commonly display a distinctive apron of thin silt/sand over adjacent mud-filled troughs on one or both the flanks of either straight-crested or interference ripples, is considered to be characteristic of tempestite (Seilacher, 1982). They form when the 'the ripple troughs were already filling with mud so that still emergent ripple sand crests could spill over the mud under continued wave action' (Seilacher, 1982). Careful observations are necessary since such ripples outwardly resemble flat-topped ripples that are associated with tidalites.

8. Seilacher (1982) has described "kinneyia" structures (Fig. 2.7.7) as a signature of storm activities. This structure usually prefers flat surface or the top of flattened ripples and contour the margins of the crests. Kinneyia consists of small ovate grooves, or pits that are uniform in size (2-12mm), regularly spaced (0.4 to 1.5 times the width), and have characteristically steep, or even overhanging, slopes (Pfluger, 1999). Occasionally, pits grade into long furrows along the same bedding plane. Outwardly they resemble either raindrop prints, or minute ripple marks. But on close scrutiny these small pits show vertical flanks that are too steep to persist at the sediment-water interface. Their preferential occurrence on flat surfaces or at the flattened tops of the ripples and almost vertical flanks of the pits suggest an origin later than ripples. Experimental observations suggest that the gas bubbles, produced by decaying organic matter and trapped below a microbial mat surface, are the primary cause for the formation of kinneyia structure (Pfluger, 1999).

9. Micro-folded (pleated) horizons (Fig. 2.7.8) with folding inclined down the paleoslope and geopetal asymmetry (with synclines being more pointed than the anticlines) have been interpreted to be due to seismic shocks (Seilacher, 1984).

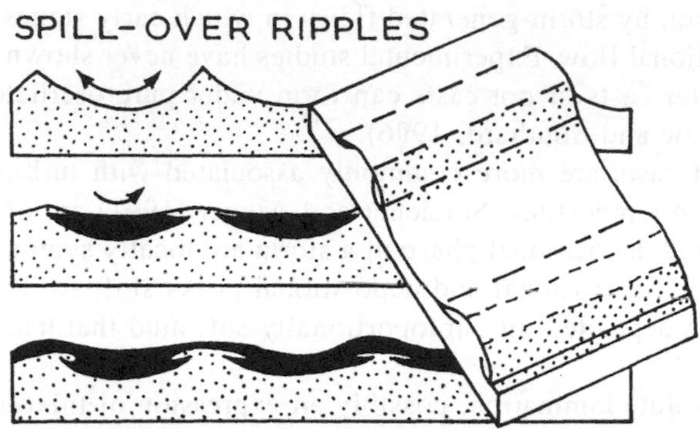

Fig. 2.7.6: Assumed origin of spill-over oscillation ripples. Crest is flattened by continuing wave activity after mud had started to settle in the ripple troughs. Spill-over aprons were later tilted down by mud compaction (Reproduced from A. Seilacher: Distinctive features of sandy tempestites, Fig. 3, page 338, in, G. E. Einsele and A. Seilacher, eds., Cyclic and Event stratification, © 1982, Springer-Verlag GmbH & Co. KG., with the kind permission).

10. Fault grading (Fig. 2.7.9) has been suggested to be another criterion for the identification of seismites in rock record. Fault grading starts with a homogenous top layer which grades downward into rubble horizon (in which displaced lumps preserve the lamination) and to micro-faulted zones with gradational contacts between these zones and the bottom, but with a sharp boundary at the top (Seilacher, 1982, 1984).

11. Convex-down stacking of bowl-shaped shells (Fig. 2.7.10) is thought to be the effect of seismic shock. Such orientation of bowl-shaped shells is clearly unstable and is unlikely to have been deposited by any current activities. Sudden seismic shocks must have thrown the sediments into suspension and at the time of re-settlement, heavy bowl-shaped shells are deposited in convex down position (Seilacher, 1984).

12. Soft-sediment deformation structures, e.g., ball-and-pillow structures, deformed cross-bedding, convolute lamination, flame structures (Fig. 2.7.11a, b), pseudo-nodules (Fig. 2.7.12) etc., besides clastic dykes and sills, are possible signatures of seismic shocks. While recognizing the fact that there are other triggering mechanisms producing soft-sediment deformation structures, (e.g., wind waves of storm origin, descent of landslides, increase in pore fluid pressure because of rapid sedimentation, impact of meteorite), the extensive lateral and vertical extent of such structures are perhaps the best criteria for the seismic shock (Allen, 1986).

The geometry of clastic dykes and sills that is tabular in geometry look similar to those of gravity-filled cavities. Thus it is necessary to differentiate the two similar-looking structures. Forcefully injected liquefied sedimentary material often shows faint internal laminations paralleling the fractured walls, while gravity-filled laminations are perpendicular to the walls and may show graded laminations. While cross cutting relationship of the clastic dykes with

Fig. 2.7.7: Kinnneya on flattened tops of interference ripples, Beduh Shales (Lr. Triassic), Sinat (N. Iraq). Kinneya structures contour ripples and are thus younger (Reproduced from A. Seilacher : Distinctive features of sanndy tempestites, Fig. 4a, page 340, in: G. E. Einsele and A. Seilacher, eds., Cyclic and Event Stratification, © 1982, Springer-verlag GmbH & Co. KG., with the kind permission).

the host rock helps in their identification in the field, it is, however, difficult to identify clastic sills in the field because of conformable relationship with the host rock. The features that are helpful in this respect are (i) presence of feeder dyke/s near-by, and (ii) presence of rip-up clasts in sills derived from overlying and underlying host rocks. Mohindra and Bagati (1996) have described various liquefaction structures, e.g., flame structures (Fig. 2.7.11), complex recumbent folds, pseudo-nodules (Fig. 2.7.12) etc., in 120 meter thick fluvio-lacustrine sediments exposed in the Sumdo area in the Lower Spiti Valley, Himachal Pradesh, India, which are laterally traceable for quite a distance. They have interpreted these structures as seismites. Similar soft-sediment deformation structure, triggered by seismic shocks, have also been reported by Anand and Jain (1987), Allen (1964, 1974, 1982), Mayall (1983), Scott and Price (1988) Stewart (1982), Weaver (1976) amongst many others.

Seismic shocks not only trigger the formation of various deformational structures, but may have a larger effect in causing a change in the architectural

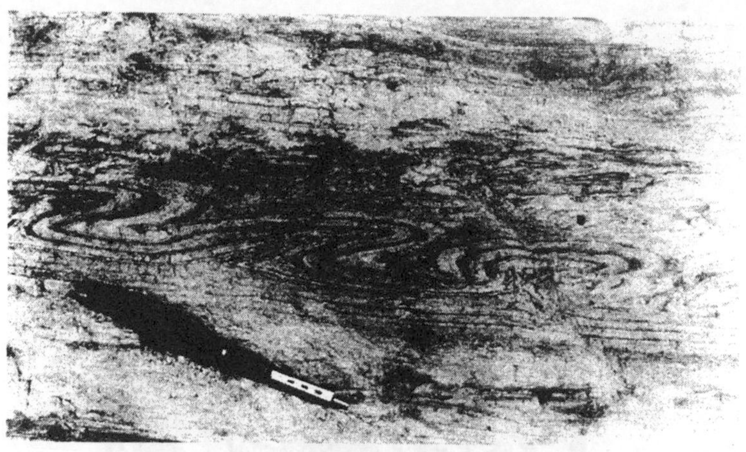

Fig. 2.7.8: Pleated laminae within the Lisan marls of South Jordan. The location of these Pleistocene lake sediments in the Jordan rift zone make a seismic origin of micro-slumps very probable. They are distributed throughout the sequence and can be followed laterally throughout the outcrop. In all of them the folds verge toward the graben axis. (Reproduced from A. Seilacher: Sedimentary structures tentatively attributed to seismic events. Marine Geology, Vol. 55, Fig. 1, © 1984, Springer-Verlag GmbH & Co. KG., with the kind permission).

pattern of the basin itself and related depositional systems. DeCelles (1988) has described from the Middle Tertiary of San Emigdio Range, southern California, three laterally-separated disturbed zones of (i) 50 m thick lens of very coarse, angular granitic breccia, (ii) several large penecontemporaneous structures, and (iii) a mixture of chaotically deformed sandstone, mass flow conglomerate and breccia with granite clasts. These disturbed zones were not only triggered by a series of earthquakes, but seismic shocks have had their effects in changing the basin architecture. DeCelles (1988) has further stressed that such disturbed zones, though may be volumetrically minor, should be closely looked into which may reveal major reorganization of the depositional basin, as revealed in the present example.

13. Presence of molar tooth structures in fine-grained carbonate rocks provides another signal of seismic activity. Recently, Fairchild et al., (1997) have described occurrence of this structure in micrite beds of the Neoproterozoic Xinmincum Formation, north China. These structures are sub-vertically oriented cracks, lensoid in shape (tapering upwards and downwards), and are 1-20 cm long, filled-in with microsparry calcite cement. These structures preferentially develop in fine-textured lithologies. Fairchild et al. (1997) have offered a mechanical explanation for their formation. Seismicity superimposed on body stresses has been suggested as most likely external forcing mechanism for the cracks to form. Cracks generated this way would tend to be elongated parallel to the propagation direction of the seismic waves (Love waves) and open perpendicular to this direction. Varying types and spatial centers of wave

activity from different earthquakes would generate cracks with variable orientation in different beds. Subsequent filling of these cracks by micro-spars, possibly with a displacive force and differential compaction during burial would tend to give rise to structures that are now known as molar tooth structures (in cross-sections they resemble molar teeth of elephant and hence the term).

14. The faunal spectrum in an individual fossiliferous storm bed may show a distorted picture. This distinction is an effect of a) repeated erosion in shallow water and subsequent deposition in deeper water, b) diagenetic distortion, c) 'mixed fauna' from different habitats, d) repeated phases of burial and re-exposure (Aigner, 1982). Such faunal distortion may help in identifying storm beds in rock record.

15. "Proximality trend" is another signature of tempestite facies. Aigner (1982, 1985) has described proximal to distal changes in "bed thickness, degree of amalgamation, grain size, bioclast and intraclast content, and in changing sedimentary structures, paleocurrents and faunas" (Fig. 2.7.13). Proximal tempestites are relatively thick, coarse-grained with profuse intra-, and extra-clasts and commonly form composite and amalgamated beds. Distal tempestites, on the other hand, are fine-grained, thinner, one-event mudstone beds. This lateral gradation in physical properties of tempestites is because of decreased intensity of both storm waves and storm-induced currents from shallow to deep offshore basins. Proximality trends can also help in paleogeographic reconstruction of storm-generated depositional systems by indicating a) the source areas of the storm sands, and b) paleobathymetric trends.

Offshore muddy tempestites and distal turbidites look alike. Careful observations are necessary to distinguish them. Their discrimination is of considerable importance for environmental analysis—paleo-depth, circulation, distance from land, sedimentation rate of hemi-pelagic and turbiditic sediments. The following criteria may help:

a. Distal turbidites are characterized by base cutout Bouma sequence and have current-formed structures towards the top of the sequence. Tempestites contain hummocky cross-stratification and other wave-induced sedimentary structures.
b. There are also differences in ichnofaunal associations. Pre-event and post-event ichnofaunal associations of tempestites are essentially the same feeding and dwelling burrows that are characteristic of Cruziana ichnofacies. This is not so in the case of pre-, and post-turbidite facies. Pre-turbidite ichnofacies are varied and highly regular patterns of open burrow systems. Food is extracted indirectly by way of farming bacteria or fungi. Post-turbidite trace fossils are episodic colonizers of newly deposited mud and are dominantly deposit feeders by sorting and ingestions. Burrows are back-filled by sorted sediment and fecal material unlike pre-turbidite facies (Einsele and Seilacher, 1991).

Fig. 2.7.9: Sections through typical seismites (upper bed: Scale 15cm) showing fault grading. Note the sharp boundary against undeformed lamination on top and gradational transition underneath. Three zones can be recognized in the graded sequence of structures (from top).

a: Soupy zone: a kind of liquefaction has wiped out all previous depositional structures. Indistinct lamination near the top indicates that the uppermost mud layer had gone into suspension.

b: Rubble zone: where compaction was somewhat more advanced, larger fragments of the original sediment survived the shock but swim with varying orientation in the soupy matrix.

c. There is difference in carbonate content between the "background" sediments (hemipelagic mud) and turbidites. Slow rates of sedimentation of hemipelagic mud below the carbonate-compensation depth leads to complete dissolution of carbonate skeletons. Carbonate turbidites, on the other hand, are not affected by dissolution because of rapid sedimentation. Hence, the colour of the carbonate turbidites is light, while that of hemipelagic mud is greenish-gray or dark-gray or brown (Hesse, 1975). A similar style of sedimentation has been reported from both modern and ancient sedimentary basins (for details, see Hesse, 1975).

However, if the turbidites are emplaced above the carbonate-compensation depth, the distinction on the basis of carbonate content is not possible; in that case bioturbation textures form the basis of discrimination - hemipelagic muds will show higher intensity of bioturbation than in turbidites. Moreover, turbiditic mud will be graded, whereas hemipelagic mud will be more homogeneous.

d. There is a difference in micro-faunal content between muddy turbidites and hemipelagic sediments. Turbidite facies contains plankton and shallow-water benthic forms compared to hemipelagic muds that show benthic agglutinants and complete lack of shallow-water forms. O'Brien et al. (1980) have suggested clay fabric as a distinguishing criterion; hemipelagic mud shows preferred orientation of the clays than in turbiditic muds. This is possibly because mud in clayey turbidites is deposited in flocculated state, while hemipelagic mud is deposited more in dispersed state under less turbulent conditions.
e. Muddy tempestites may have layers of reworked flat pebble conglomerates composed of mud that alternate with layers of the same lithology having similar petrographic composition, rare in biogenic silica and organic matter (unless emplaced in oxygen-free environment). Signatures of a) shallow water organisms emplaced into the deep-sea, b) beds with silt laminae that show decrease in grain size and thickness upward (graded siltstone), c) presence of escape burrows, d) restriction of bioturbation to the top of the bed and e) very thick massive mud beds (of uniform petrography and without bioturbation) are other signatures of fine-grained turbidites (Hesse, 1975; Piper, 1972,1978; Piper et al., 1991; Stow et al., 1984).

Fig.2.7.9 (contd.)

c: Segmented zone: still more coherent older layers only broke along antithetic step faults of miniature scale. Their effect decreases with depth before they die out, leaving fewer and fewer faults at larger, but still fairly regular intervals. Thus the seismite has no defined base and certainly no basal slip surface.

d: Undisturbed sediment: laminations are left undeformed, while major faults may cross the beds at distances of a few meters (Reproduced from A. Seilacher: Fault-graded beds interpreted as seismites. Sedimentology, vol. 13, © 1969, Plate 1, Blackwell Science with the kind permission).

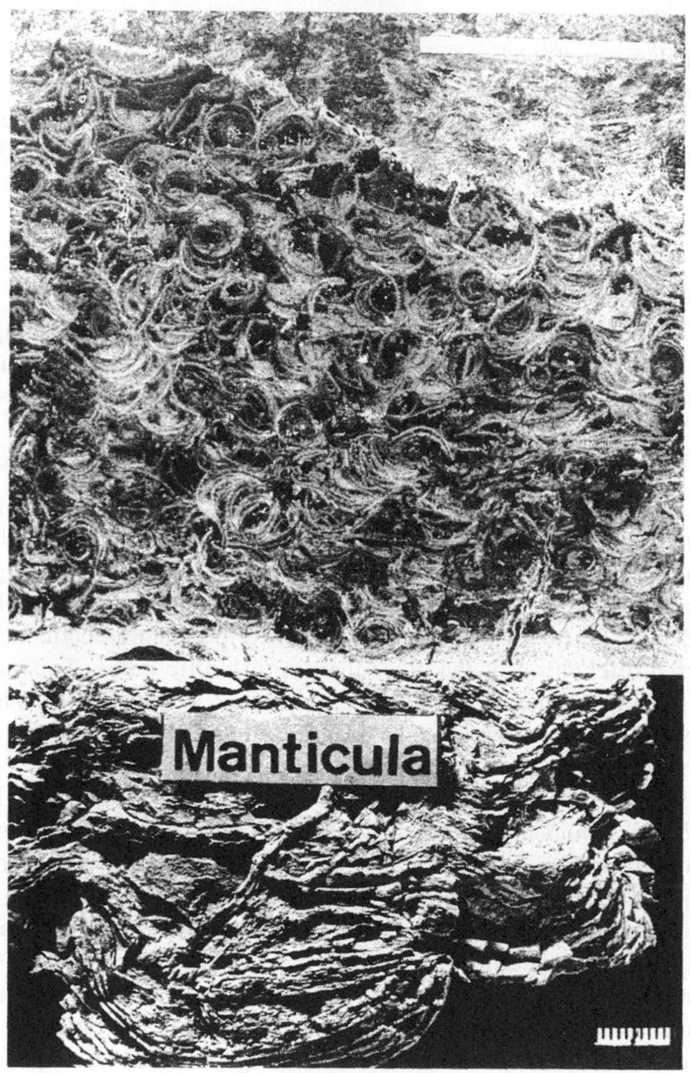

Fig. 2.7.10: Monospecific shell beds in geosynclinal pelagic carbonate muds commonly show a uniform structure with an unusually high degree of stacking. The predominance of convex-down stacks indicates the water had already quieted when the shell sank down as the coarsest grade of a reworked sediment layer. This and the probable bathymetry suggest a tsunami rather than storm wave origin. Lithospheric after-shocks could have increased the stacking of the already settled shells.

The example on top (mid-Jurassic Bositra shells from Tithonian of Stua Alm, northern Italy) shows sharp erosional base and an overlying layer of typical "filament limestone", in which stacking is less obvious because cementation has delayed and allowed the shells to be compactionally flattened. The boundary between the two shell layers is highly stylolitized.

The lower picture (Upper Triassic Nugget Point, southern New Zealand) shows the convex-down stacking in a much larger pelecypod species, but in a similar facies. Scale 1 cm. (Reproduced from A. Seilacher: Sedimentary structures tentatively attributed to seismic events. Marine Geology, vol. 55, Fig. 4, © 1984, Springer-Verlag GmbH & Co. KG., with the permission).

Fig. 2.7.11 a: Hydroplastic intrusion of siltyclay material exhibiting flame structures. Note undeformed lower strata. Pencil (circled) is 16 cm. (Reproduced from R. Mohindra and T. N. Bagati: Seismically induced soft-sediment deformation structures (seismites) around Sumdo in the lower Spiti (Tethys Himalaya). Sediment. Geology, 101, p. 69-83, Fig. 6, ©1996, Elsevier Science - NL, Sara Burgerhartstraat 25, 1055 KV Amsterdam, The Netherlands, with the kind permission).

16. The facies association is another positive way of discriminating tempestite from turbidite. Tempestites are commonly associated with rocks of shallow-water facies (shoreface, foreshore and even beach sand and gravel). Being associated with shallow-water facies, the thicknesses of tempestites are also restricted and have lower preservation potential in the stratigraphic record. Turbidites, on the other hand, may attain a thickness of hundreds to thousand

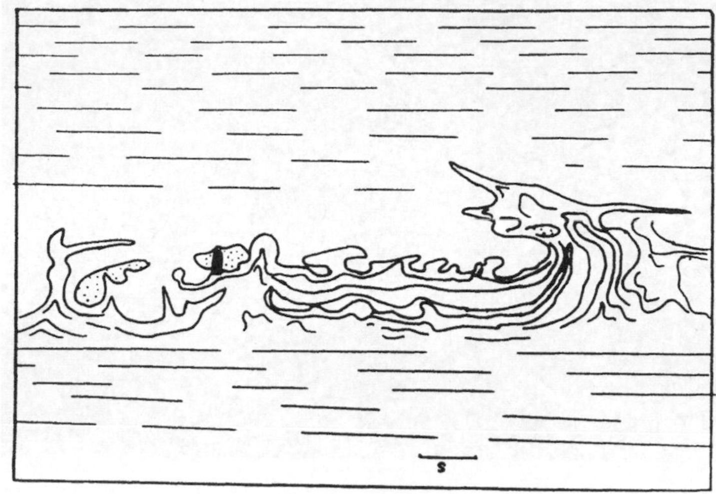

b: Photograph and line drawing of complex flame structures. Scale bar 13 cm (Reproduced from R. Mohindra and T. N. Bagati: Seismically-induced soft-sediment deformation structures (seismites) around Sumdo in the lower Spiti (Tethys Himalaya), Sediment. Geology, vol. 101, p. 69-83, Fig. 8, ©1996, Elsevier Science-NL, Sara Burgerhartstraat 25, 1055 KV Amsterdam, The Netherlands, with the kind permission).

of meters and are associated with deep-water marine facies with higher preservation potential.

17. Storm-associated sediments are usually well sorted, bimodal with coarser particles concentrating near the bottom. Turbidites are usually poorly sorted and have graded beds; proximal turbidites may show inverse grading and are usually associated with debris flows and mud flows.

18. Abundant mud infiltration fabrics, such as, shelter porosity, micrograded sediment perched on individual shell, and sediment screening (Fig. 2.7.14) are other excellent features of periodic high-energy events in an otherwise low-energy setting (Kreisa, 1981; Kreisa and Bambach 1982). These features form at the waning phases of storms. [Sediment screening occurs where fine sediment is partially blocked from infiltrating into the voids of a grain-supported fabric of the shell-lag deposits resulting in decreased infiltration downward]. Kreisa and Bambach (1982) noted "infiltration fabrics found in relatively thin shell beds interbedded with mudrock appear to be excellent evidence of episodic high-energy events in a normally low energy environment". While shelter porosity can also be generated through winnowing by continuous wave/tidal currents, their repetitive occurrences in couplets of coarse and fine sediments are highly unlikely other than storm surges (Kreisa, 1981).

Specht and Brenner (1979), in their study on bioclastic carbonates in Upper Jurassic epicontinental mudstones of east central Wyoming, have suggested storm-wave origin for the bioclastic limestone on the basis of certain textures

Fig. 2.7.12: Photograph and line drawing showing deformation features. Diapiric intrusion of clayey silt material deforms the strata into flame-like recumbent folds with pseudo-nodules (cycloids) of coarse sand. Scale bar is 13 cm. (Reproduced from R. Mohindra and T. N. Bagati: Seismically-induced soft-sediment deformation structures (seismites) around Sumdo in the lower Spiti (Tethys Himalaya), Sediment. Geology, vol. 101, p. 69-83, Fig. 10, © 1996, Elsevier Science-NL, Sara Burgerhartstraat 25, 1055 KV Amsterdam, The Netherlands, with the kind permission).

and fabrics. Amongst those are i) unabraded grains, ii) sheltered mud patches with winnowed fabric, and their extreme variability, iii) fining-upward grading of skeletal non-matrix grains, iv) general upward trend of increasing grain support and v) upward reduction in matrix material (showing an increased winnowing effect). The above features are similar to recent graded storm deposits reported by Kumar and Sanders (1976).

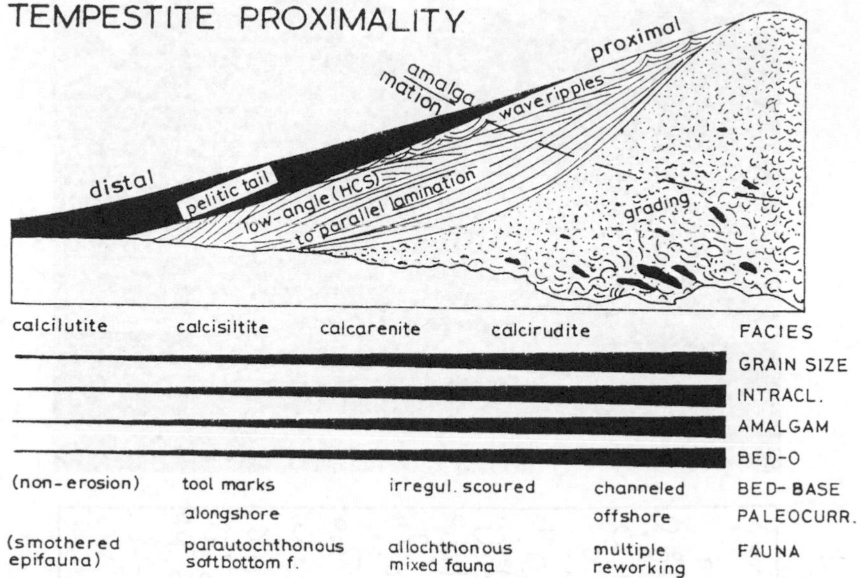

Fig. 2.7.13: Generalized trends in vertical and lateral variation of "ideal" tempestites: proximality as a guide for environmental interpretations (Reproduced from T. Aigner: Storm Depositional Systems: Dynamic Stratigraphy in Modern and Ancient Shallow Marine Sequences. Lecture Notes in Earth Sciences, Fig. 49, page 174, ©1985, Springer-Verlag GmbH & Co. KG. with the kind permission).

19. The couplets of coarse lag deposits and fine sand/silt form a good signature of storm deposits. The presence of infiltration fabric further strengthens the interpretation (Kreisa, 1981). It is important, however, to note that the formation of such couplets is dependent upon the availability of an appropriate grain-size distribution within the basin of deposition. Therefore, one may come across storm deposits without such couplets, with either coarse lag deposit or fine laminated sands only (Kreisa, 1981).

20. Extreme variability in thickness (pinch-and-swell parallel lamination) and lateral discontinuity are important features of storm beds, in contrast to turbidites that are characteristically uniform in thickness and laterally continuous for a considerable distance. Extreme variability in thickness indicates rapid deposition from suspension without significant transport by traction (Kreisa, 1981).

21. Shallow-water gravity slide/slump structures can be products of storm surges, particularly when slump/slide deposits are in close association with hummocky cross-stratification. Myrow and Hiscott (1991) have described similar associations of structures from the Chapel Island Formation, southeast Newfoundland, Canada. They suggested "fluctuations in bottom pressure associated with the passage of large surface waves create cyclical shear stresses that cause

Fig. 2.7.14: Shell bed showing infiltration fabrics. Note shelter porosity beneath shells (now filled with bladed calcite cement); coarser peloidal sediment and skeletal debris perched on larger shells and grading upward to micrite under sheltered voids; increase in before-cement porosity caused by screening action of larger shells at bottom of section. Scale: width of the field is about 1 cm. (Reproduced from R. D. Kreisa and R. K. Bambach: The role of storm processes in generating shell beds in Paleozoic shelf environment, *in*, G. Einsele and A. Seilacher, eds., Cyclic and Event Stratification , Fig. 2, page 203, ©1982, Springer-Verlag GmbH & Co. KG, with the kind permission).

an increase in pore fluid pressure and associated decrease in strength, leading to failure and downslope movement" (Myrow and Hiscott, 1991).

22. Moore and Moore (1988) have drawn attention to another type of event deposits triggered by huge submarine slides. They have described deposits of coarse-grained gravels at Lanai, Hawaii, formed 105,000 years ago, caused by the generation of successive giant waves triggered by slides. They have suggested that failure and downward movement of the huge Lanai submarine landslide (40 km wide at the top) created an ocean disturbance, producing giant

waves (wave length 10–30 km and a minimum initial velocity of 140 m/sec), known as backfill waves. Moore and Moore (1988) explain that backfill waves form when an underwater slide moves downward and forms a depression in the water surface above the head of the slide; the water that rushes in to fill the depression overshoots and propagates as a packet of waves that radiate outwards. Side scan sonar system has actually identified this huge submarine slide, known as Lanai slide.

23. Tsunamis are long period sea waves, having a wavelength between 100–200 km and an amplitude of few centimeter to meter with a travel speed of several hundred kilometers/hour. These extremely powerful sea waves are generated by the seafloor faulting, submarine slides along mid-ocean volcanic islands/passive continental margins, or volcanic eruption. When tsunamis hit the shorelines, their wave heights are further enhanced by local topographic effects, causing resonance phenomena, as well as from interference and focusing of the incident waves in bays and estuaries along the coastline (Harbitz, 1992, cited in Driscoll et al., 2000). Since tsunamis generated in nature are much larger than produced singly by either of the processes, it is believed that a combination of process is responsible for their generation.

When such a high velocity, low amplitude tsunami hits the shoreline, its amplitude is greatly enhanced, as indicated above, with attendant erosion and transport of large quantity of sediments, either landward or seaward. Naturally, therefore, the most commonly described sedimentary product of tsunamis is a sheet of sand with large rip-up clasts at the bottom that overlies an erosional surface. The sand is normally graded (attesting to rapid deposition from suspension) with a decrease in thickness and grain size landward (Atwater and Moore, 1992; Bourgeois et al., 1988; Dawson et al., 1988; Minoura and Nakaya, 1991; Minoura et al., 1996).

Recently, Bondevik et al., (1997) have described a tsunami deposit in shallow marine basins and coastal lakes, western Norway, where normally graded basal sandstone with large rip-up clasts overlie an erosional base. The sand bed is overlain by mixed organic sediments, and contains rip-up clasts of peat, lacustrine gyttja and plant fragments. The sequence thins and decreases in grain size landward. The repeated recurrence of sequence of sand and organic deposits has been interpreted as pulses of erosion and re-deposition.

Moore (2000) has also described a late Pleistocene conglomerate deposit as a product of tsunami from the south coast of Molokai, Hawaii. Generally, the characteristics that identify a tsunami deposit from a storm deposit are: i) landward fining in grain size, ii) compositional trend, iii) greater landward extension of the deposit. "Landward fining occurs in many tsunami deposits because the grains are commonly carried in suspension from a restricted source" and the coarser particles, being higher in settling velocity, settle down nearer the source on flow deceleration (Moore, 2000). Had this been a product of multiple storm surges or small tsunamis, there should not be i) any grain size trend at all or a separate fining trend for each shoreline, ii) an age trend for different deposits, and iii) a change in the type of matrix or degree of cementation.

2.8 Discontinuity Surface

Barring the cases of faults and synsedimentary slides, a discontinuity surface within a sedimentary succession is usually represented in the field as a surface across which (from bottom to top) one or more attributes of the rocks vary, not gradually, but abruptly. The attributes that are tangible and significant in this respect are: 1. Texture, 2. Diagenetic fabric, 3. Lamina/bed thickness 4. Sedimentary structure, 5. Lithology, 6. Sedimentary facies, 7. Structural attitude, 8. Fossil content (Fig. 2.8.1). Abrupt nature of the change in any of the above attributes across a surface implies a length of time, however small, for which no record is present in the succession either due to non-deposition or erosion (Barrel, 1917; Bates and Jackson, 1982; Clari et al., 1995). For example, a lamina is bounded by discontinuities manifested by subtle but abrupt textural variation, representing the smallest time gap in a sedimentary succession (Fig. 2.8.1).

The discontinuity surfaces that are much smaller in extent as compared to the dimension of a sedimentary basin usually represent short-term breaks reflecting variability of the physico-chemical conditions of sedimentation such as hydrodynamic fluctuations; these surfaces are useful in identifying the depositional mechanism and environment. For example, abrupt transition of texture, lamina/bed thickness, sedimentary structure or variation of lamina/bed lithology from sandstone to shale within a heterolithic lithosome are usually reflective of velocity fluctuations in the depositing flow. On the other hand, the discontinuity surfaces that are spatially extensive, generally termed unconformities, represent a break of prolonged period (Fig. 2.8.1). These surfaces are of stratigraphic significance because they reflect some specific combinations, during a time interval, of the relative magnitudes of the rates of eustatic change in sea level or base level, basin subsidence (a function of tectonics) and sedimentation (a function of tectonics and climate) — the three fundamental variables influencing evolution of a sedimentary basin fill (Posamentier et al., 1988 a, b).

Although unconformity surfaces are formed over a long period of time and are virtually diachronous over the space of the basin, the rocks lying above an unconformity everywhere, are in general younger than the rocks lying below it and here lies the utilitarian significance of these surfaces: the rock succession between two unconformities apparently represents a continuum of sedimentation ignoring the short-term breaks (Posamentier et al., 1988b). Thus, unconformities are best surfaces to divide a sedimentary succession into a

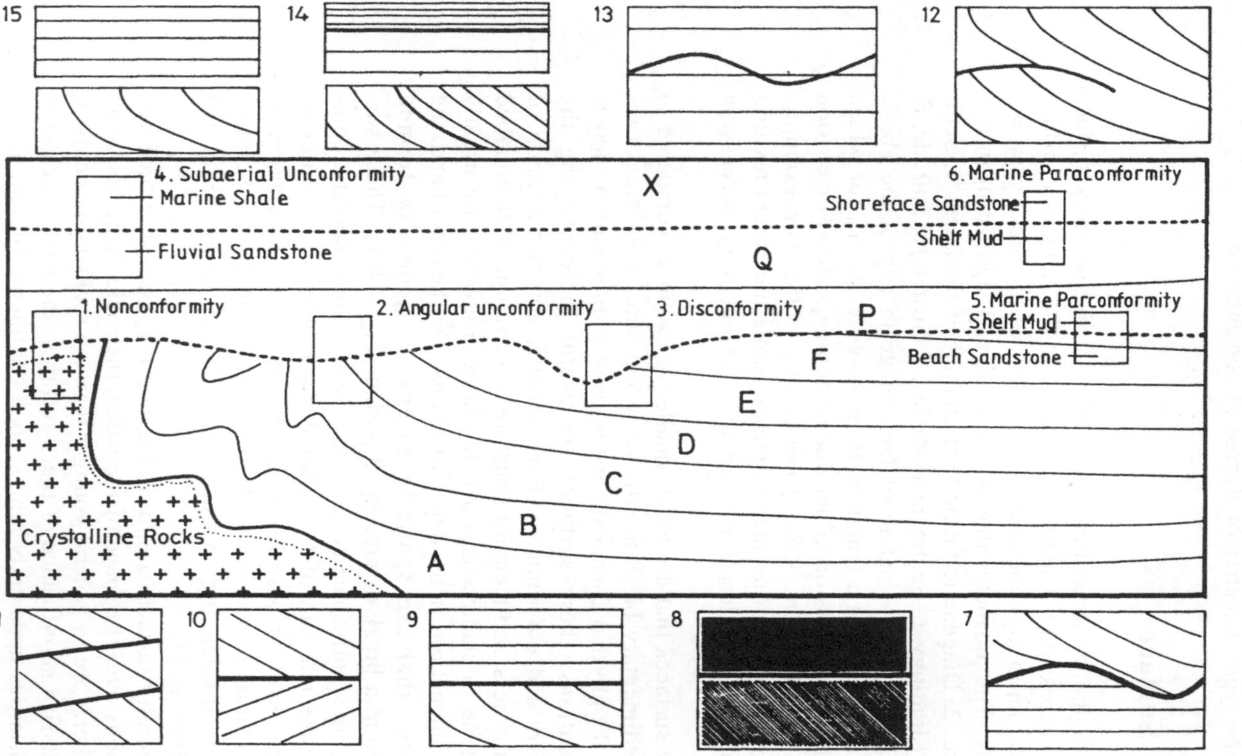

Fig.2.8.1: Diagrammatic representation of different orders of discontinuity surfaces that may be present in a sedimentary succession. The central box represents a hypothetical cross-section of a sedimentary basin showing the basement and the overlying lithostratigraphic units (A,B,C ....etc.). The smaller boxes within it mark the different types of unconformities: 1. Nonconformity; 2. Angular unconformity; 3. Disconformity marked by erosional truncation of underlying beds; 4. Subaerial unconformity without recognizable erosional truncation but marked by facies contrast; 5 & 6. Marine paraconformity - marked by facies contrast. The boxes outside the central box represent different discontinuities that have little stratigraphic significance but representing hydrodynamic fluctuations in the depositing flow; 7. Discontinuity marked by erosional truncation and abrupt variation sedimentary structure; 8. Discontinuity marked by abrupt change in lithology; 9. Discontinuity represented by sharp change in sedimentary structure; 10,11,12,13. Discontinuities characterized by erosional truncation; 14. Discontinuity marked by abrupt change in stratal thickness; 15. Discontinuity between laminae marked by subtle but abrupt textural variation.

manageable number of successive time-slices each recording the complete history of continuous sedimentation during an interval of geological time.

Recognition of unconformities in the field involves three steps: i. identification of a surface across which one or more tangible attributes change abruptly, ii. adjudgement of the prolonged nature of the time gap across the surface and iii. the spatial extent of the surface. In some specific cases, long period time gap across a discontinuity appears self-evident. For example, the surface between the crystalline basement and the overlying sedimentaries is a discontinuity reflecting an abrupt change in the lithological attribute (Fig. 2.8.1). The intensity of the lithological variation is so abrupt that the rock type changes from endogenous to exogenous indicating a prolonged time gap. It is thus an unconformity of stratigraphic significance and is traditionally known as nonconformity (Dunbar and Rodgers, 1957; Pirsson, 1915). Again, a surface, across which the structural attitudes of beds and/or deformation styles change abruptly (Figs. 2.8.1, 2.8.2), demonstrates stratigraphically significant time gap between the times of formation of the underlying and overlying beds and is also an unconformity, traditionally termed angular unconformity (Dunbar and Rodgers, 1957; Hutton, 1788). Moreover, in both the cases, the unconformity surface is essentially a surface of subaerial erosion (Figs. 2.8.1, 2.8.2).

However, there may be unconformity surfaces within a sedimentary succession showing no structural discordance but representing a period of erosion or non-deposition (subaerial or submarine); they are known as disconformity (Grabau, 1913) and paraconformity (Dunbar and Rodgers, 1957) respectively. If the fossil contents of the rocks change abruptly across such a surface or there is abrupt facies transition across it violating the Walther's Law, it may be designated as a disconformity or paraconformity (subaerial or submarine) depending on whether the surface is associated with erosion or non-deposition (Fig. 2.8.1).

However, there may be disconformity or paraconformity surfaces in a sedimentary succession implying prolonged time gap but without any discernible change in any of the tangible attributes, so far mentioned, in the over and underlying rocks on outcrop scale. Presence of such unconformities is best revealed by the variation in the arrangement of "stratal surfaces" across the unconformity surface as summarized succinctly by Doglioni et al. (1990). However, "stratal surfaces" are essentially time boundaries (Vail et al., 1977); they do not necessarily follow the lithological boundaries and are represented by the seismic reflection lines. So identification of unconformities from the pattern of stratal arrangements is conveniently done from seismic sections and is not quite feasible in the field. In order to recognize such unconformities in the field, other indirect clues are to be sought for as summarized below following Krumbein (1942) and Krumbein and Sloss (1963).

Fig. 2.8.2: Angular unconformity between the Ordovician strata (right) and the Silurian strata (left), New York.

Features Indicative of Subaerial Disconformity / Paraconformity

1. The cause of subaerial exposure of the shelf is often related to a fall in relative sea level (forced regression, Posamentier et al., 1992; Posamentier and James, 1993). Relative sea level fall may result in a drop in non-marine base level causing the rivers to incise valleys. Presence of such fluvial valley fills is a clear indication of a subaerial disconformity. However, valley incision may also take place due to increased discharge and stream power or decreases in stream load independent of relative sea level fluctuation. The features of incised valleys indicating a relative fall in base level are: i. valley-fills are characteristically wider and thicker, ii. deep valleys with terraces, iii. lag deposits over incised valleys are commonly much coarser than the overlying valley-fill deposits.

2. Subaerial surfaces of erosion (disconformity) are usually undulatory in shape and truncate the underlying beds (Fig. 2.8.1). In many cases the wavelength of the undulatory erosion surface is too large to be perceived on outcrop scale. However, undulatory geometry, as well as, truncation of beds can be decipherd by careful mapping.

3. Igneous intrusions which terminate abruptly at a surface of contact, and which do not show evidences of thermal alteration of the overlying beds, mark unconformities caused by subaerial erosion. Pebbles of intrusive bodies may also be found in the immediately overlying beds.

4. The occurrence of conglomerates, characterized by angular, and weathered rock fragments of the underlying formations, termed basal conglomerates, indicate a subaerial disconformity.

5. If a surface remains subaerially exposed for a prolonged period without suffering significant erosion or deposition, the process called pedogenesis sets in (see Section 2.6). So presence of well-developed, thick paleosols within a sedimentary succession reveals a subaerial unconformity (paraconformity). Criteria of similar significance are the presence of land plants, roots and rootlets, colour mottling, illuvial clay-rich layers, concretions and other soil structures.

Subaerial exposure of a surface followed by a prolonged depositional break may result in the development of duricrust. Duricrusts are hard cemented crusts, a striking feature of many tropical and subtropical landscapes, of soil surfaces formed through precipitation of various types of cements. This represents a product of terrestrial processes within the zone of weathering (Goudie, 1973). The caliche is a particular type of duricrusts. Duricrusts within a sedimentary succession provide a strong signal of terrestrial weathering zone, and hence an unconformity. Silcretes and ferricretes are other two common types of duricrusts having the same geological significance. Bauxite, laterite, pisolitic horizons are also indicative of subaerial exposure with a depositional break and hence an unconformity.

Concentration of iron oxides is known to be associated with unconformities. Sharp (1940) has reported concentration of iron oxide below the epi-Algonkian surface of weathering in the Grand Canyon region. Local concentrations of iron- oxide may also take place in the formation of lateritic soils. Since iron-oxide zones also occur in some conformable sequences, this criterion may be only suggestive; association of other features is required for confirmation.

6. Leith (1925) has described many erosion surfaces, which are capped by chert. Such silicified erosion surfaces are believed to have formed by the concentration of residual chert during the weathering of limestones. This indicates subaerial weathering surfaces. During weathering of phosphatic limestone, relatively more soluble calcite may be leached, leaving a residual concentration of calcium phosphate near the surface. Such surfaces along with phosphatic nodules may indicate possible erosion surfaces.

7. Limestone exposed subaerially frequently develop porous zones due to development of cavernous porosity and karstic topography from dissolution. These more porous zones in limestones may indicate unconformities. Similarly, porous zones may develop along surfaces in terrigenous rocks, when they are subaerially exposed to acidic meteoric water; unusually high porosity and permeability develop beneath such erosional surfaces from the dissolution of unstable framework grains and cements.

8. In desert environment, where wind does the major mechanical work, lag deposits of coarser material may develop through winnowing of finer fractions. Coarser fractions may show wind faceted dreikanters and einkanters, identify-

ing them as products of aeolian abrasion. Occurrence of such lag deposits within a sedimentary succession indicates an unconformity.

9. Silicification of fossils during weathering of limestone may develop beekite rings on the surface of silicified shells. Beekite rings or small dough-nut-like circlets are bluish-to-white opaque-to-translucent quartz. Such beekite rings have been claimed to be good indicators of an erosional interval (Howell, 1931).

10. Zones of residual, weathered chert, known as tripolitic chert, are usually associated with stratal discontinuity surfaces. Such chert zones may be chalky white in colour and extremely porous. The porous nature of such chert is visible in thin sections (using blue dyed epoxy). White weathering rims are common. Weathered chert may occur both above and below stratal discontinuity surfaces, and forms a good signature of proximity of discontinuity surfaces.

Features Indicative of Submarine Disconformity/Paraconformity

1. The major physico-chemical changes in sedimentary environment, responsible for the interruption of sedimentation, results in diagenetic changes. The most common evidence for submarine discontinuities are: (a) presence of rock grounds- seafloor surfaces which, regardless of their origin and characteristics, are hardened before deposition starts anew; evidences of hardening with a depositional break include erosional truncation of rigid bodies such as shells, grains and cements, boring and encrusting organisms along the surface, neptunian dykes in the underlying rocks; (b) staining, crusts and nodules of authigenic minerals, (c) contrast between compactional features in the rocks across a surface (e.g. stylolites in the underlying rocks versus fittted fabrics and/or dissolution seams in the overlying rocks (Bathurst, 1987; Buxton and Sibley, 1981) suggests that only the underlying rocks suffered an early cementation phase and hence an unconformity.

2. Glauconites are marine authigenic green minerals ranging in composition from glauconitic smectite to glauconitic mica. These minerals are reported from a) shallow open marine environments where rate of sedimentation is very low, and b) occur above a major hiatus produced by contour currents at water depths of 1,600–2,500 m (Odin, 1985). Since occurrence glauconite is also reported from conformable sequences, other evidences are necessary to establish an unconformity surface. Glauconite minerals in deep marine facies may suggest erosion by contour currents; however, their occurrence in shallow marine sediments indicate period of slow deposition (condensed sequence, Shanmugam, 1988). Maximum marine flooding at the end of transgression is characterized by very slow sedimentation rates producing condensed sequence, which may be treated as a bundle of short-period submarine discontinuities that together constitute a stratigraphic break equivalent to an uncorformity. In addition to

glauconite, condensed section deposits include chert band, limestone band and organic rich radioactive shale.

3. Occurrence of phosphatic nodules and pellets in conjunction with glauconite minerals are considered to indicate submarine unconformities - surfaces of non-deposition rather than erosion or exposure. Pettijohn (1926, cited in his book "Sedimentary Rocks", 1975) has interpreted zones rich in phosphatic granules or "pebbles" as the residum on a "corrosion surface" or diastemic plane caused by submarine solution.

4. Concentrations of manganiferous compounds have been noted in association with unconformities. Manganese nodules are commonly associated with erosional surfaces produced by strong bottom currents in the deep sea. The strong currents keep the manganese nodules from being covered with sediments, allowing the nodules to grow (Watkins and Kennett, 1972). Manganese hardgrounds in rock record may be used to recognize submarine erosional surfaces created by bottom currents (Tucholke and Embley, 1984).

5. Stratigraphic breaks of greater or lesser magnitude may develop in connection with coral reefs. The reef is built to the surface of the sea, whereupon it may spread laterally. If the sea level later rises the reef is built upwards again. In section, therefore, there may be irregular lateral spreads of the reef, marking successive stands of the sea. According to Twenhofel (1932), such lateral spreading, usually marked by a shaly or sandy zone through the reef, is an indication of a stratigraphic break.

6. Ravinement surfaces marked by transgressive lag deposits formed due to shoreface retreat indicates submarine disconformity.

7. Drowning unconformity results due to shut down of carbonate factory following rapid rise in sea level and is recognized by a sharp change from shallow-water carbonate to deep shelf, slope or basinal deposits.

3. PALEOGEOGRAPHIC SETTINGS

Introduction

Terrigenous clastic sediments are transported and brought to a sedimentary basin from the source-land by means of moving air, water or ice, under the influence of gravity. In contrast, carbonate sediments are produced within the basin, if supply of terrigenous clastics is limited and the basin is subaqueous (marine/lacustrine).

At a particular time a sedimentary basin is typified by a particular spatial array of depositional systems defining its geography and physiography. A depositional system is a geographic domain of sediment accumulation in a sedimentary basin, characterized by a unique combination of climatic, tectono-geomorphologic, fluid-dynamic and biologic processes. A depositional system may consist of several, more localized domains representing the architectural elements—the building blocks of the system (e.g. Miall, 1985, 1988; see Chapter 4). The processes operating in a system leave their signatures in the resulting sedimentary deposit that help recognizing specific depositional systems from the rock record (see Walker, 1992; Reading, 1996).

The strata occurring in a sedimentary succession are representative of different depositional systems that prevailed in the basin during its life term. In a sedimentary succession the products of the depositional systems are stacked vertically. The paleogeography of an ancient sedimentary basin at a particular time can be reconstructed from a sedimentary succession by knowing the set of depositional systems that existed in the space of the basin contemporaneously. How can we know, from a sedimentary succession, which depositional systems existed in the basin adjacent to each other at a particular time? As stated in Walther's law: from a package of sediments representing a continuum of sedimentation, the ancient array of depositional systems (i.e., the paleogeography) can be reconstructed by spreading out each depositional system (occurring vertically in the package) like a deck of cards (Fig. 3.1). Reconstruction of paleogeography using Walther's law is, however, applicable only to a progradational vertical succession where the boundaries between the products of each depositional system are gradational. If in a vertical succession the products of the depositional systems are stacked with sharp or erosional boundaries or the succession is retrogradational, there is no certainty that the depositional systems existed adjacent to each other contemporaneously.

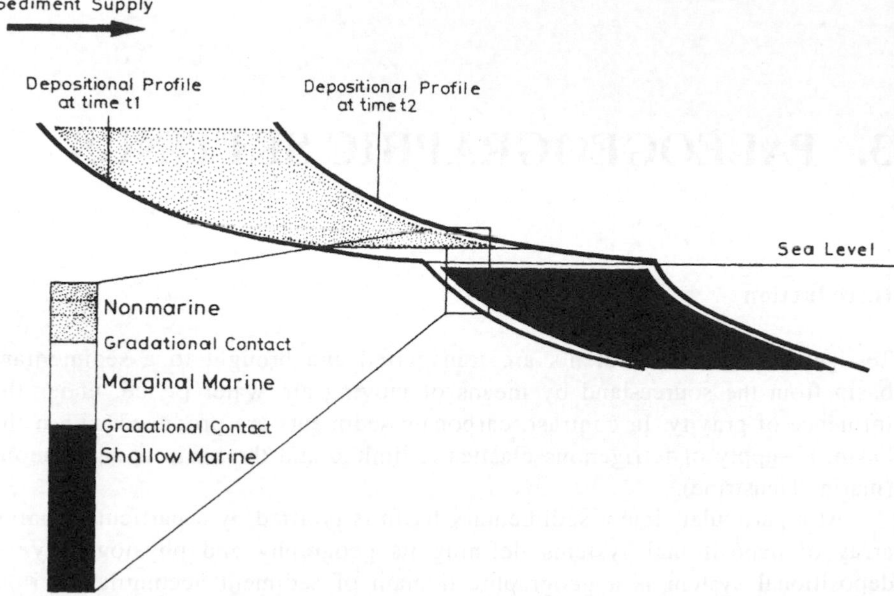

Fig. 3.1: Schematic diagram representing Walther's law of sedimentary succession. In response to sediment supply and deposition, the depositional profile shifts progressively down the basin inducing superposition of relatively up-dip depositional systems over the down-dip depositional system. The resulting sediment package at a location is represented by a progradational succession of depositional systems, essentially with gradational boundaries. Such a package occurring in ancient sedimentary succession can be utilized to understand the lateral distribution of contemporaneous depositional systems in the basin at a particular time.

In course of evolution of a sedimentary basin, its geographic setting does not remain fixed, but changes with time in response to changes in relative base level, source land tectonics and climate. In this chapter we discuss the different models of geographic setting (i.e., the array of different depositional systems that may occur adjacent to each other at a particular time) that may characterize sedimentary basins under different combinations of relative rates of base level change, source-land tectonics and climate, in order to reconstruct the paleogeographic evolution of a sedimentary basin from a sedimentary succession. Usually, at a particular time, a sedimentary basin is consisted of either siliciclastic or carbonate depositional systems. However, basins characterized by both siliciclastic and carbonate depositional systems, prevailing in different parts of the basin, are not uncommon.

Siliciclastic Depositional Settings

A survey of present-day siliciclastic depositional systems reveals that they are limited in variety (Table 3.T1). In low latitude areas under arid climatic con-

Table 3.T1: Siliciclastic Sediment Dispersal Systems

Alluvial Fan System	Gravity Flow Dominated, Stream Flow Dominated	***Subdivisions:*** Upper Fan, Middle Fan, Lower Fan
Fluvial Systems	Straight and anastomosing rivers (suspended load, muddy), High-sinuosity channel systems (meandering rivers, mixed load, sand plus mud), Low sinuosity channel systems (braided rivers, bed load, gravel plus sand)	***Subdivisions:*** Upstream, Midstream, Downstream. ***Sub-environments:*** Channel, Levee, Crevasse splay, channel, and Flood Plain
Delta Systems	River (meandering) Delta (river, wave or storm dominated), Braid Delta, Fan Delta	***Subdivisions:*** Delta Plain, Delta Front, Prodelta. ***Sub-environments:*** Distributory Channels, Distributory Mouth Bars, Crevasse Channels, Crevasse Splays, Levees
Marginal Marine Systems	Estuary, Tidalflat, Beach, Barrier Bar, Lagoon, Chenier Plain	
Shallow Marine Systems	**Shoreface:** Wave, Storm or tide-dominated **Inner Shelf:** Wave, storm or tide-dominated **Outer Shelf:** may or may not be storm or tide-dominated	
Deep Marine Systems	Sub-marine Channel, Sub-marine Fan, Basin Plain	

dition detritus from the upland areas is reworked by wind into erg — the desert depositional system. In polar arid areas or during ice ages transport and deposition of sediments remains dominated by continental glaciers. Lacustrine system may develop independently under the influence of local tectonics. Alluvial fans are common along linear mountain fronts. Rivers are passages through which detritus is transferred from upland areas into the sea or lake via deltas and estuaries. The hydrodynamic processes in the shore and shelf again redistribute the sediments introduced to the coast among different depositional systems such as tidal flat, barrier bar, beach, lagoon, chenier plain shoreface, deeper shelf. Sediments may further be transferred to deep marine basin plain from the shelf under the influence of gravity through the slope, by deep oceanic currents and due to settlement from suspension clouds.

Although many of the above depositional systems may occur independently in isolation, an ideal, complete siliciclastic sediment dispersal complex comprises of the following array of depositional systems down the basin: alluvial fan, alluvial plain fluvial delta, shelf, slope and basin plain (Table 3.T2). Sediment dispersal through these systems leads to the development of a depositional profile of the basin comprising the graded river profile, graded shelf profile and the graded slope profile (Fig. 3.2). The development of the equilibrium profile

Table 3.T2: Models of siliciclastic geographic setting

ALLUVIAL PLAIN, DELTAIC SHORELINE

Alluvial Fan	Braid plain	Meandering river	River Delta	Shelf	Slope	Basin Plain
Alluvial Fan	Braid plain	Braid Delta	Shelf	Slope	Basin Plain	
Alluvial Fan	Fan Delta	Shelf	Slope	Basin Plain		

ALLUVIAL PLAIN, SHORELINE AND ESTUARINE SYSTEMS

Alluvial Fan	Braid Plain	Straight or meandering river	Bay Head Delta	Estuary Basin	Shore-face	Shelf	Slope	Basin plain

TIDAL SHORELINE WITHOUT RIVERS

Backshore/ Marsh/ Swamp	Tidalflat	Shoreface (Tide dominated)	Shelf (Tide dominated)	Slope	Basin Plain

TIDAL SHORELINE WITHOUT RIVERS AND WITH LONGSHORE SUPPLY OF SAND

Backshore/ Marsh/ Swamp	Tidalflat	Lagoon with flood–tidal delta	Barrier-bar Beach	Shoreface with ebb tidal delta	Shelf	Slope

WAVE-DOMINATED SHORELINE WITHOUT RIVERS AND WITH LONGSHORE SUPPLY OF SAND

Backshore/Marsh/ Swamp	Bay lagoon, Storm wash-over	Barrier-bar beach	Shoreface	Shelf	Slope	Basin Plain

WAVE-DOMINATED SHORELINE WITHOUT RIVERS AND WITHOUT LONGSHORE SUPPLY OF SAND OR MUD

Backshore/Marsh/Swamp	Beach	Shoreface	Shelf	Slope	Basin Plain

WAVE-DOMINATED SHORELINE WITHOUT RIVERS AND WITH LONGSHORE SUPPLY OF MUD

Backshore/Marsh/Swamp	Chenier Plain	Shoreface	Shelf	Slope	Basin Plain

ALLUVIAL PLAIN, EPEIRIC SEA

Alluvial Fan	Braidplain	Meandering River	Peritidal Flat

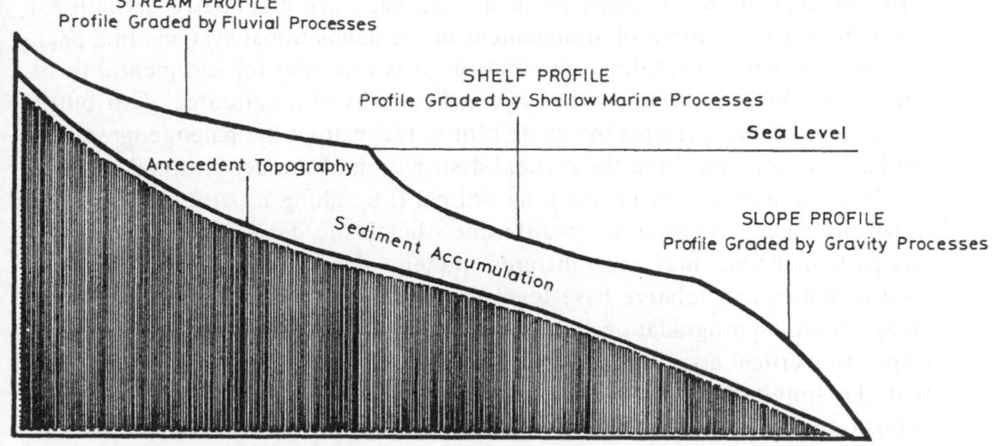

Fig 3.2: Schematic diagram illustrating the concept of graded profiles. The profiles develop on an antecedent topography as a result of sediment deposition and reworking by fluvial, and shelfal currents and by gravity processes. See text for further explanation.

results from the interplay between sediment supply and fluvial, wave, storm and tidal energy in the basin. Sediments entering the non-marine domain are transported by and deposited from traction and suspension currents to attain and maintain the graded river profile. The sediments entering the shelf setting through river systems are re-distributed by tide, storm and wave-driven currents or fluvial currents that try to attain and maintain the graded shelf profile. Beyond the shelf, the basin is too deep (below storm wave base) for the tide and storm processes to rework the sediments. It is when gravity processes operate to move sediments further deep into the basin (sediment gravity flows), until a stable gradient is achieved (i.e., the graded slope profile such as the profile exhibited by the present-day continental slopes).

The development of the slope, however, depends on the tectonic character of the basin. Passive margin and marine extensional basins show well-developed slope systems, whereas continental sag, foreland ramp, and epeiric basins generally lack the slope system (see Chapter 5).

In siliciclastic settings, as dispersal of sediments continues, the different depositional systems shifts longitudinally down the basin—a process called progradation. As a result, the deposits of a depositional system are progressively buried by the one occurring longitudinally up the basin (Fig. 3.1). Therefore, if we have the vertical section at any location of a sedimentary basin we would find a package of sediment beds in which the deposits of progressively up-the-basin depositional systems are stacked successively one above another — as depicted in the Walther's law (Fig. 3.1). Such a package essentially represents a continuum of sedimentation and the boundaries between the deposits of different depositional systems in such a package are essentially gradational. Since the vertical pattern of arrangement of the depositional systems in a package proxies their longitudinal distribution, it is essential for sedimentologists, working in the field, to know the different patterns of longitudinal distribution of the depositional systems to enable him to reconstruct the paleogeography of the basin at any time from the vertical distribution of the depositional systems.

In a sedimentary basin, the longitudinal (i.e., along a particular direction down the basin) distribution of different siliciclastic depositional systems, at any particular time, may show different patterns depending on climate, sourceland tectonics and relative base level (Fig. 3.3; Table 3.T2). Each pattern of array would, on progradation, generate a package of sediments characterized by a specific vertical array of depositional systems (Fig. 3.3). It should be realized that the longitudinal distribution pattern may vary within a basin as the depositional systems may be distributed differently in the transverse direction. Moreover, the proximal (such as alluvial fan), distal (such as deep-marine system) and some of the medial depositional systems may or may not be present in the series shown in Table 3.T2 depending on the geomorpho-tectonic attributes of the basin.

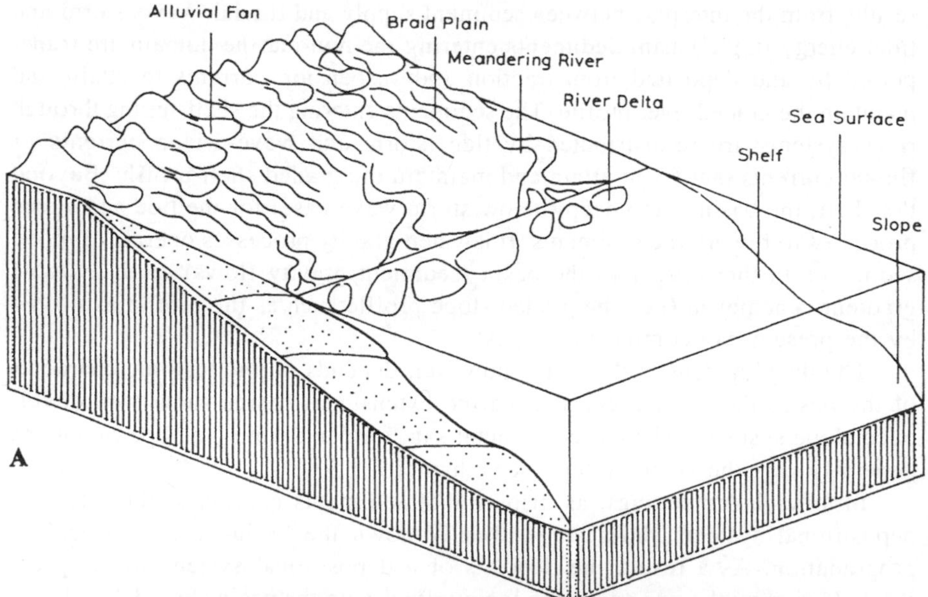

Fig. 3.3 a: Block diagram showing a siliciclastic geographic setting comprising alluvial fan, braidplain, meandering river plain, river delta, shelf and slope depositional systems. In longitudinal profile, a sediment package, formed in a single, continuous depositional episode is shown. Note superposition of relatively updip depositional systems over the downdip depositional systems in the sediment package. As variants of this complete sediment dispersal complex, there may be a number of other geographic settings : 1. alluvial fan-braid delta-shelf slope; 2. alluvial fan-fandelta-shelf slope; 3. alluvial fan-slope, submarine fandelta.

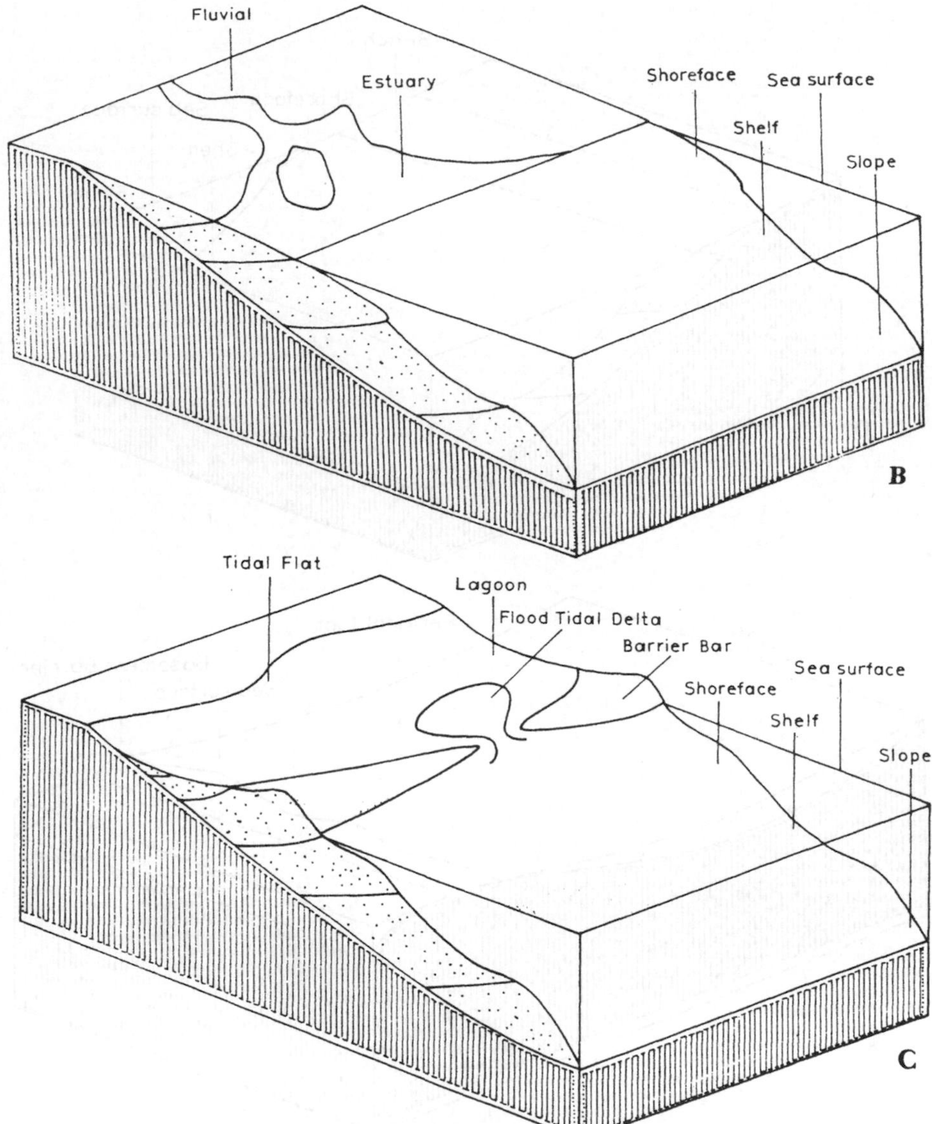

Fig.3.3 b: Block diagram showing a siliciclastic geographic setting comprising fluvial, estuarine, shelf and slope depositional systems. In longitudinal profile, a sediment package, formed in a single, continuous depositional episode is shown. Note superposition of relative up-dip depositional systems over the down-dip depositional systems in the sediment package.

c: Block diagram showing a siliciclastic geographic setting comprising lagoon, barrier bar-beach, shoreface, shelf and slope systems. Alongshore supply of sand builds the barrier bar, while dominance of tidal effect results into flood tidal deltas and tidal flat; ebb tidal deltas may also be present in the setting. In longitudinal profile, a sediment package, formed in a single, continuous depositional episode is shown. Note superposition of relatively up-dip depositional systems over the down-dip depositional systems in the sediment package.

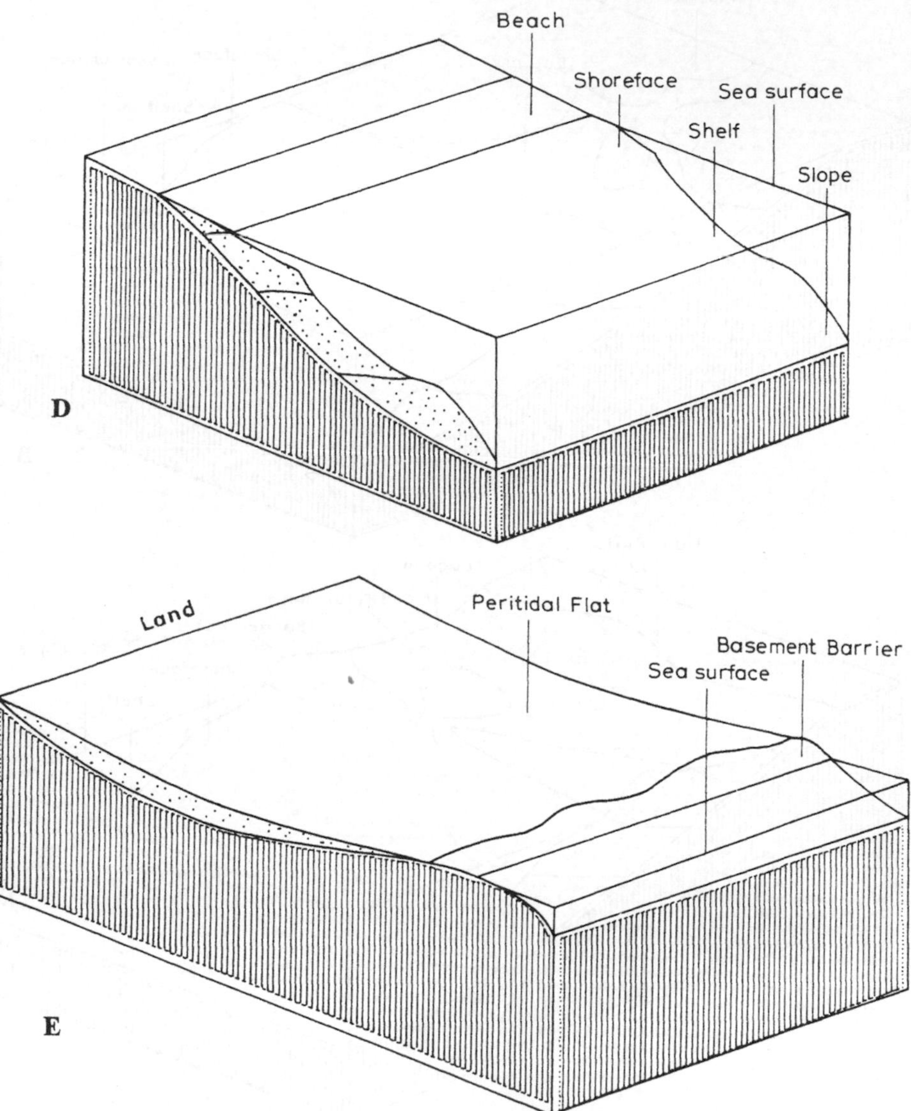

Fig.3.3 d: Block diagram showing a siliciclastic geographic setting comprising beach, shoreface, shelf and slope depositional systems. This setting is characteristic of microtidal, wave-dominated coasts with limited alongshore supply of sand. The beach will be replaced by chenier plain if there is alongshore supply of mud instead of sand, and by tidal flat if the coast is tide-dominated. In longitudinal profile, a sediment package, formed in a single, continuous depositional episode is shown. Note superposition of relatively updip depositional systems over the downdip depositional systems.

e: Block diagram showing the geographic setting of epeiric seas. Note occurrence of a single depositional system all over the sea; as a result the sediment package formed by a single, continuous depositional event is essentially aggradational as shown in the longitudinal profile.

Carbonate Depositional Settings

In contrast to terrigenous clastic sediments, carbonate sediments are not delivered to a basin but are largely produced within the basin under the condition of limited siliciclastic input (James and Kendall, 1992). The domain of marine carbonate production is called the carbonate factory—the shallow sub-tidal region of the basin. The productivity of the carbonate factory depends on latitude, temperature, salinity, water depth, sun light intensity, turbidity, water circulation, pCO_2, and nutrient supply.

Carbonate sediment particles either directly precipitate from water (i.e., non-skeletal) or are produced as organic skeletons (i.e., skeletal). The sediments may be loose (such as carbonate mud, ooid, peloid, aggregates, clasts and organic skeletons) or may build up rigid, wave/current resistant bodies, such as organic reefs, rapidly lithified sandy shoals. Loose carbonate particles can be acted upon by the prevailing hydrodynamic processes (such as storm and tide) and suffer re-distribution similar to siliciclastic sediments. A portion of the sediments generated in the sub-tidal carbonate factory may be transported from their place of origin, both towards the shore and the deep interior of the basin and are trapped in the near-shore and deep marine depositional systems (Fig. 3.4). The rest, however, remain in place in the sub-tidal domain. The planktonic microorganisms, living in the upper part of the seawater, upon death, are swept off and preferentially accumulate in deeper water as suspension fallout (pelagic deposits). Sediment gravity flows originating on steep slopes of carbonate build-ups (e.g., reefs) also aid in transporting sediments to the deep-water domain. The major carbonate depositional systems are shown in Table 3.T3.

The domain of carbonate sediment production and accumulation is called carbonate platform. The ability of carbonate sediments to be generated and accumulated, *in situ,* as i) organic build-ups, ii) to be early cemented, iii) to have interlocking texture due to non-spherical morphology of skeletal grain, iv) to be transported and deposited as clastic particles, result in a wide variety of

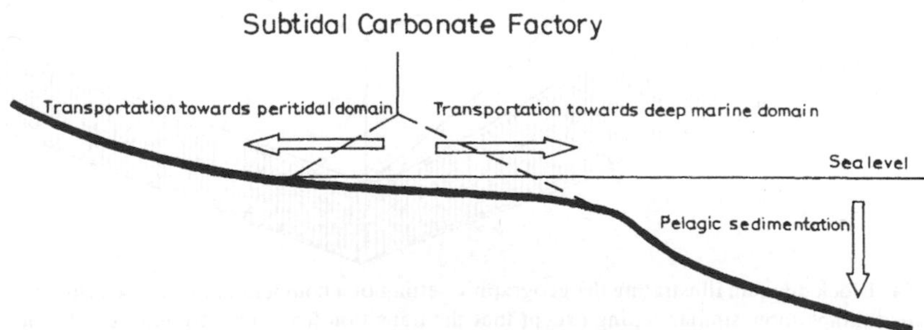

Fig. 3.4: Diagram illustrating dispersal of carbonate sediments from the subtidal domain - the marine carbonate factory.

Table 3.T3: Carbonate Depositional Systems

Coastal and Nearshore Systems	1. Coastal dunes, Beaches, Shoreface 2. Lagoon, Barrier Islands, Tidal Inlets, Tidal Deltas 3. Tidal Flats, Tidal Channels and Sabkhas 4. Evaporite Salinas
Platform Interior Systems	1. Open, shallow to deep storm-dominated shelf 2. Shelf Lagoon
Platform Edge Systems	1. Reef 2. Lithified Grain Shoal
Slope and Basin Plain Systems	1. Upper Slope Talus 2. Lower Slope Mass Flows 3. Toe-of-slope Talus and Mass Flows 4. Pelagic and Hemipelagic Suspension Fallouts

platform morphology often with slopes exceeding the angle of repose (Read, 1985; Burchette and Wright, 1992; Fig. 3.5; Table 3.T4). The antecedent topography exerts a major control on the platform morphology and in the course of basin evolution, one morphology may develop into another under the influence of rate of sediment production, tectonics and eustasy. Homoclinal ramps develop on gentle regional paleoslopes in continental passive margin and fore-

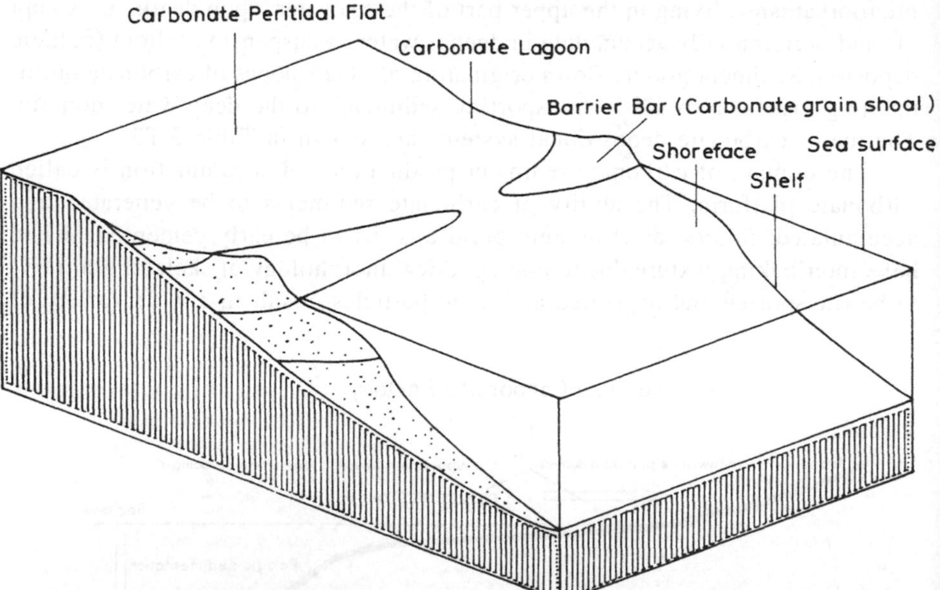

Fig.3.5a: Block diagram illustrating the geographic setting of a homoclinal carbonate ramp. Distally steepened ramps show similar setting except that the transition from inner to outer shelf domain is marked by sharp increase in slope. In longitudinal profile, a sediment package, formed by a single, continuous depositional episode is shown. Note superposition of relatively updip depositional systems.

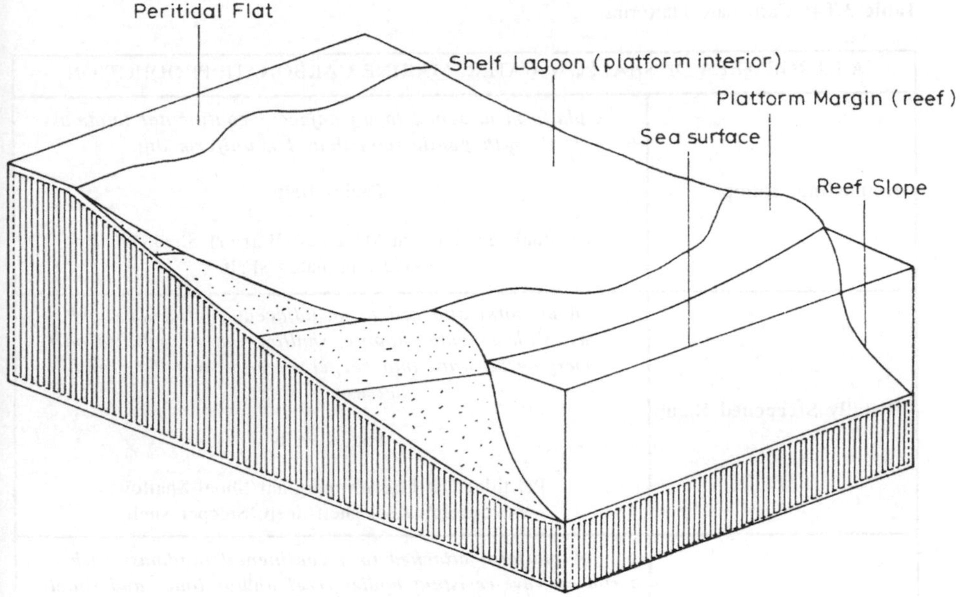

Fig. 3.5 b: Block diagram illustrating the geographic setting of a rimmed carbonate platform. In the longitudinal profile, a progradational sediment package, formed by a single continuous depositional episode is shown.

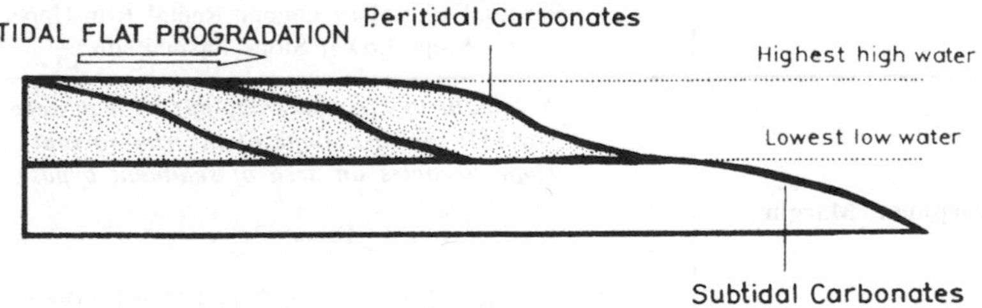

Fig. 3.6: Diagram illustrating progradation of tidal flat over the subtidal carbonate factory producing a shallowing-upward sedimentary succession and progressively diminishing the extent of a carbonate factory.

land basins. Distally steepened ramps may develop where faulting or flexuring steepens the outer part of the ramp or if the ramp develops on drowned rimmed shelves. Isolated platform may develop on horsts or volcanoes. Rimmed shelf typically develops from ramps with high carbonate production localized along the edge of the platform.

As long as the carbonate factory remains operative, transportation of clastic carbonates to the peritidal domain allows the peritidal depositional systems to prograde over the sub-tidal depositional system producing a conformable shallowing-upward succession (Fig. 3.6). Eventually, the whole setting becomes

Table 3.T4: Carbonate Platforms

A LARGE AREA OF SHALLOW-WATER, MARINE CARBONATE PRODUCTION	
Homoclinal Ramp	***A platform attached to an adjacent continental landmass with gentle (less than 1^0), uniform dip.*** Facies Belts Peritidal Flat-Lagoon-Marginal (Barrier) Shoal-Shallow, wave-dominated shelf
Distally Steepened Ramp	*A platform attached to an adjacent continental land mass with a gentle sloping, shallow water landward part merging seaward into deeper water with a discernible increase in slope.* Facies Belts Peritidal flat-Lagoon-marginal Shoal-Shallow, Wave-dominated, shelf-deep, Steeper shelf
Rimmed Shelf	***A platform attached to a continental landmass with a rim of wave-resistant bodies (reef and/or lime sand shoals) on the seaward side dividing the domain into a broad, shallow platform on the landward side and a steep but accretionary platform (up to 35^0) on the seaward side.*** Facies Belts Peritidal Flat-Shelf-Lagoon-Reefal Rim-Upper Slope-Lower Slope-Basin Plain
Escarpment Margin	***A variant of rimmed shelf but with a steep (>35^0) seaward slope of the rim so that slope becomes an area of sediment bypass.*** Facies Belts Peritidal flat-Shelf Lagoon-Reefal Rim-Lower Slope-Toe-of-Slope-Basin Plain
Isolated Platform	***A platform detached from the land, characterized by shallow-water carbonate accumulation surrounded by deep water, with or without a rim.*** Facies Belts Similar to any of the above
Epeiric Platform	***A flat platform attached to an adjacent continental landmass, characterized by very shallow water.*** Facies Belts Peritidal to Shallow Subtidal Flat

peritidal and as a result, production of carbonate sediment ceases. Sediment production can resume if and when the setting becomes sub-tidal in response to a rise in relative sea level. Similarly, the growth of organic build-ups (reefs) continues till they reach the sea level; for further growth, a rise in relative sea level is necessary.

4. SEQUENCE STRATIGRAPHY

Introduction

"Stratigraphy is the study and description of rock successions. It leads to interpretation and fuller understanding of rock successions as sequences of events in the geological history of the Earth" (Whittaker et al., 1991). The fundamental procedure of stratigraphy is to divide a sedimentary succession into units, which can be correlated over the basin, correlation being the demonstration of correspondence between geological units in some defined property and in relative stratigraphical position (Whittaker et al., 1991). Accordingly, lithostratigraphic units are described and defined using their gross lithological characteristics and by their interrelationships with adjacent units. However, lithostratigraphic units are often time-transgressive (diachronous) and are not suitable for revelation of sequential changes of a sedimentary basin (Owen, 1987).

Sedimentary basins are the geographic domains of sediment accumulation and preservation under the influence of tectonism, eustasy and climate. The evolution of sedimentary basins is a function of several dynamic variables such as rate of base level fluctuation, rate of sedimentation/supply, and rate of basin floor movement (Jervey, 1988). Sequence stratigraphy is a stratigraphic approach that analyses the evolution of a sedimentary succession in response to temporal changes in the relative magnitudes of these variables (Miall, 1995b; Posamentier et al., 1988b; Posamentier and Vail, 1988b; Van Wagoner et al., 1990, Emery and Myers, 1996). Thus, whereas in lithostratigraphy a sedimentary succssion is divided into units based on lithological variation, the sequence stratigraphic units are those that developed during a time interval when the relative magnitudes of the rates of base level fluctuation, sedimentation/supply and basin floor movement remained steady and uniform at the scale of stratigraphic resolution (Fig. 4.1). A chronostratigraphic implication is, therefore, an inherent attribute of sequence stratigraphic units (Catuneanu et al., 1998). In other words, the bounding surfaces of sequence stratigraphic units are essentially laterally correlatable time planes. Under certain circumstances, however, the boundaries of lithostratigraphic and sequence stratigraphic units may coincide.

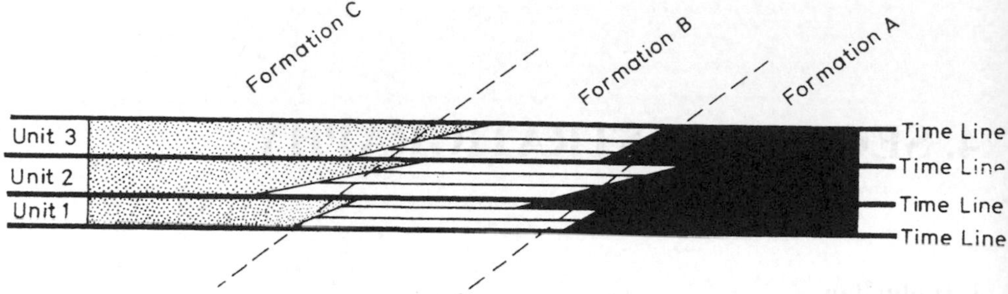

Fig. 4.1: Distinction between lithostratigraphy and sequence stratigraphy. Lithostratigraphic units are defined by lithological uniformity, whereas sequence stratigraphic units are those that formed during a specific time interval and may consist of several lithologies.

Principles and Methodology

Sediments can accumulate in a sedimentary basin, if there is *accommodation space.* Accommodation space is the space between the basin floor and the base level surface (Fig. 4.2). In non-marine, alluvial basins, the base level is the graded stream profile, whereas in marine or lacustrine basins sea level or lake level represents the base level. A sedimentary basin undergoes non-deposition or erosion if the available accommodation space is zero or negative (Fig. 4.3). There are two definitions of base level (cf. Shanley and McCabe, 1994): 1) stratigraphic base level (Barrel, 1917; Sloss, 1962; Wheeler, 1964) which is the '... potential energy surface that describes the direction in which a stratigraphic system is likely to move, toward sedimentation and stratigraphic preservation or sediment bypass and erosion'; and 2) geomorphic base level (Bates and Jackson, 1987; Powell, 1875; Schumm, 1993), which is sea level.

The basin floor and the base level surface are dynamic entities in the sense that they can move up and down relative to a datum e.g., the centre of the earth (Fig. 4.2). The basin floor can move up and down relative to the centre of the earth as a result of tectonic upliftment or subsidence. The base level surface may move up and down with respect to the centre of the earth due to extra-basinal factors. For example, the marine base level i.e., the sea surface moves up and down with respect to the centre of the earth due to eustatic rise and fall of sea level. Displacement of non-marine base level in response to relative sea level changes varies depending on the relative steepness of the stream and shelf profiles (Fig. 4.4). Non-marine base level surface may also be displaced relative to the centre of the earth due to climatic factors or source-land tectonics independent of sea level fluctuations.

From Figure 4.2 it is apparent that in a sedimentary basin

$$dA/dt = dA_{bl}/dt - dB_d/dt$$

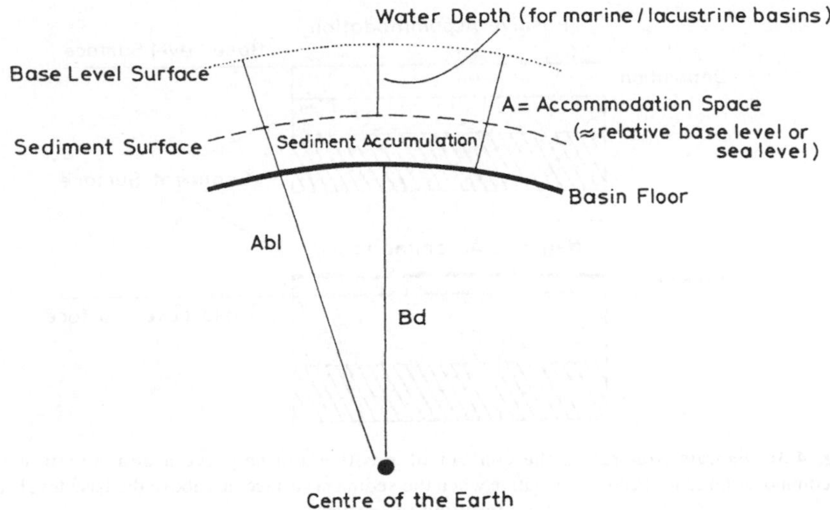

Fig. 4.2: Schematic diagram illustrating disposition of different surfaces characterizing a sedimentary basin. Abl- absolute base level, Bd- distance between the basin floor and the centre of the earth. A- accommodation space.

Where, dA/dt = rate of change of accommodation space (positive for increase and negative for decrease) and dB_d/dt = rate of basin floor movement (positive for upliftment and negative for subsidence), dA_{bl}/dt = rate of base level fluctuation relative to the centre of the earth, due to extrabasinal factors (positive for rise and negative for fall).

The above equation reveals that the space for accumulation of sediments in a sedimentary basin is created and destructed due to the combined effect of subsidence or upliftment of the basin floor and extrabasinal factors such as eustatic sea level fluctuation, climate and sourceland tectonics.

The accommodation space, available at any time in a sedimentary basin, is utilized by being progressively filled with sediments supplied to or produced in the basin. A sedimentary succession records the history of creation/destruction and utilization (i.e., filling) of accommodation space in a sedimentary basin through time.

In a sedimentary basin, sediments are not introduced (in case of siliciclastic basins) or produced (in case of carbonate platforms) continuously, but episodically. In a marine basin, as long as sicliclastic sediment supply remains on, the short-term rate of sedimentation outpaces the short-term rate of change of accommodation space and a progradational sediment package is developed obeying Walther's law (Fig. 4.5; see Chapter 3). Similarly, in carbonate platforms as long as the carbonate factory produces sediments, a progradational or aggradational sediment package is formed (see Chapter 3). When supply of

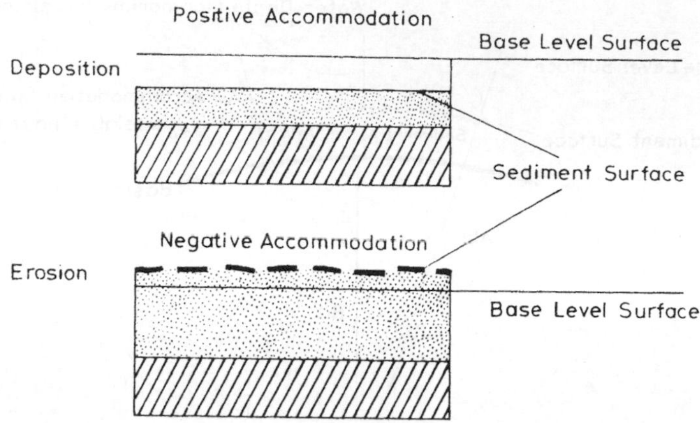

Fig. 4.3: Diagram illustrating the concept of positive and negative accommodation space. Accommodation space becomes negative when the sediment surface lies above the base level surface resulting in erosion.

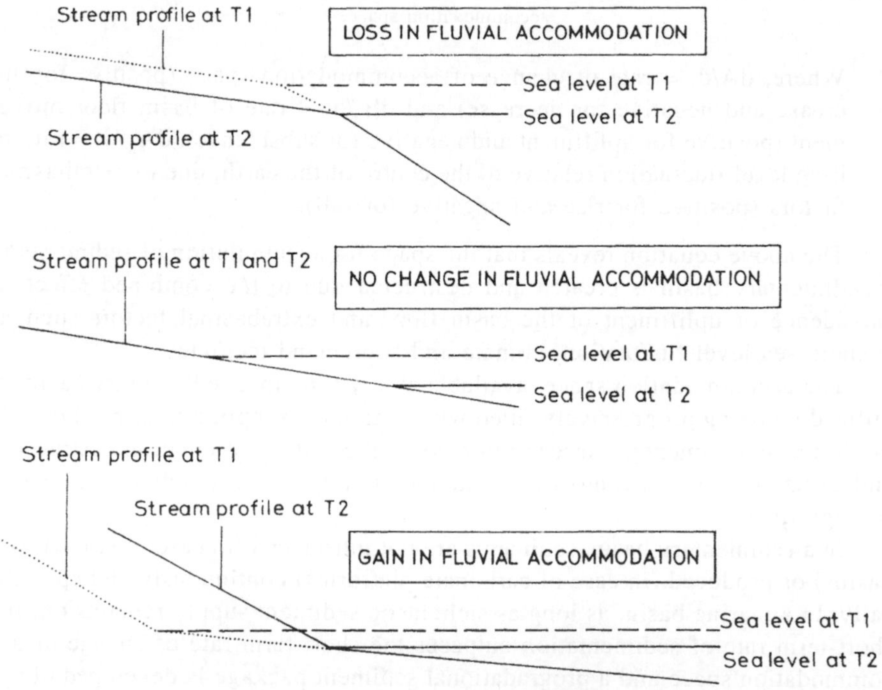

Fig. 4.4: Changes in fluvial accommodation space in response to fall in relative sea level for different relative steepness of the stream and shelf profiles.

silicilastics or production of carbonate sediments ceases, the event of creation or destruction of accommodation space takes place. Creation and destruction of accommodation space only shift the locii of the depositional systems up and down the basin respectively, without net sedimentation (Fig. 4.5). Studies on the sedimentary successions have revealed that on the finest level of stratigraphic resolution, sedimentary basins evolve through alternate phases of continuous sedimentation and creation or destruction of accommodation space (Emery and Myers, 1996; Fig.4.5). In addition to phases of continuous sedimentation and creation/destruction of accommodation space, a sedimentary basin in its lifetime also passes through periods when there is zero or negative accommodation space so that the basin suffers non-deposition or sub-aerial erosion, irrespective of sediment supply, producing unconformities (Fig. 4.3). In carbonate platforms, a rapid rise in relative sea level accompanied by deteriorating changes in nutrient supply, siliciclastic input, salinity, oxygenation or predation may also lead to shut down of the carbonate factory, to result in a submarine unconformity called drowning unconformity (Schlager, 1988). Such unconformities do not, however, indicate zero or negative accommodation space.

Therefore, in addition to the basal unconformity surface (either an angular unconformity or a non-conformity on which the basin initiates; (see Section 2.8) the other basic components of a sedimentary succession are:

1. *The progradational sediment package representing a continuum of sedimentation.* This is termed as **parasequence** in the nomenclature of sequence stratigraphy (Figs. 4.5, 4.6; Van Wagoner et al., 1988). A parsequence is essentially a conformable sedimentary succession and may consist of a vertical array of depositional systems, indicating progradation obeying Walther's law that prevailed adjacent to each other contemporaneously defining a particular paleogeography (see Chapter 3). However, at some locations of the basin, a parasequence may also be represented by the deposit of a single depositional system (Fig. 4.7); in that case, the progradational nature of the package is reflected in the systemic upward variation in grain size (mostly upward coarsening, upward fining in case of tidal flat system). At the present level of understanding, parasequences are well-defined in shallow marine setting (Fig. 4.7; Van Wagoner et al., 1990). Characteristics of parasequences in alluvial, deep marine and other settings are yet to be understood.

2. *The surface across which there was creation of accommodation space.* This is ideally a paraconformity surface (see Section 2.8) and is termed as **flooding surface** in the nomenclature of sequence stratigraphy (Fig. 4.8; Van Wagoner et al., 1988, 1990; Helland-Hansen and Gjelberg, 1994; Helland-Hansen and Martinsen, 1996). These surfaces occur as parasequence boundaries.

3. *The surface across which there was destruction of accommodation space.* This is ideally a paraconformity surface (see Section 2.8). This is termed as **regressive surface** in the nomenclature of sequence stratigraphy (Fig. 4.9; Posamentier et al., 1992). These surfaces also occur as parasequence boundaries.

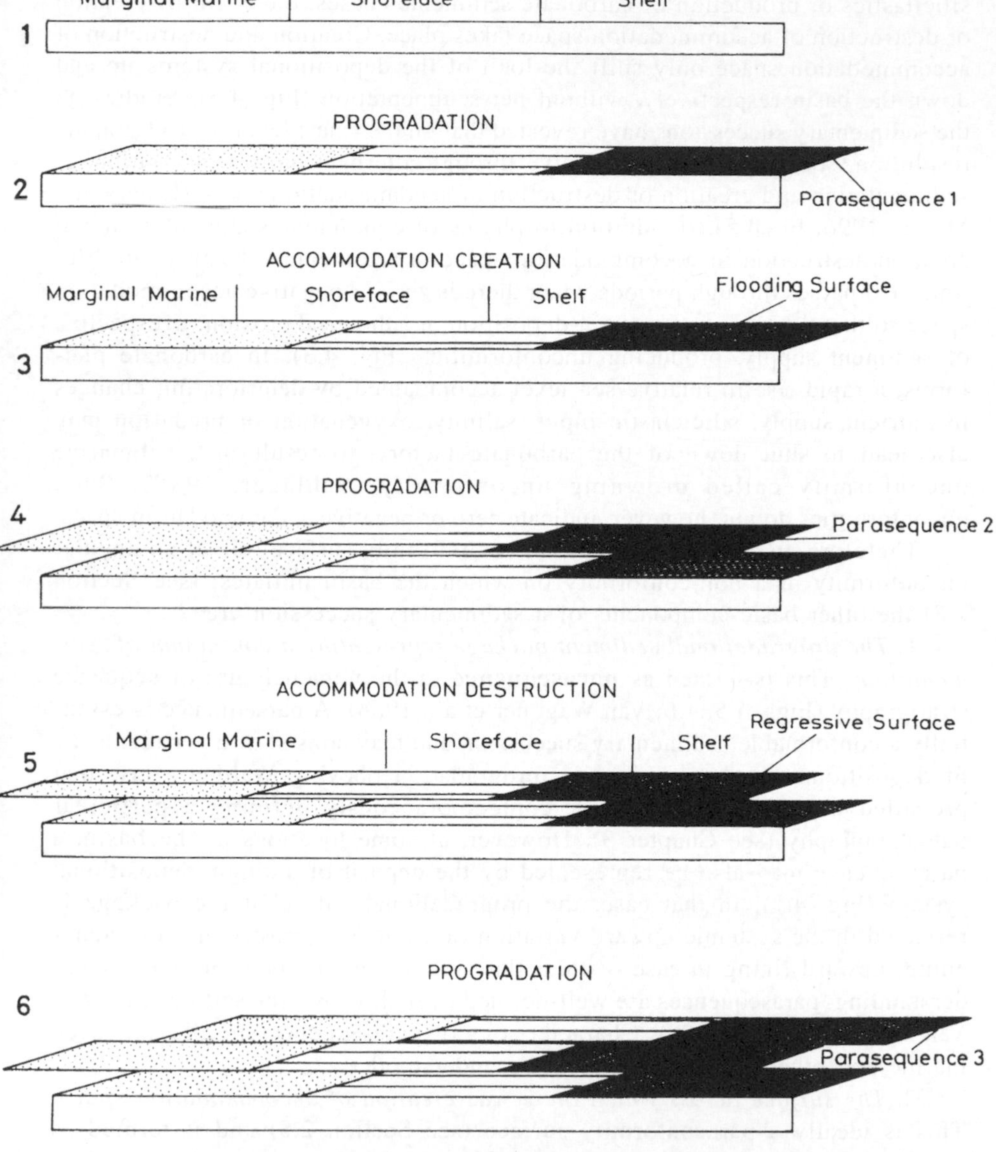

Fig. 4.5: Schematic diagram illustrating development of sedimentary succesison through alternate phases of accommodation space creation or destruction and progradation of co-existing depositional systems. See text for further explanation.

Fig. 4.6 a: Successive coarsening-up parasequences, each showing gradational, upward transition from innershelf to shoreface deposits.

b: Successive fining-up parasequences, each showing gradational upward transition from shoreface to intertidal flat deposits. Note sharp nature of the parasequence boundaries (a is reproduced from R. W. C. Arnott, The parasequence definition - are transgressive deposits inadequately addressed? J. Sediment. Research, Vol. B62, No. 1, Fig. 1, © 1995, Society for Sedimentary Geology (SEPM), Tulsa, Oklahoma, U.S.A. with the kind permission).

4. *The surface bearing evidences of subaerial exposure and erosion.* This is a subaerial unconformity (disconformity) surface representing periods of zero or negative accommodation space in the basin (Fig. 4.10; see Section 2.8). The unconformity surface, however, may be correlated with a conformity basinward termed as **correlative conformity** (Embry, 1995).

Fig. 4.6 b

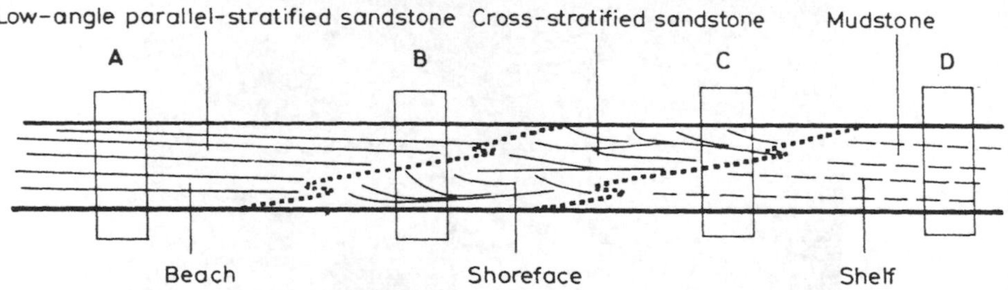

Fig. 4.7: Schematic diagram illustrating variable manifestations of a parasequence, formed by progradation of beach, shoreface and shelf systems, at different locations. At A, the parasequence is represented only by beach deposit. At B, the parasequence consists of beach deposit gradationally overlying shoreface deposits, whereas the parasequence at C, shoreface deposit gradationally overlies shelfal mudstones. At D, the parasequence is represented only by shelfal mudstones.

5. *The* ***drowning unconformity*** *indicating shut down of the carbonate factory following a rise in relative sea level.* This is exclusive to carbonate platforms (Schlager, 1988; see Section 2.8).

The sedimentary succession occurring between two subaerial unconformity surfaces is termed a **sequence** that developed between two base level falls. The building blocks of a sequence are parasequences. The parasequences within a sequence may be arranged vertically in different ways depending on the relative

Fig. 4.8: Shelfal mudstones sharply overlying shoreface sandstones indicating abrupt increase in water depth. The surface separating the litho units is a flooding surface.

magnitudes of the *long-term* rate of accommodation space creation (dA/dt) and rate of sediment supply (in siliciclastic basins, dJs/dt) or rate of carbonate sediment production (in carbonate platforms, dG_c/dt- growth rate of reef or rigid carbonate bodies at platform edge and dGc/dt- production rate of clastic carbonates at the platform interior) as discussed below (Schlager, 1993).

Fig. 4.9: Shoreface sandstone sharply overlying shelf mudstones indicating abrupt fall in water depth. The surface separating the litho units is a regressive surface.

Stacking Patterns of Parasequences

Siliciclastic Basins

1. $dA/dt > dJs/dt > 0$

If the long-term rate of accommodation space creation exceeds that of sediment supply, a retrogradational parasequence set will develop in which each parasequence is separated from the other by flooding surfaces (Fig. 4.11).

2. $dJs/dt = dA/dt > 0$

If the long-term rate of accommodation space creation equals that of sediment supply, an aggradational parasequence set will develop in which each parasequence is separated from the other by flooding surfaces (Fig. 4.11).

3. $dJs/dt > dA/dt > 0$

If the long-term rate of accommodation space creation is less than that of sediment supply, a progradational parasequence set will develop in which each parasequence is separated from the other by flooding surfaces (Fig. 4.11).

4. $dJs/dt > 0 > dA/dt$

If the long-term rate of accommodation space creation is less than that of sediment supply and is also less than zero, a forced-progradational parasequence set will develop in which each parasequence is separated from the other by regressive surfaces (Fig.4.11).

Fig. 4.10: Shoreface sandstones sharply overlying deepwater limestone. The top of the limestone show evidences of karstification indicating subaerial exposure. The surface separating the litho units is, therefore, a sequence bounding subaerial unconformity (disconformity).

Carbonate Platforms

1. $dA/dt > dGc/dt = dGr/dt > 0$

This will result in a carbonate ramp with a retrogradational parasequence set in which each parasequence is separated from the other by flooding surfaces similar to siliciclastic basins.

2. $dGc/dt = dGr/dt = dA/dt > 0$

This will result in a carbonate ramp with an aggradational parasequence set in which each parasequence is separated from the other by flooding surfaces similar to siliciclastic basins.

3. $dGc/dt = dGr/dt > dA/dt > 0$

This will result in a carbonate ramp with a progradational parasequence set in which each parasequence is separated from the other by flooding surfaces similar to siliciclastic basins.

4. $dGc/dt = dGr/dt > 0 > dA/dt$

This will result in a carbonate ramp with a forced progradational parasequence set in which each parasequence is separated from the other by regressive surfaces similar to siliciclastic basins.

5. $dGr/dt > dA/dt > dGc/dt > 0$

This will result in a rimmed platform with an accretional margin; the platform interior will develop a retrogradational parasequence set in which each parasequence is separated from the other by flooding surfaces similar to siliciclastic basins.

6. $dGr/dt > dGc/dt > dA/dt >0$

This will result in a rimmed platfrom with an accretional margin; the platform interior will develop a progradational parasequence set in which each

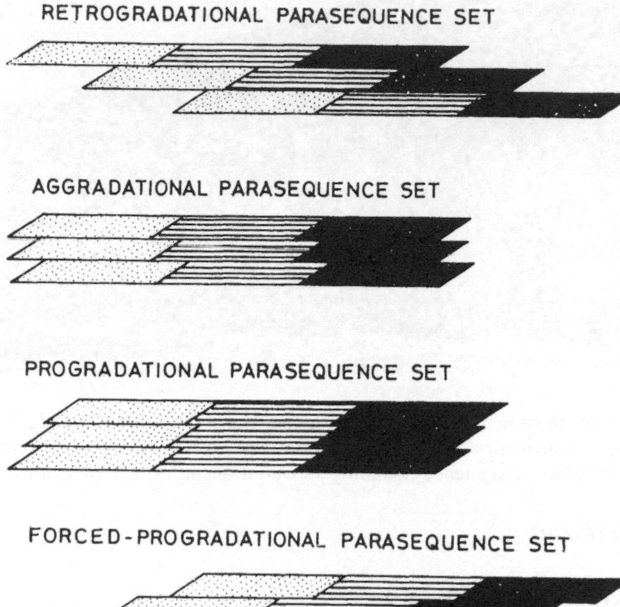

Fig. 4.11: Different stacking patterns of parasequences reflecting different long-term, relative rates of accommodation space variation and sedimentation. Different symbols represent different depositional systems; downdip direction of the basin towards right.

parasequence is separated from the other by flooding surfaces similar to siliciclastic basins.

7. dA/dt > dGr/dt > dGc/dt > 0

This will result in a rimmed platform with an escarpment margin; the platform interior will develop a retrogradational parasequence set in which each parasequence is separated from the other by flooding surfaces similar to siliciclastic basins.

8. dGr/dt > dA/dt > dGc/dt = 0

This will result in a rimmed platform with an accretional margin; the platform interior will remain starved.

9. dGr/dt > dGc/dt = dA/dt > 0

This will result in a rimmed platform with an accretional margin; the platform interior will develop an aggradational parasequence set in which each parasequence is separated from the other by flooding surfaces similar to siliciclastic basins.

10. dGr/dt > dGc/dt > 0 > dA/dt

This will result in a rimmed platform with an accretional margin; the platform interior will develop a forced progradational parasequence set in which each parasequence is separated from the other by regressive surfaces.

In a sequence, one or all types of parasequence sets may be present. If there are multiple sets, they may show a variety of vertical arrangements (Fig. 4.12). In a sequence, the surface separating a retrogradational parasequence set below from an aggradational to progradational parasequence set above is termed **maximum flooding surface** (Fig. 4.12; Van Wagoner et al., 1988). The surface separating a progradational parasequence set below and forced-progradational parasequence set above is termed **basal surface of forced regression** (Catuneanu et al., 1998).

Systems Tracts

The strata occurring in a sedimentary succession are representative of different depositonal systems that prevailed in the basin during its lifetime. The geography of a sedimentary basin at a particular time is characterized by a particular set of depositional systems prevailing in the basin adjacent to each other contemporaneously. The different geographic settings that may characterize sedimentary basins have been discussed in chapter 3. In course of evolution of a sedimentary basin, the set of contemporaneous depositional systems prevailing in the basin may vary with time. In other words, the paleogeographic setting (i.e., a particular set of contemporaneous depositional systems) of a basin may change with time. For example, a sedimentary basin characterized by an alluvial plain, delta and shelf at a particular period may evolve into a paleogeography

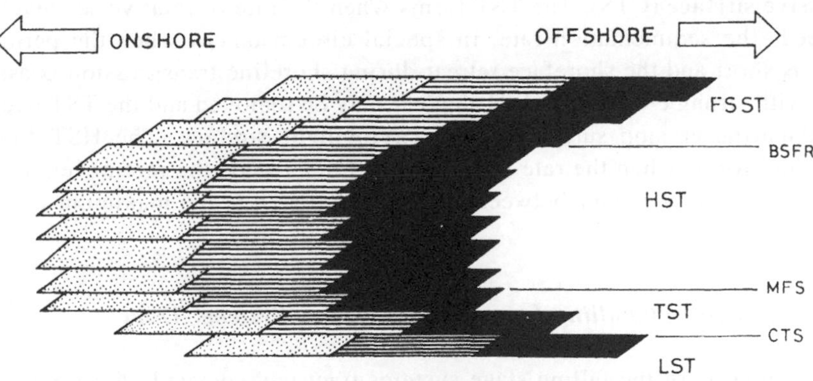

Fig. 4.12: Schematic diagram showing an idealized arrangement of parasequence sets and systems tracts reflecting progressive increase in accommodation space followed by a progressive decrease in a marine setting. The different shades represent different depositional systems. Note the maximum flooding surface (MFS) between the transgressive (TST) and highstand systems tracts (HST), conformable transgressive surface (CTS) between the lowstand (LST) and TST and the basal surface of forced regression between the HST and falling stage systems tract (FSST).

characterized by beach, shoreface and shelf or, a carbonate ramp may develop into a rimmed platform. The products of different depositional systems are usually represented in the sedimentary succession as a group of parasequences. The group of parsequences, representing a particular set of depositional systems, i.e., a particular paleogeography different from that represented by the groups of parsequences occurring above and below, constitutes what is known as **systems tract**.

The temporal variation in paleogeography is a virtual reflection of the temporal change in the relative magnitudes of the long-term rates of accommodation space variation and sedimentation. As a result, the parasequences in a systems tract are usually stacked in any of the four patterns: retrogradational, aggradational, progradational or forced-progradational (Fig. 4.12). Accordingly, four types of systems tracts are recognised: i) **Lowstand systems tract (LST)** characterized by progradational parasequence set; ii) **Transgressive systems tract (TST)** characterized by retrogradational parasequence set; iii) **Highstand systems tract (HST)** characterized by aggradational to progradational parasequence sets; and **iv) Falling stage systems tract (FSST)** characterized by forced-progradational parasequence set. However, under special circumstances a systems tract may consist of more than one type of parasequence sets.

The FSST forms during periods of relative sea level fall, which may eventually lead to formation of an unconformity. The LST forms when the relative sea level fall stabilizes or is slowly rising so that the sedimentation rate exceeds the rate of rise. The LST may overlie an unconformity or a FSST whence its lower boundary is a **correlative conformity**. The upper boundary of the LST is marked by the first major marine flooding surface termed **conformable transgressive surface** (CTS). The TST forms when the rate of relative sea level rise exceeds the sedimentation rate. In special circumstances when the period of LST is short and the shoreface retreat during shoreline transgression is associated with strong erosion, the LST deposit may be removed and the TST overlies a submarine erosion surface termed **ravinement surface**. The HST follows TST and forms when the rate of sedimentation exceeds the rate of relative sea level rise. The boundary between the TST and HST is the maximum flooding surfaced (MFS).

Characteristics of Falling Stage and Lowstand Systems Tract

The formation of the falling stage systems tract is associated with progressive destruction of accommodation space due to fall in relative base level ($dA/dt < 0$). In marine settings, the systems tract is associated with a fall in relative sea level — termed forced regression (Posamentier et al., 1992). Therefore, a forced-progradational parasequence set would form in areas of positive accommodation space, till the accommodation space remains positive despite progressive reduction of space. However, continuation of relative sea level fall would eventually lead to a situation of zero and negative accommodation space. As a result, subaerial sedimentary bypass and erosion would set in to form a subaerial unconformity — the sequence boundary.

In shallow marine setting, sediments are transported over the exposed shelf through fluvial systems, and if the shelf gradient is greater than that of the fluvial profile, rivers may also incise valleys on the exposed shelf. The depth of incised valleys would depend on the relative difference in the gradients of the fluvial and shelf profile.

The sediments crossing the exposed shelf may form slope and submarine fans if a slope system exists. In a ramp setting, sediments bypassing the exposed shelf would form a forced-progradational parasequence set characterized by deltaic and wave-dominated shoreface-shelf depositional systems.

In carbonate platforms, a fall in relative sea level may result in exposure, shrinking and shifting of the carbonate factory depending on the morphology of the platform (Fig. 4.13; Sarg, 1988; James and Kendall, 1992). The exposed carbonates suffer subaerial diagenesis resulting in a subaerial unconformity surface characterized by karstification (in warm humid climate) or calichification (in arid climate), provided the duration of exposure is sufficient. In platforms attached to a landmass the karst or caliche profiles may be veneered by prograding fluvial-deltaic or eolain siliciclastics. During periods of falling relative sea level, platform margin carbonates may suffer enhanced re-sedimentation in deeper water due to slope failures forming slope and submarine fans. In ramps, on the other hand, a forced-regressive parasequence set would develop.

The depositional systems that comprise the LST are mainly deltaic systems prograding onto the shelf with significant reworking by wave.

Characteristics of Transgressive Systems Tract

The formation of the transgressive systems tract is associated with accommodation space creation (dA/dt > 0) at a rate faster than that of sediment supply or production. Consequently, a retrogradational parasequence set develops. The siliciclastic marine setting is characterized by estuarine, shoreface and shelf depositional systems. The tidal influence is pronounced due to increased shelf width and tidal deltas may be common. The succession is mud dominated. Culmination of transgression promotes deposition of deep-water fine-grained sediments at a very slow rate forming condensed section (see Section 2.8).

In carbonate platforms, rise in relative sea level during transgression may drown the platform and carbonate production may cease, forming drowning unconformity (James and Kendall, 1992; Sarg, 1988). However, drowning is seldom a consequence of sea-level rise alone as has been discussed earlier. If other factors remain favourable, a sea level rise may not be able to cease

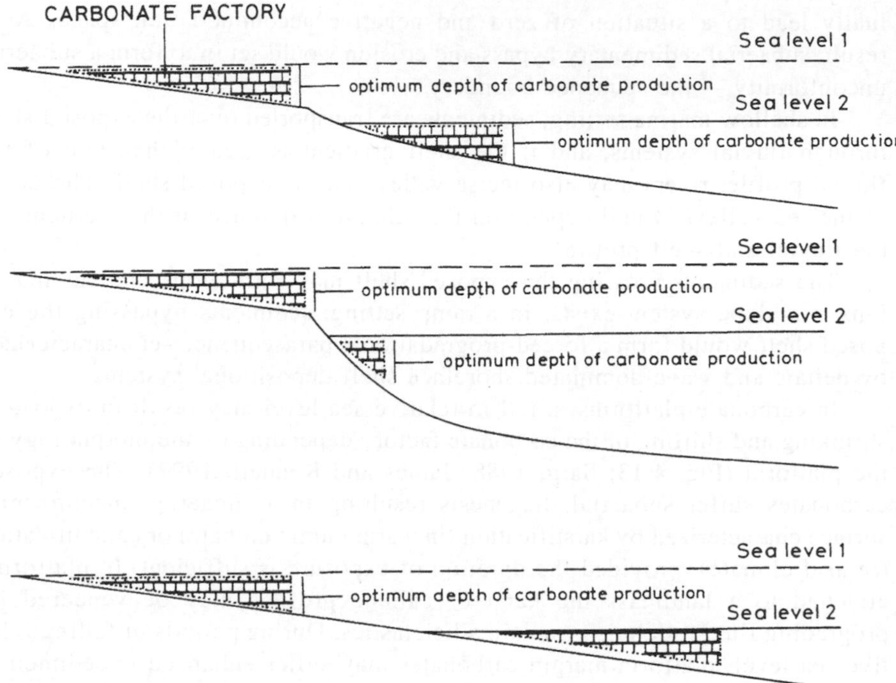

Fig. 4.13: Shrinking and shifting of the carbonate factory in response to fall in relative sea level. Note that the change in the dimension of the carbonate factory following the fall in relative sea level depends on the platform morphology. In distally steepened ramps the factory shifts location and shrinks (top); in rimmed platforms the factory shifts and shrinks more than the distally steepened ramp (middle); in horizontal platforms the factory shifts but does not change its dimension.

carbonate production. In such a situation, subtidal carbonates, stratigraphically thick reefs dominate the succession (James and Kendall, 1992).

Characteristics of Highstand Systems Tract

The formation of the highstand systems tract is associated with accommodation space creation (dA/dt >0) at par or lower rate than that of the sediment supply. As a result, an aggradational to progradational parasequence set develops.

The siliciclastic shallow-marine domain is characterized by a decrease in tidal influence with an increase in wave effect. Deltaic and wave-dominated shelf depositional systems characterize a sand-dominated succession.

Carbonate platforms are dominated by peritidal carbonates; reefs show lateral spreading rather than vertical growth (James and Kendall, 1992; Sarg, 1988). Parasequence boundaries are often characterized by meteoric diagenetic features. Enhanced export of sediments from the platform margin gives rise to thick wedges of slope deposits of mass flow origin.

Stacking of Systems Tracts

An ideal sequence formed between two base level falls, usually consists of a lowstand systems tract at the bottom, followed by the transgressive systems tract, highstand systems tract and falling stage systems tract (Fig. 4.12). The boundary between the transgressive and highstand systems tract is the maximum flooding surface. A sedimentary basin fill usually consists of many sequences. Therefore, while analysing a sedimentary succession, first it is to be divided into sequences by identifying the subaerial disconformity surfaces. The next step is to identify parasequences and their stacking patterns within a sequence to distinguish different types of parasequence sets followed by delineation of maximum flooding surface and systems tracts. All these data may then be utilized to interpret the sequence in terms of relative rates of accommodation space creation/destruction and sediment supply or production through time. When this is done for all the sequences, pattern of stacking of the sequences can be taken into consideration (Fig. 4.14) to understand the tectono-sedimentary evolution of the basin.

In summary, the succession preserved in a sedimentary basin may be conceived of as being composed of sediment packages bounded by discontinuities of varying magnitudes (Fig. 4.14). The package bounded by subaerial unconformities is the sequence. The sequence again consists of systems tracts bounded by their respective discontinuity surfaces of lower magnitude than the subaerial unconformity (one of which is the maximum flooding surface). The systems tracts can further be split into parasequences bounded by discontinuities of the lowest stratigraphic magnitude (flooding surface and regressive surface).

Alluvial Architecture and Sequence Stratigraphy

The methodology of sequence stratigraphy described above was developed for marginal marine and marine strata and its straightforward application to non-marine alluvial successions is limited because the key surfaces such as flooding surface, regressive surface, parasequences and their stacking pattern are difficult to define in the non-marine domain (Shanely and McCabe, 1994). Moreover, the non-marine accommodation space may change independently of relative sea level changes particularly in alluvial basins located far away from the contemporary shoreline. In such basins accommodation space is affected by local tectonics and climate. Therefore, the system of sequence stratigraphy for such basins is likely to be different from that of marine basins. In addition, the description of alluvial strata and architecture requires a different methodology

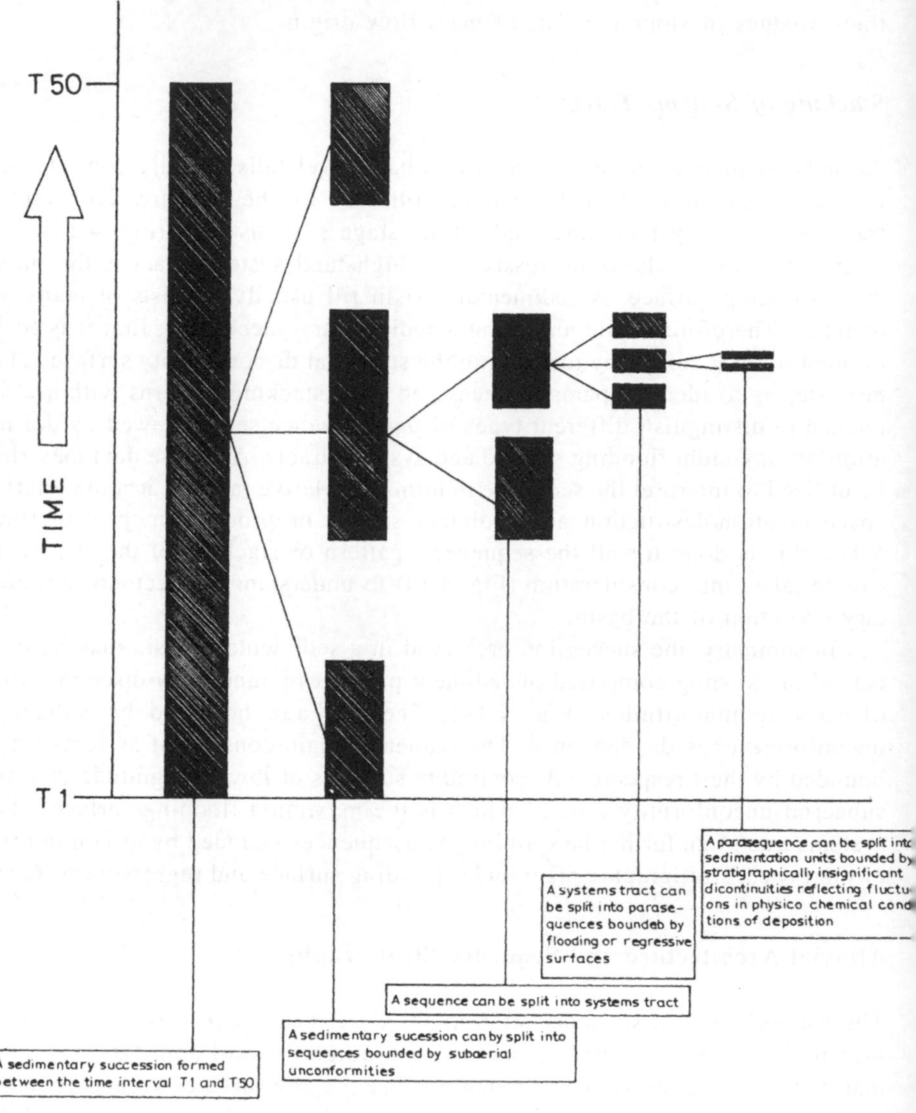
T 50
TIME
T 1
A sedimentary succession formed between the time interval T1 and T50
A sedimentary sucession can by split into sequences bounded by subaerial unconformities
A sequence can be split into systems tract
A systems tract can be split into parase-quences boundeb by flooding or regressive surfaces
A parasequence can be split into sedimentation units bounded by stratigraphically insignificant dicontinuities reflecting fluctu ons in physico chemical cond tions of deposition

Fig. 4.14: Resolution of a sedimentary succession into sequence stratigraphic units of varying magnitudes.

from that applicable to marine strata (Bridge, 1993; Miall, 1985, 1987, 1995). Alluvial sequence stratigraphy begins with characterizing the alluvial architecture.

Alluvial Architecture

An alluvial plain is the domain of sediment accumulation bordered by topographic barriers and its deposits consist basically of two suites of strata: i) Channel-belt coarser sedimentaries and ii) Overbank finer sedimentaries (Table 4.T1). The channel-belt is the conduit through which major discharge of sediment and fluid takes place, whereas overbank areas flanking the channel-belt, receive sediments only during periods of floods. Overbank sediments are deposited as levees on the edge of the channel-belt, from crevasse channels spurting out breaching levees, or as splays, from small, ephemeral flood plain channels or in the broad low-lying areas of the flood plain termed flood basin, following sheet floods. A channel-belt may be occupied by a single channel or by multiple channels (Table 4.T1). The single channel may be straight or meandering. Multiple channels in the alluvial tract may show a braided pattern of arrangement or may be anastomosed.

Three principal architectural elements of alluvial deposits are microform, mesoform and macroform. Microforms include ripples and plane beds. Dunes are regarded as mesoforms. Macroforms are essentially compound bedforms or bars upon which micro and/or mesoforms are superimposed. Accretion of the macroform produces inclined strata termed macroform strata that internally consist of micro and/or mesoform strata. The azimuths of the macroform strata may be orthogonal, oblique or parallel to that of the associated micro and mesoform strata. The first type of compound stratification is commonly known as epsilon cross-stratification or lateral accretion bedding and the parent macroform is termed laterally accreting bars e.g. point bar (Fig. 4.15). The third type of compound cross-stratification popularly goes by the name of downcurrent-dipping cross-stratification and the associated macroform is called foreset macroform or downstream accreting bars (Fig. 2.2.19).

Another type of macroform of the fluvial system is a channel that is filled dominantly by vertical accretion producing plano-concave sediment bodies (Fig. 4.16). Channels in many cases are filled with sandy sediments; however, conglomeratic as well as muddy channel fills are also common. The principal diagnostic features of channel-fill sandstone are the concave-up, erosional contact with the underlying beds (Fig. 4.16). The channel-fill sediment bodies produced by vertical accretion usually acquire the channel morphology and in transverse section the channel form can be observed (Fig. 4.16). The channel fill may be a) massive, b) parallely stratified either horizontally or conformal to the cross-sectional channel geometry, c) stacked sets of massive strata, cross-strata and horizontal strata, d) debris flow or turbiditic sandstone in case of submarine channels.

A channel-belt sediment body is produced by superimposition of bars and channel-fills (Fig. 4.17). The building blocks of ancient channel-belt sediment bodies are storeys; a storey is defined as the deposit of a single channel bar and

Table 4.T1: Chart summarizing the essential features of alluvial systems for their characterization from the ancient record.

<table>
<tr><td colspan="4">ALLUVIAL PLAIN (several kms wide)
A domain bordered by topographic barriers; may consist of one or more channel belts with flanking floodplain</td></tr>
<tr><td colspan="4">CHANNEL BELT (10s of m – a few km wide)</td></tr>
<tr><td>Single channel
(high sinuosity)</td><td>Single channel
(low sinuosity)</td><td>Multiple channel
(braided)</td><td>Multiple channel
(anastomosed)</td></tr>
<tr><td>Meandering, mobile (laterally migrating) channel, characterized by point bars, chute and neck cut-offs</td><td>Characterized by alternate bars</td><td>Mobile (laterally switching) channels, characterized by foreset macroforms</td><td>Fixed channels</td></tr>
<tr><td colspan="4">CHANNEL ORDERS IN A MULTIPLE-CHANNEL BELT</td></tr>
<tr><td>Channels between composite macroforms (stabilized bars/ islands)</td><td colspan="2">Channels between simple macroforms</td><td>Channels dissecting macroforms</td></tr>
<tr><td colspan="4">AVULSION</td></tr>
<tr><td colspan="2">Channel-belt avulsion (1st order)
Local or Regional</td><td colspan="2">Anabranch avulsion- channel switching within a multiple-channel belt (2nd order)</td></tr>
<tr><td colspan="4">INDIVIDUAL CHANNEL-BELT SEDIMENT BODY
(10s of m – a few km wide, 1m – 10s of m thick)</td></tr>
<tr><td colspan="2">Single storey</td><td colspan="2">Multistorey/Multilateral</td></tr>
<tr><td colspan="2">A storey is defined as the deposit of a single channel bar and adjacent channel fill</td><td colspan="2">Results when the rate of channel migration within the aggrading channel-belt is large enough to result in superposition of channel bars and fills in any kind of channel-belt or due to anabranch avulsion and/or deposition in multiple-channel braided or anastomosed channel-belts</td></tr>
</table>

adjacent channel-fill (Bridge, 1993). In vertical profiles, a channel-belt deposit may consist of single or multiple storeys. Multistorey channel-belt deposit results when the rate of channel migration within the aggrading channel-belt is large enough to result superposition of channel bars and fills in any kind of channel-belt or due to channel switching in multiple-channel braided or anastomosed channel-belts (Table 4.T1).

Characterization of channel-belt sediment bodies are best done by identifying different orders of surfaces within the bodies (Table 4.T2; Fig. 4.17; Miall, 1988, 1995). The first order surface represents the boundaries between sets of micro/mesoform strata. The boundaries of storeys are fifth order sur-

Table 4.T1: contd.

ALLUVIAL ARCHITECTURE			
Dispositional pattern of channel-belt sediment bodies and floodplain depósits			
Randomly distributed, isolated channel-belt sediment bodies encased in fllodplain mudstone	*Superimposed channel-belt sediment bodies* (caused by regional river-belt avulsion)		
	Vertically superposed	*Laterally superposed*	*Both vertically and laterally superposed*

SHAPE DESCRIPTORS OF SEDIMENT BODY UNDER CONSIDERATION				
Tabular Sheet *(w/t >15>100)* rectangle in 2D	***Elongate Sheet*** rectangle in 2D	***Wedge*** triangle, rectangle, trapezoid in 2D	***Lobe*** convexo-planar lens, triangle, trapezoid in 2D (asymmetry: a/b- 0-1)	***Ribbon (w/t<15)*** plano-cocave lens, rectangle, triangle in 2D

faces. The sixth order surface represents the boundary between channel-belt and overbank deposits or between two different channel-belts.

It is useful to describe the three-dimensional shapes of alluvial sediment bodies and their appearances in differently oriented vertical sections (Table 4.T1; Bridge, 1993). The convention is to describe the upper bounding surface first, followed by the lower bounding surface. For example plano-concave or plano-convex etc.

A channel-belt moves downward (vertical incision) or upward (aggradation) in response to decrease and increase in base level respectively. The channel-belt may also shift laterally (migrate) with or without a vertical component. The width of a channel-belt sediment body approximates the lateral extent of the alluvial plain within which the channel-belt could shift position by lateral migration. The thickness, on the other hand, depends on the depth of channels and amount of aggradation and incision. The width of a channel-belt deposits of a multiple-channel braided river is virtually larger than that of a channel-belt characterized by single channel, meandering river. Consequently, the channel-belt sediment body produced by braided rivers are typically sheet-like with high width/thickness ratio. Again, a meandering channel-belt would tend to produce sheet-like deposits if it is free to migrate laterally (Collinson, 1978), in contrast to ribbon-like channel-belt bodies produced by less-mobile meandering rivers or fixed channels of an anastomosed channel-belt (Bridge and Leeder, 1979).

The coarser fraction of the fluvial sediment load usually accumulates in the channel-belt, whereas, the finer fraction is preferentially preserved in the neighbouring overbank areas. The rock record reveals that alluvial successions are represented by an alternation of channel-belt and overbank deposits (Fig. 4.18). The vertical stacking of channel-belt and overbank sediment bodies in an allu-

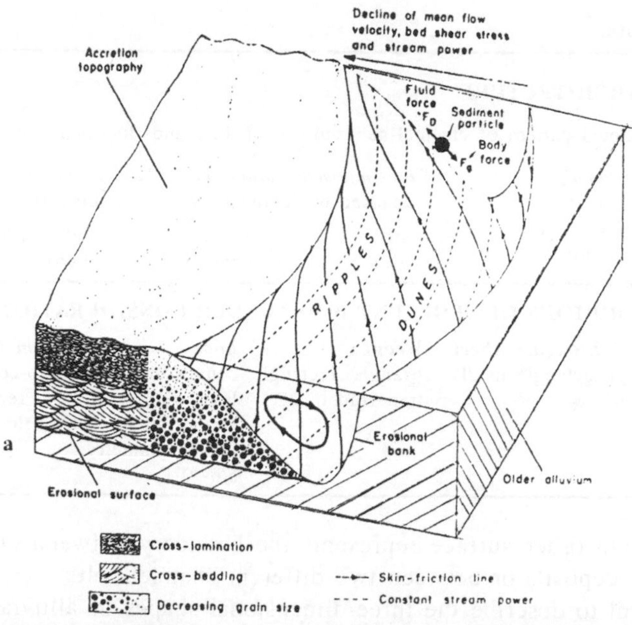

Fig. 4.15 a: Diagram illustrating lateral accretion of meandering channel. Note: (1) Lateral accretion surfaces in the deposit left by the laterally migrating channel in cross-section and in bedding plane; (2) the stratification style in the deposit bounded by the accretion surfaces; (3) orthogonal relationship between the orientations of the lateral accretion surfaces and the current directions revealed by the stratification formed by bedforms; (4) the vertical variation in sedimentary structure and texture. (Reproduced from J. R. L. Allen, 1970, Physical Processes of Sedimentation: An Introduction. London, George Allen & Unwin Ltd., Fig. 4.5, p. 133, with the kind permission of the author).

b: Natural example of lateral accretion surfaces. Paleoflow was across the plane of the section.

vial succession is assumed to be the consequence of repeated, regional channel-belt avulsion i.e. sudden abandonment of the whole channel-belt in favour of a new tract, in an aggrading alluvial plain (Bridge and Leeder, 1979).

For a constant aggradation rate, higher avulsion frequency leads to a greater degree of superposition and interconnection of channel-belt sediment bodies. The resulting alluvial succession thus consists of amalgamated channel-belt

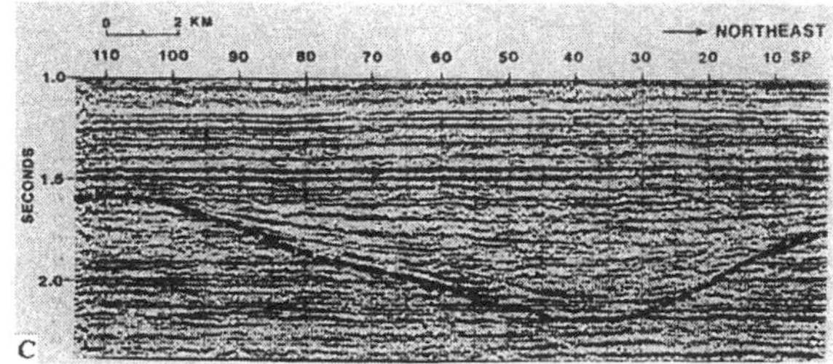

Fig. 4.16: Photographs (**a, b and c**) showing channels in cross-section (Fig. 4.16 c is reproduced from G. Shanmugam and R. J. Moiola: Submarine fans: characteristics, models, classification and reservoir potential, Earth Science Reviews, Vol-24, Fig.12, p.398. © 1988, Elsevier Science-NL, Sara Burgerhartstraat 25, 1055 KV Amsterdam, The Netherlands, with the kind permission).

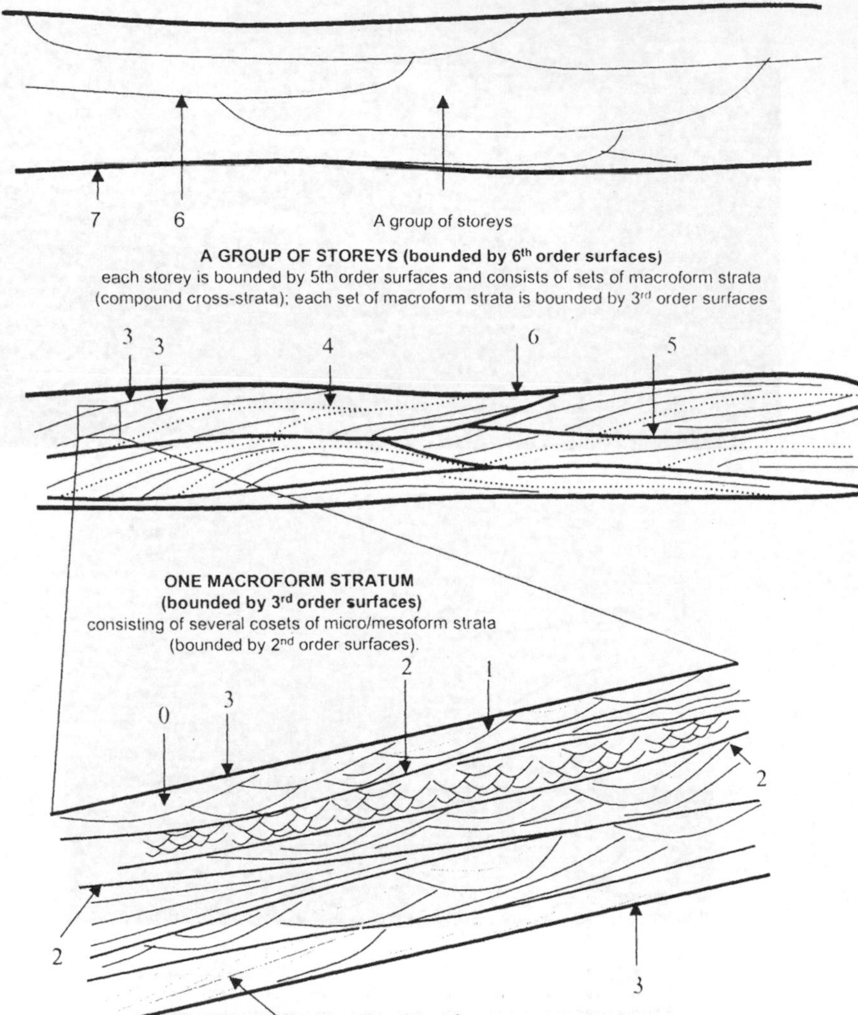
A SANDSTONE BODY (bounded by 7th order surfaces)
consisting of several groups of storeys, each group is bounded by 6th order surfaces
7
6
A group of storeys
A GROUP OF STOREYS (bounded by 6th order surfaces)
each storey is bounded by 5th order surfaces and consists of sets of macroform strata (compound cross-strata); each set of macroform strata is bounded by 3rd order surfaces
3
3
4
6
5
ONE MACROFORM STRATUM
(bounded by 3rd order surfaces)
consisting of several cosets of micro/mesoform strata
(bounded by 2nd order surfaces).
2
1
3
0
2
2
3
Intraset reactivation surface

Fig. 4.17: Internal architecture of a channel-belt sandstone body (adapted from Miall, 1988, 1995 and Bridge, 1993).

deposits with little overbank fines (Fig. 4.19). On the other hand, for a constant avulsion frequency higher aggradation rate tends to produce a succession of alternating poorly connected, isolated, channel-belt sediment bodies and overbank fines (Fig. 4.19, Bridge and Leeder, 1979).

The principal objectives of architectural analysis of alluvial successions

Table 4.T2: Hierarchy of Bounding Surfaces in a Channel Sediment Body (A bounding surface is a surface that bounds sediment bodies of consistent characteristics defined by the user)

Rank	*Description*
0	Boundaries between micro/mesoform (ripple, dune, plane bed) strata > 0 Reactivation surfaces within a set of micro/mesoform strata
1	Boundaries between sets of micro/mesoform strata
2	Boundaries between cosets of micro/mesoform strata
3	Boundaries between macroform (compound bedform- frontally, laterally, obliquely, vertically accreting bars) strata (a stratum represents one *event* of accretion of a macroform)
4	Boundaries between sets of macroform strata. A set represents one *episode* of accretion of a macroform (bar)
5	Boundaries between cosets of macroform strata. A coset represent the whole or remnant of a macroform (bar or channel). Also termed *storey*
6	Boundaries between groups of storeys. Each group of storeys represents the individual channel complexes of a channel-belt.
7	Boundaries between the channel-belt sandstone body and the overbank mudstone.

Fig. 4.18: Field photo showing vertical alternation of channel belt sandstone body and floodplain finer sedimentaries. (Reproduced from W. M. Stear: Morphological characteristics of ephemeral stream channel and overbank splay sandstone bodies in the Permian Lower Beaufort Group, Karoo Basin, South Africa. IAS Spec. Pub. No.6, Fig.3, p.408, © 1983, Blackwell Science Limited with the kind permission).

1. Identification of the effects of intrabasinal or autogenic processes in the sedimentary package such as changes in the fluvial style and processes, channel-belt dimension, avulsion frequency and sedimentation rate.
2. Identification of the effects of extrabasinal or allogenic processes such as tectonics, basin subsidence and climate.

Alluvial Sequence Stratigraphy

The basic premise of sequence stratigraphy is to analyze a sedimentary succession in terms of the rates of accommodation space creation and filling. This also applies to alluvial successions (Currie, 1997) with the difference that the base level is not the sea level but a non-marine, stratigraphic base level that may fluctuate independent of sea level changes. This modification is necessary because in many alluvial successions the relationship with equivalent shoreline strata is difficult to establish (Martinsen et al., 1999).

In alluvial basins the long-term rate of sediment supply (S) usually exceeds or equals the long-term rate of accommodation space creation (A) either by subsidence or upward movement of base level. Therefore, the long-term rate of sedimentation proxies the geological rate of subsidence and the A/S ratio will always have values between 0 and 1. If the available accommodation space is negative, the channel-belts incise valleys signifying disconformities. Unconformity without channel incision may result if the available accommodation space is zero and there is a lack of sediments. Assuming constant sediment flux, periods of higher subsidence will be characterized by higher A/S value

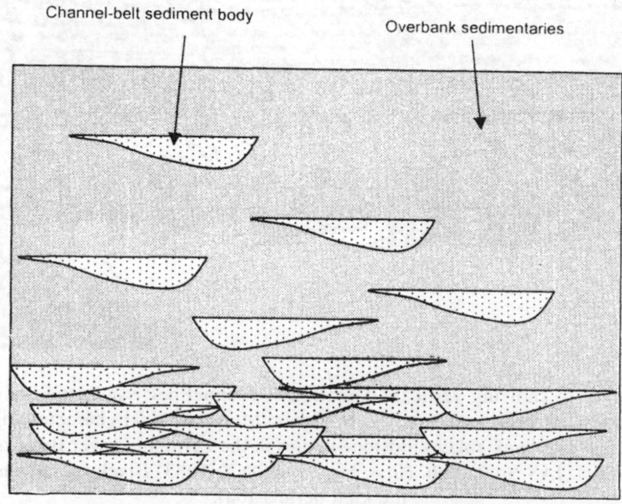

Fig. 4.19: Different patterns of stacking of channel-belt and floodplain sedimentary bodies due to variable avulsion frequencies at a constant aggradation rate.

and vice versa. It is likely that the lower and higher A/S values will have different expressions on the alluvial architecture and there will be surfaces within the alluvial succession marking abrupt changes in the A/S value (Martinsen et al., 1999; Olsen et al., 1995). Barring the unconformity surfaces that bound an alluvial sequence, the other surfaces commonly recognized in alluvial successions are:

1. Surfaces across which the stacking pattern of channel-belt and overbank sediment bodies changes but the fluvial style remains same. Periods of lower subsidence rate are characterized by amalgamated channel-belt sediment bodies with little or no overbank fines whereas periods of higher subsidence rate are reflected in an alluvial succession showing isolated channel-belt bodies encased in overbank fines (Fig. 4.19; Burns et al., 1997; Martinsen et al., 1999).
2. Surfaces across which dimensions and flow directions of channel-belt change coupled with a change in fluvial style. Channel-belt dimensions increase during periods of lower subsidence rate with a dominant flow transverse to the basin axis and the fluvial style is usually braided. During periods of higher subsidence the alluvial plain may be characterized by meandering or anastomosed fluvial systems flowing roughly parallel to the basin axis (see also Chapter 5). The factors that influence change in fluvial style are poorly understood. However, mean channel gradient and the proportion of fluid and sediment discharge is perhaps a major controlling factor. Braided channel-belt requires a larger gradient than meandering or anastomosed channel-belt.
3. Higher discharge capacity of rivers and lower sediment load results into anastomosed network of channels, whereas lower discharge capacity and higher sediment load leads to a braided channel pattern (Orton and Reading, 1993)
4. Surfaces across which frequency, type and maturity of paleosol change (Wright and Marriot, 1993; Krauss, 1999; see also Section 2.6). Extremely low rate of sedimentation or non-deposition is a prerequisite for pedogenesis. That is why paleosols mark major unconformities signifying long periods of non-deposition and non-erosion and landscape stability. These paleosols are strongly developed, thick and solitary and are often laterally correlatable over long distances and are essentially allogenic.

Sedimentation in the alluvial floodplain is virtually episodic with intervening periods of non-deposition or erosion. Consequently, in the alluvial domain soils preferentially form on the floodplain deposits, which are essentially autogenic. At a relatively slower rate of pedogenesis than sedimentation, compound paleosols develop showing alternation of paleosol profile and unaffected deposit. With increasing rate of pedogenesis, the proportion of unaffected sediments decreases and composite paleosols result. In situations, where sedimentation and pedogenesis go hand in hand, but the rate of pedogenesis is high cumulative paleosols form leaving no deposit unaffected. In conditions, when the water table rises and comes closer to the surface, hydromorphic paleosols develop. In contrast, under conditions of falling water table, well-drained paleosols form. Therefore, periods of high subsidence and low sediment flux, are characterized by development of hydromorphic paleosols. On the other hand, well-drained paleosols would form during periods of low subsidence and sediment flux. Hydromorphic and well-drained paleosols may be considered to represent the two end members of a continuous spectrum and each of them may be compound, composite or cumulative depending on the relative rates of pedogenesis and sedimentation.

5. TECTONIC SIGNATURES IN THE BASIN FILL

General

A sedimentary succession is the content of a sedimentary basin. Strata are preserved in a basin due to synsedimentary tectonic subsidence. Rises in absolute base level also aid in accumulation and preservation of sediments; however, the cumulative magnitude of the rises over the life span of a sedimentary basin is far less than the thickness of sedimentary successions represented in different ancient sedimentary basins (>1 km). Hence, rise in absolute base level alone cannot produce significantly thick sedimentary successions and synsedimentary tectonic subsidence is the key characteristic of sedimentary basins. Without tectonic subsidence there would be hardly a sedimentary basin (Ingersol and Busby, 1995).

Subsidence of sedimentary basins is induced by two basic mechanisms: a) flexure of the crust due to loading and b) down-faulting due to stretching and/or thinning, with isostasy being the essential component in both the processes (Fig. 5.1; Angevine et al., 1990; Blundell, 1991). Supracrustal thrust loading in orogenic belts induces flexuring of the crust producing asymmetric sedimentary basins in the foreland termed *foreland basins* (Fig. 5.1; Allen and Homewood, 1986; Miall, 1995a). Subcrustal loading due to injection or flow of high-density material underneath the crust and/or thermal contraction of the crust also induces downward flexuring producing oval-shaped sedimentary basins termed *sag basins* (Fig. 5.1; Sleep et al., 1980; Klein, 1995) Stretching of the crust, on the other hand, results in the development of normal faults in the upper level of the crust producing narrow, linear, graben-type basins termed *rift basins* (Fig. 5.1; Coward et al., 1987; Leeder, 1995; Mckenzie, 1978; Bott, 1995; Olsen, 1995; Sengor, 1995). Displacements along crustal scale strike-slip faults are usually associated with loacalized down-faulting and sagging producing basins known as *strike-slip basins* (Fig. 5.1; Balance and Reading, 1980; Christie-Blick and Biddle, 1985; Nilsen and Sylvester, 1995). The ultimate aim of studying sedimentary successions is to reconstruct the tectono-sedimentary evolution of the basin. This requires understanding of the subsidence mechanism and the tectonic setting of the basin by identifying signatures of tectonic activity from

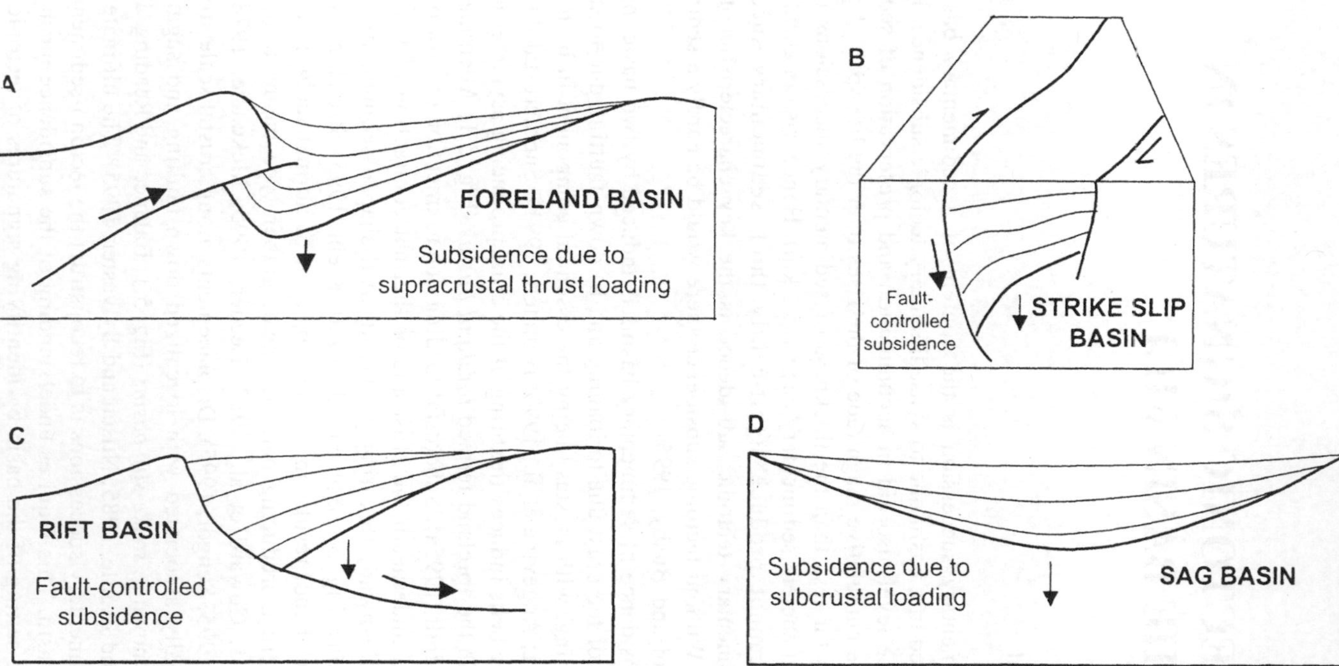

Fig. 5.1: Diagram illustrating configuration, subsidence mechanism, basin-fill geometry and stratal architecture of different types of sedimentary basins.

within the succession (Allen and Allen, 1990; Busby and Ingersol, 1995; Einsele, 1992; Frostick and Steel, 1993a; Miall, 1990).

The geometry and thickness distribution of the basin fill, the nature of the contact between the basin-fill and the adjacent country are of primary significance. The basin fills of rift, strike-slip and foreland basins are typically asymmetric due to localization of tectonic activity and therefore of maximum subsidence along one basin margin (Fig. 5.1). However, in rift basins the active margin is characterized by normal faults, in contrast to strike-slip and thrust faults that prevail along the margins of strike-slip and foreland basins respectively. The basin fills of sag basins are typically symmetric with maximum thickness occurring at the center and the basin margins are not usually faulted (Fig. 5.1).

In order to decipher the tectonic setting from a sedimentary perspective, it is necessary to understand whether: (i) the basin was of ramp type without a slope system containing only continental or shallow marine systems or both; (ii) the basin possessed both shelf and slope systems with a distinct shelf-slope break; and (iii) the width of the shelf was narrow or wide (see Reading, 1996).

The commonly encountered depositional systems in rift basins are alluvial fan, fluvial, deltaic and lacustrine (Leeder and Gawthorpe, 1987; Schlische, 1991). Rifting of crustal segments below sea level may eventuate development of marine systems if there is a connection with the sea. The depositional systems of foreland basins range from alluvial to deep marine. However, in some foreland basins developed on thick crusts the deep marine system may be absent with widespread development of shallow marine systems (Cant and Stockmal, 1993). Sediments in sag basins tend to be continental to shallow marine (Klein, 1995).

Carbonate platforms develop when subsidence is continuously or episodically slow so that siliciclastic input is low, gradients are slight, and basin water depths are relatively shallow (Burchette and Wright, 1992). The associated tectonic regimes are, therefore, characterized by gentle flexural subsidence, such as in foreland basins, the margins of sag basins, and the hanging-wall dip-slopes of rift basins.

Changes in paleogeographic setting of sedimentary basins through time is essentially a reflection of the (a) variation in the subsidence rate relative to the rate of sediment supply, (b) changes in the direction of tilt and the location and size of the depocentre (i.e., the site of maximum subsidence), (c) changes in the character and orientation of the basin margins and in overall basin size and shape, (d) changes in the gradient both at margins of the basin through faulting and within the basin through tilting of the basin floor, (e) deflection of the sedimentary systems where tectonically controlled morphological changes act as barriers to sediment transfer both into the basin from the hinterland and between various areas within the basin and (f) changes in the rate of sediment supply due to uplift or erosion of the supplying hinterland (Frostick and Steel, 1993b).

Tectonism exerts a significant influence on sedimentation, not only on the type and thickness of sediment, but also on the areal distribution of depositional systems within the basin. The role of tectonism in determining system distribution is especially important in asymmetrically subsiding basins, such as half grabens, foreland basins, and some strike slip basins (Alexander and Leeder, 1987; Cant and Stockmal, 1993; Nilsen and McLaughlin, 1985). Recent depositional models suggest that periods of rapid subsidence in asymmetrical basins result in deposition of fine-grained sediment, such as lacustrine or fine-grained, longitudinal fluvial systems, directly above the locus of maximum subsidence and close to the margin of the uplifted terrane (Blair and Bilodeau, 1988; Mack and Seager, 1990). Because denudation is significantly slower than the rates of uplift and subsidence, coarse-grained alluvial fan or braidplain deposit is restricted to a narrow zone directly adjacent to the uplifted terrain. However, as the rate of subsidence and uplift diminishes erosion rate eventually surpasses subsidence rate and a sheet of coarse sediment spreads across the basin (Fig. 5.2; Burns et al., 1997). Thus, near the locus of maximum subsidence, finer-grained facies are syn-tectonic and coarser grained facies are post-tectonic.

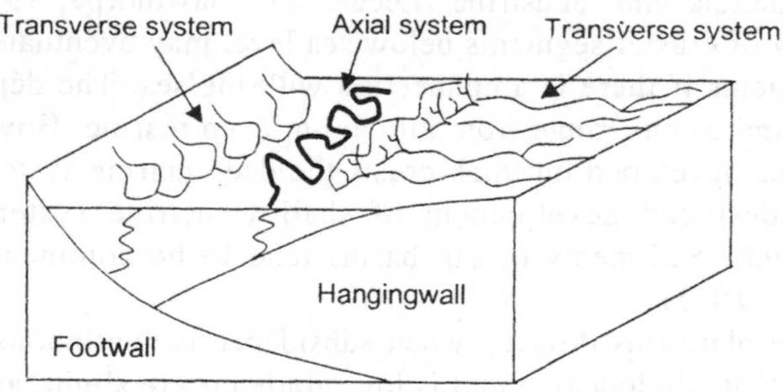

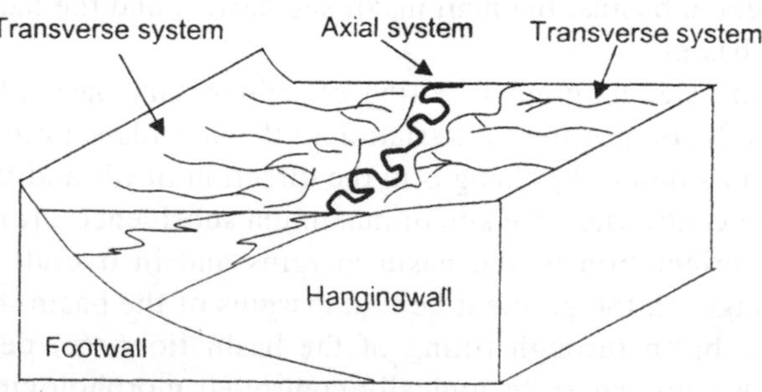

Fig. 5.2: Diagram illustrating disposition and spread of axial and transverse fluvial systems during periods of rapid and slow subsidence in a rift basin.

Change in the subsidence rate relative to the rate of sediment supply is reflected in the variation in stacking patterns of parasequences in a marine succession. Periods of higher subsidence rate are characterized by widespread lacustrine environment in continental setting and deeper shelf environment in marine setting with alluvial systems occurring along the margins. In contrast, periods of lower subsidence rate promote widespread development and progradation of alluvium. Basin-wide sandy alluvial or shallow marine sediments both suggest filling of the basin and diminution of the effects of tectonic barriers (Frostick and Steel, 1993a). Intensive aeolian reworking of alluvial deposits can indicate that the basin was in a period of relative stasis, i.e., almost filled with sediment with little or no tectonism (Frostick and Steel, 1993b). The volume of rift basins progressively increases since initiation (Schlische, 1991). Therefore, if sediment supply remains constant the early basin fill is fluvial and as the basin widens, sediment influx is spread over a wider area and basin topography may increase, leading to deposition of lacustrine sediments; as strain rate decreases fluvial deposition may recur producing a package of fluvial-lacustrine-fluvial deposits.

Sub-aerial unconformities in the sedimentary succession indicate periods of zero or negative accommodation space in the basin. These unconformities result due to (a) lower rate of subsidence than the rate of sediment supply, (b) a fall in relative base level arising from uplift or (c) a fall in absolute base level. However, discontinuity surfaces may also result from lack of sediment supply or production (in case of carbonate sediments). Unconformities in rift basins develop on intrabasinal and basin margin highs and are rarely basin-wide unless formed by a fall in absolute base level. In foreland basins, unconformities typically develop near the basin margin and disappear basinwards. Moreover, upwards in the succession, forward propagation of thrusts into the evolving sedimentary pile in foreland basins progressively shifts the zone of erosion towards the center of the basin. As a result, unconformities tend to be located more towards the basin interior with time. In sag basins, unconformities are typically basin-wide.

Intraformational unconformities signifying angular discordances between over and underlying strata clearly indicate synsedimentary tectonism (Riba, 1976).Syn-sedimentary tectonic activity also results in diastrophic deformation of strata. In foreland basins, the strata tend to be folded near the thrust margin (Anadon et al., 1986). In rift and strike slip basins, fault-controlled subsidence produces tilting, folding and rupturing of strata on the hangingwall (Fig. 5.3; Dula, 1991; Gibbs, 1987). The other signatures of synsedimentary tectonism are penecontemporaneous deformation features, such as slides, slumps etc (see Section 2.2).

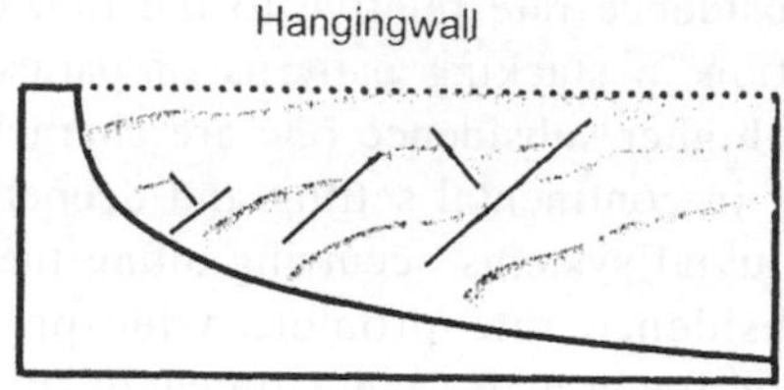

Fig. 5.3: Schematic sketch showing deformation features of hanging-wall strata in a rift basin.

Sedimentary Signatures of Rift Basin

Rifts are elongated, sediment-filled depressions (30-60 km wide, some times longer) bounded by crustal-scale normal faults formed as a result of tectonic stressing. Field observations indicate that majority of rift basins consist of half-grabens with the active fault at one margin (Fig. 5.1; Frostick and Reid, 1987). The bordering topographic highs are termed rift shoulders. Along the length of the rift, there may be a number of half-graben segments and the active faults bounding each half-graben usually show alternate reversal of dip direction along the length of the rift and flip position from one side of the rift to the other (Fig. 5.1; Frostick and Reid, 1987; Hamblin and Rust, 1989; Leeder, 1995; Rosendahl, 1987). Individual segments are joined by smaller faults termed as accommodation zone, transfer zone or transfer fault that stand at higher elevations than the intervening basins (Jackson and Leeder, 1994). Accommodation space for sediment accumulation is created by synsedimentary fault movement that involves: 1) rotation and offset of fault blocks or 2) hanging wall deformation forming rollover and/or antithetic faults (Fig. 5.3; Dula, 1991; Gibbs, 1987; Schlische, 1991). The basin is thus highly asymmetric with the finite magnitude and rate of subsidence increasing towards the border fault. Consequently, the basin-fill is typically wedge-shaped, increasing in thickness towards the active margin. Differential subsidence causes the strata to dip towards the active border fault and synsedimentary tectonism is indicated by gradual increase in dip of successively older strata and their progressive onlap into the hanging wall (Fig. 5.1).

Removal of the hanging-wall load from the footwall due to lateral displacement or block rotation offset during fault movement leads to isostatic and kinematic uplift of the footwall including the rift shoulders (Leeder and Gawthorpe, 1987; Wernicke and Axen, 1988). Footwall upliftment and hanging-wall subsidence produce steeper footwall scarp slope and gentler hanging-wall dip slope. The footwall area serves as the major sediment source for the basin with additional contributions from the shoulder of the hanging wall block. The transverse sediment input (not necessarily orthogonal to basin margin) from the footwall may feed a major sediment transporting system oriented along the basin axis (axial or longitudinal system). The axial system is preferentially located near the footwall at the site of maximum subsidence and follows the regional tectonic gradient (Alexander and Leeder, 1987; Frostick and Reid, 1989, 1990; Leeder and Alexander, 1987).

In terrestrial settings, the steeper footwall ramp is the locus of alluvial fans and fan delta spreading towards the interior of the basin where finer grained fluvial or lacustrine systems prevail (Leeder and Gawthorpe, 1987; Frostick and Steel, 1993c). The hanging-wall ramp is also characterized by alluvial fan and fan deltas but being broader and gentler, the systems gradients are normally gentler than footwall-sourced fans. During periods of higher subsidence, footwall-linked depositional systems cannot prograde and the sediments are preferentially trapped close to the margin; on the other hand, the fine-grained, longitudinal fluvial system or a lacustrine system occupy the locus of the maximum subsidence and the hanging-wall-linked systems gain a wider spread (Fig. 5.2; Blair and Bilodeau, 1988; Frostick and Steel, 1993c; Leeder and Gawthorpe, 1987; Mack and Seager, 1990;). In contrast, during periods of relative quiescence, the footwall-linked alluvial systems define a larger drainage basin area with basin wide progradation and concomitant shifting of the axial system towards the hanging wall dip slope and shrinking of the hanging-wall linked alluvial systems (Fig. 5.2). However, relative size of the drainage basin in the footwall and hanging-wall also depends on the volume of sediment shed which is a function of the relative rates of uplift of the footwall and hanging-wall (Mack and Seager, 1990). Successive occurrences of narrow and wide packages of transverse alluvial deposits interlayered with longitudinal alluvial deposits thus have the potential to unravel the tectonic history of the basin.

Subsidence rates in active extensional basins are high, with the development of marked topography, which inhibits carbonate ramp development, but can promote the growth of rimmed shelves and isolated buildups. Due to footwall uplift the rate of siliciclastic sediment input may be high, particularly in humid climates. Carbonate platforms of all types are thus restricted to drowned rifts with low clastic supply, arid settings where clastic input is only periodic, or to intrabasinal highs isolated from marginal clastic sediment input (Burchette and Wright, 1992). Locally carbonates may be intercalated with shallow-marine fan-delta sediments (Burchette, 1988).

The disposition of different depositional systems and their preservation are, however, controlled by the relative rates of hanging wall subsidence and sedimentation (Ravnas and Steel, 1998). When hangingwall subsidence is faster than sedimentation, the fault becomes a significant morphological step, which affects the depositional systems and the sediment thickness on the two margins (Fig. 5.4). On the other hand, if the subsidence rate does not exceed the sedimentation rate, the same system is maintained across the fault (Fig.5.4). In the latter case the growth fault will only control the differential thickness between the two walls of the fault, because the sediments are continuously leveling the fault scarp (Doglioni et al., 1998). The occurrence of lateral facies variations within syntectonic deposits is mainly related to superficial effects of a growing

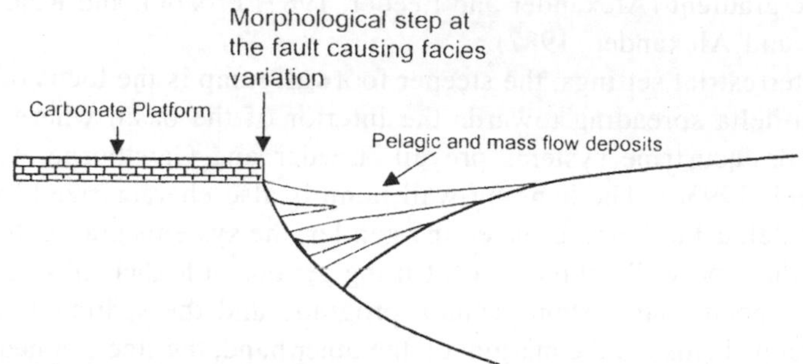

Fig. 5.4: Disposition of different systems in a rift basin when the rate of fault-controlled subsidence is greater than the rate of sedimentation (modified from Doglioni et al., 1998).

tectonic structure. Marine rift basins in arid climates commonly become restricted (e.g. Tertiary of Red Sea), a characteristic accentuated by their compartmentalization, so that ramps in such settings may be dominated by hypersaline facies. When the sedimentation rate is lower than the total subsidence rate, submarine depositional environments may develop a carbonate platform in the footwall and pelagic sedimentation in the hanging-wall (Fig. 5.4). A subaerial environment with negative total subsidence may experience erosion of the footwall and alluvial or lacustrine sedimentation in the hanging-wall. Narrow facies distribution and boundaries characterize subaerial settings of grabens where sedimentation does not compensate the tectonic offset. Where the sedimentation rate is higher than the subsidence rate, the topographic gradient is levelled by sedimentation, both in subaerial and submarine environments. As the sedimentation exceeds the tectonic rates, the topography becomes less and less irregular. The sedimentation rate controls the surface morphology during the growth of different tectonic structures.

Many rift basins pass through an initial phase of domal uplift due to upwelling of hot mantle plumes (active rifts, Olsen and Morgan, 1995). Rifts develop following doming due to stretching of the extrados as well as due to isostatic compensation (Cloos, 1939; White and Mckenzie, 1988). Early doming leads to diversion of river drainage away from the site of rifting, extensive volcanic activity and pronounced erosion (Frostick and Steel, 1993c). The basin fill of this type of rift is thus likely to be floored by an erosional unconformity and the succession is dominated by organic-rich lacustrine deposit and volcaniclastics with little or no siliciclastics (Cohen, 1989) and may contain a basal deposit recording radial drainage pattern (Cox, 1989; Frostick and Read, 1989). In other kind of rifts, fault-controlled subsidence takes place first, followed by magmatic intrusion and doming (passive rifts, Olsen and Morgan, 1995). The basin-fill succession of these rifts shows a basal siliciclastic package reflecting centripetal drainage followed by an erosional unconformity that forms during doming (Frostick and Steel, 1993c). Both the types of rifts, however, at the terminal phase of their history suffer regional subsidence due to cooling and contraction termed post-rift subsidence. The strata that accumulate during post-rift subsidence drape the whole basin. Most ramps associated with extensional basins develop during post-rift subsidence stage when subsidence becomes largely flexural; high subsidence rates are restricted to the basin center, and relief is reduced or infilled

Sedimentary Signatures of Strike-Slip Basins

Movement of crustal blocks along strike slip faults may induce localized subsidence along the length of the fault in areas of step-over and fault-bend, forming sedimentary basins (Fig. 5.1; Mann et al., 1983). Subsidence, in most cases, is associated with normal faulting, although flexural sag may also accompany (Nilsen and Sylvester, 1995). Strike slip basins are generally small and short-lived compared to those that develop in other plate tectonic settings. Basins range in size from small sag ponds to rhombochasms as wide as 50 km (Nilsen and Sylvester, 1995). Basin length-to-width ratios are typically 4:1 with a range from 1 to 10:1. The basins are asymmetric in profile, with their structurally deepest parts close to and sub-parallel to the syndepositionally most active strike slip margin. The basins contain very thick sedimentary successions compared to their real dimensions. The basin fill is characterized by high rates of sedimentation and is derived from multiple basin margin sources that change through time as a result of continuous lateral movement along basin margin faults. Syntectonic unconformities are common in the basin fill along with abundant synsedimentary slumping and deformation, possibly in response to basin wide shaking from earthquakes along basin margin faults. The basin is dominantly filled by axial drainage, sub-parallel to the principal displacement zones. The depocenteres migrate in the same directions as source terranes along the principal displacement zones; these migration directions are generally opposite to the directions of axial sediment transport. The basin fills are typically non-marine, characterized by diverse depositional systems, including talus, landslide, alluvial fan, braided and meandering fluvial, deltaic, fan delta, shoreline, shallow and deep lacustrine. Small debris-flow dominated alluvial fans containing coarse sedimentary breccia and conglomerate form along the syndepositionally active principal displacement zones. Along the inactive or less active margins are larger stream-flow dominated alluvial fans and fluvial deposits that contain finer-grained conglomerates and little breccia.

The sedimentary fill of strike-slip basins show repetitive, basin wide, upward coarsening successions. These successions seem to be tectonically controlled by basin wide changes in base level that induce progradation of marginal and axial coarse grained facies over axial finer-grained lacustrine and related facies.

The different criteria (Nilsen and Sylvester, 1995) that may be used to recognize strike-slip basins are:

1. Mismatches across basin margins signifying fault displacement contemporaneous with basin filling rather than post-depositional displacement.
2. Longitudinal and lateral basin asymmetry
3. Episodic rapid subsidence.
4. Abrupt lateral facies changes and local unconformities.
5. Marked contrasts in stratigraphy, facies geometry and unconformities among different basins in the same region.

Sedimentary Signatures of Foreland Basin

Foreland basins are asymmetrical moats that form in front of an orogenic belt in response to flexural subsidence of the crust as a result of supracrustal thrust loading (Allen and Homewood, 1986; Miall, 1995a). Narrow and deep basins develop on a weak crust, whereas strong crust produces basins that are wide and shallow (Cant and Stockmal, 1993). The basin may have discrete compartments or sub-basins (Lucchi, 1986; Fig. 5.5). Sediments are shed from the evolving orogenic belt.

The foreland basin fill strata show evidences of deformation (folding and/or thrusting, Doglioni and Prosser, 1997; Massari et al., 1993) close to the active margin that die out basinward. Blind thrusts may underlie anticlines shown by the basinal strata. Synsedimentary tectonism leads to development of syntectonic progressive and angular unconformities within the basin fill (Fig. 5.6; Riba, 1976, Anadaon et al., 1986) and disconformities form on the uplifted crests of anticlines (Doglioni and Prosser, 1997). Cumulative wedge systems or progressive unconformities indicate a depositional surface that is tilted by uplift of one side and the subsidence of the other, with no interruption of sedimentation. Such features would be generated over an anticlinal flank and would especially well develop on the flanks of overfolds and overthrusts (Riba, 1976).

Coarse-grained alluvial fans characterize the active margin of the basin, that passes into sand-dominated fluvial systems or fan delta basin-ward and in the basin interior marine depositional systems prevail. During periods of

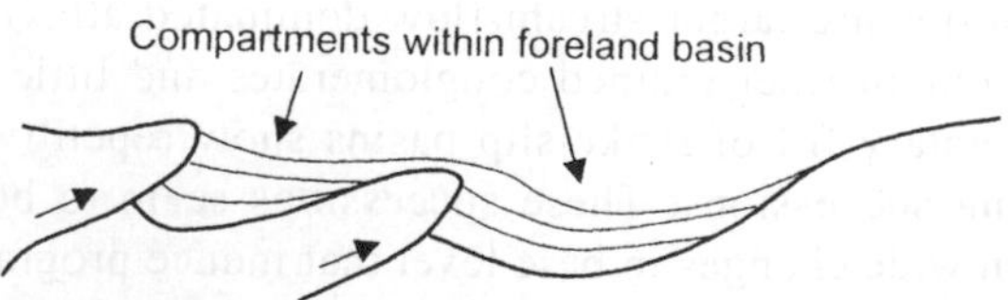

Fig. 5.5: Generation of compartments within a foreland basin due to periodic thrust propagation (modified from Lucchi, 1986).

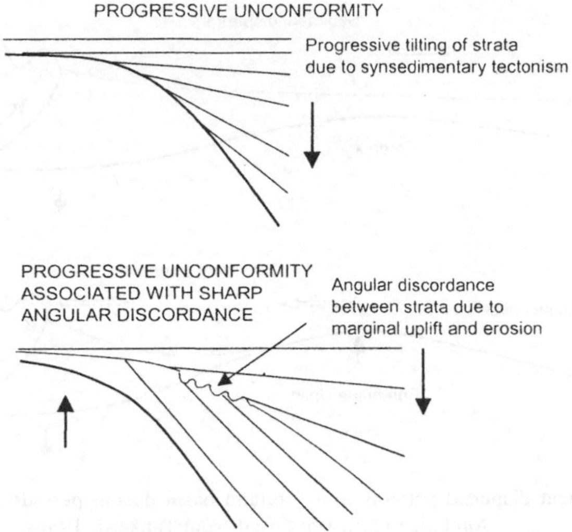

Fig. 5.6: Diagram showing a stratal package in which successive older strata are progressively more inclined indicating synsedimenatry tectonism. The whole package defines what is known as progressive unconformity or cumulative wedge system (modified from Riba, 1976).

relative quiescence, foreland basins receive a large sediment supply from the fold-thrust belt and transverse drainage extends across the entire basin (over-filled stage). In contrast, axial drainage may occupy the deeper part of the basin fed by transverse drainage systems during periods of high tectonic subsidence (under-filled stage).

Flexural subsidence of the foreland is also associated with bulging of the crust far away from the orogenic belt and is termed as forebulge. Forebulge also supplies sediments to the basin from the opposite direction, if uplifted sufficiently to form positive areas (Fig. 5.7). In the case of a subaerial, fluvial basin low gradient alluvial systems are usually present along the margin of the forebulge that supplies coarse clastics to the basin interior transversely. In the case of a subaqueous basin, the forebulge may serve as a shoal area, with deposition of offshore barrier-type islands, or as a region of enhanced wave and tide reworking, or as a carbonate bank. Detailed study of the stratigraphic onlap /offlap, wedging and pinchout relationships within and adjacent to the forebulge can provide a great deal of information with regard to basin history.

In marine foreland basins, carbonate ramps commonly form as linear belts along the peripheral bulge where slope gradient and subsidence rates are lowest. This area, being isolated by the foredeep from the main siliciclastic source area in the thrust front, experiences relatively low rates of terrigenous sediment supply. The forebulge experiences episodes of uplift and subsidence in response to thrust emplacement. As a result, the ramps associated with the

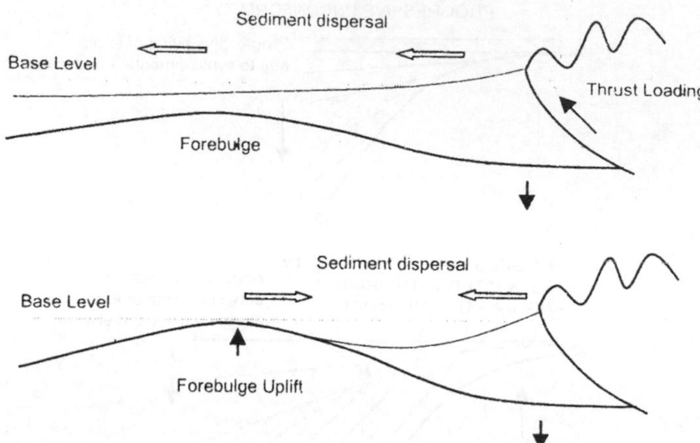

Fig. 5.7: Sediment dispersal patterns in a foreland basin during periods of active thrusting and forebulge uplift (modified from Tankard, 1986).

developing bulge may be subject to repeated episodes of uplift and drowning producing complex karsted unconformities (Burchette and Wright, 1992).

Foreland basin fill usually shows successive cycles of gradual shoaling from deep marine to shallow marine and non-marine. Transgressive-regressive cycles and inter-tonguing of marine and non-marine strata are related to periodic thrust loading. During periods of thrusting and enhanced subsidence rate, sedimentation from marine or axial fluvial systems prevail over wider areas of the basin, while the transverse fluvial systems are localized along the margin. As thrusting ceases, sediments from the newly uplifted orogenic belt are shed into the basin through widespread transverse fluvial systems. Erosional removal reduces the lithospheric load and as a consequence, the basin undergoes isostatic uplift and erosion producing unconformities.

REFERENCES

Ager, D. V., 1963, Principles of Paleoecology. New York, McGraw Hill Book Co. Inc., 371 p.

Aigner, T., 1982, Calcareous tempestites: Storm-dominated stratification in Upper Muschelkalk Limestones (Middle Trias, SW-Germany), p. 180-198. *In*: Cyclic and Event Stratification, (G. Einsele and A. Seilacher, eds.). Berlin-Heidelberg-New York, Springer-Verlag, 536 p.

Aigner, T., 1985, Storm Depositional Systems: Dynamic Stratigraphy in Modern and Ancient Shallow-marine Sequences. Lecture Notes in Earth Sciences-3. Berlin-Heidelberg-New York, Springer-Verlag, 174p.

Alam, M. M., Crook, W. A. K. and Taylor, G., 1985, Fluvial herringbone cross-stratification in a modern tributary mouth bar, Coonamble, New South Wales, Australia. *Sedimentology*, **32**:235–244.

Alam, M. S., Keppens, E. and Paepe, R., 1997, The use of oxygen and carbon isotope composition of pedogenic carbonates from Pleistocene paleosols in NW Bangladesh, as paleoclimatic indicators. *Quaternary Res.*, **16**:161–168.

Alexander, H. S., 1932, Pothole erosion. *J. Geology*, **40**:305–337.

Alexander, J. and Leeder, M. R., 1987, Active tectonic control on alluvial architecture. Tulsa. Oklahoma, p. 243-252. *In*: Recent Developments in Fluvial Sedimentology, (R.M. Ethridge, R.M. Flores, and M.D. Harvey, eds.). Tulsa, Oklahoma, SEPM Spec. Publ., 39.

Allen, J. R. L., 1964, Studies in fluviatile sedimentation: six cyclothems from the Lower Old Red Sandstone, Anglo-Welsh Basin. *Sedimentology*, **3**:163–198.

Allen, J. R. L., 1966, On bedforms and paleocurrents. *Sedimentology*, **6**:153–190.

Allen, J. R. L., 1970, Physical Processes of Sedimentation. London, George Allen & Unwin, 248 p.

Allen, J. R. L., 1971, Mixing of turbidity current heads, and its geologic implications. *J. Sediment. Petrology*, **41**:97–113.

Allen, J. R. L., 1973a, Features of cross-stratified units due to random and other changes in flow. *Sedimentology*, **20**:189–202.

Allen, J. R. L., 1973b Compressed structures (patterned ground) in Devonian pedogenic limestones. *Nature*, **243**:84–86.

Allen, J. R. L., 1974a, Sedimentology of the Old Red Sandstone (Siluro-Devonian) in the Clee Hill area, Shropshire, England. *Sediment. Geology*, **12**:73–167

Allen, J. R. L., 1974b, Studies in fluviatile sedimentation: implications of pedogenic carbonate units, Lower Old Red Sandstone, Anglo-Welsh outcrop. *Geol. Jour.*, **9**:181–208.

Allen, J. R. L., 1980, Sandwaves: a model of origin and internal structures. *Sediment. Geology*, **26**:281–328.

Allen, J. R. L., 1982, Sedimentary Structures: Their Character and Physical Basis. Vols. 1 & 2. Amsterdam, Elsevier Pub. Co., 593 and 663 p.

Allen, J. R. L., 1985, Principles of Physical Sedimentology. London, George Allen & Unwin, 272 p.

Allen, J. R. L., 1986, Earthquake magnitude-frequency, epicentral distance and soft-sediment deformation in sedimentary basins. *Sediment. Geology*, **46**:67–75.

Allen, P.A. and Homewood, P., Eds., 1986, Foreland Basins. IAS Spec. Publ., 8, Oxford, Blackwell Scientific Publications, 453 p.

Allen, Philip A and Allen, J. R., 1990, Basin Analysis: Principles and Applications. Oxford-London-Edinburgh-Boston-Melbourne-Paris-Berlin-Viena, Blackwell Scientific Publications, 450 p.

Anadon, P., Cabrera, L., Colombo, F., Marzo, M. and Riba, O., 1986, Syntectonic intraformational unconformities in alluvial fan deposits, eastern Ebro basin margins (NE Spain) p. 259–271: IAS Spec. Publ. 8, Oxford, Blackwell Scientific Publications.

Anand, A. and Jain, A. K., 1987, Earthquakes and deformational structures (seismites) in Holocene sediments from the Himalayan-Andaman Arc, India. *Tectonophysics*, **133**:105–120.

Angevine, C. L., Heller, P. L. and Paola, C., 1990, Quantitative Sedimentary Basin Modeling. Am Assoc. Petrol. Geol., Continuing Education Course Note Series No. 32, 133 p.

Arndorff, L., 1993, Lateral relations of deltaic paleosols from the Lower Jurassic Ronne Formation on the Island of Bornholm, Denmark. *Paleogeo. Paleoclamato., Paleoecology*, **100**:235–250.

Arnott, R.W.C., 1993, Quasi-palnar laminated sandstone beds of the Lower Cretaceous Bootlegger member, north-central Montana: evidence of combined-flow sedimentation. *J. Sediment. Petrology*, **63**:488–494.

Arnott, R.W.C., 1995, The parasequence definition – are transgressive deposits inadequately addressed? *J. Sediment. Research*, **B62**, No.1, p. 1–6.

Ashley, G.M., 1990, Classification of large-scale subaqueous bedforms: A new look at the old problem. *J. Sediment. Petrology*, **60**:160–172.

Astin, T. R., and Rogers, D. A., 1991, 'Subaqueous shrinkage cracks' in the Devonian of Scotland reinterpreted. *J. Sediment. Petrology*, **61**:850–859.

Atwater, B. F. and Moore, A. L., 1992, A tsunami about 1000 years ago in Puget Sound, Washington. *Science*, **258**:1614–1617.

Balance, P. F. and Reading, H. G., 1980, Sedimentation in Oblique-Slip Mobile Zones. IAS Spec. Publ. No.4, Oxford, Blackwell Scientific Publications, 265 p.

Banks, N. L., 1973, The origin and significance of some downward-dipping cross-stratified sets. *J. Sediment. Petrology*, **43**:423–427.

Barrell, J., 1917, Rhythms and the measurements of geologic time. *Geol. Soc. Am., Bull.*, **28**:745–904

Bates, R. L. and Jackson, J. A., 1984, Dictionary of Geological Terms, 3rd edition. Anchor Press / Doubleday, 571 p.

Bathurst, R. G. C., 1966, Boring algae, micrite envelops and lithification of molluscan biosparite. *Geol. Jour.*, **5**:15–32.

Bathurst, R. G. C., 1987, Diagenetically enhanced bedding in argillaceous platform limestone: stratified cementation and selective compaction. *Sedimentology*, **34**:749–778.

Bathurst, R. G. C., 1991, Pressure-dissolution and limestone bedding: the influence of stratified cementation, p. 450–463. *In*: Cycles and Events in Stratigraphy, (G.Einsele, Werner Ricken and Adolf Seilacher, eds.). Berlin-Heidelberg-New York, Springer-Verlag, 955p

Beach, D.K. and Ginsburg, R.G., 1980, Facies succession of Pliocene-Pleistocene carbonates, northwestern Great Bahama Bank. *Am. Assoc. Petrol. Geologists, Bull.*, **64**:1634–1642.

Beeunas, M. A. and Knauth, L. P., 1985, Preserved stable isotopic signature of subaerial diagenesis in the 1.2 b.y., Mescal Limestone, Central Arizona: implications for the timing and development of a terrestrial plant cover. *Geol. Soc. Am., Bull.*, **96**:737–745.

Bertrand-Sarfati, J. and Moussine-Pouchkine, A., 1983, Pedogenetic and diagenetic fabrics in Upper Proterozoic Sarnye're' Formation (Gourma, Mali). *Precambrian Res.*, **20**:225–242.

Bertrand-Sarfati, J. and Raaben, M. E., 1971, Comparison des ensembles stromatolitiques du Precambrien uperiour du Sahara Occidental et de l'oural. *Bull. Soc. Geol. France, Ser.* **12(2)**:364–371.

Bertrand-Sarfati, J. and Trompette, R., 1976, Use of stromatolites for intrabasinal correlation: example from the late Proterozoic of the northwestern margin of the Taoudenni Basin. *In*: Stromatolites, Developments in Sedimentology, 20, (M. R. Walter, ed.). Amsterdam-Oxford-New York, Elsevier Sci. Pub. Development Co., 790 p.

Bestland, E. A., 1997, Alluvial terraces and paleosols as indicators of early Oligocene climate change (John Day Formation, Oregon). *J. Sediment. Research*, **67A**:840–855.

Bestland, E. A., Retallack, G. J., and Swisher III, Carl C., 1997, Stepwise climate change recorded in Eocene-Oligocene paleosol sequences from Central Oregon. *J. Geology*, **105**:153–172.

Bhattacharya, A. K., 1997, On the origin of non-tidal flaser bedding in point bar deposits of the river Ajay, Bihar and West Bengal, N.E. India. *Sedimentology*, **44**:973–975.

Blair, T. C. and Bilodeau, W. L., 1988, Development of tectonic cyclothems in rift, pull-apart, and foreland basins: Sedimentary response to episodic tectonism. *Geology*, **16**:517–520.

Bluck, B.J., 1965, The sedimentary history of some Triassic conglomerates in the Vale of Glamargon, South Wales. *Sedimentology*, **4**:225–245.

Blundell, D. J., 1991, Some observations on basin evolution and dynamics. *J. Geol. Soc., London*, **148**:789–800.

Bondevik, S., Svendsen, J. I., and Mangerud, J., 1997, Tsunami sedimentary facies deposited by the Storegga tsunami in shallow marine basins and coastal lakes, Western Norway. *Sedimentology*, **44**:1115–1131.

Bott, M. H. P., 1995, Mechanisms of rifting: Geodynamic modeling of continental rift systems, p. 27–92. *In*: *Continental Rifts: Evolution, Structure, Tectonics*, (K.H. Olsen, ed.). Elsevier Scientific Publ.

Bouma, A. H., 1962, Sedimentology of Some Flysch Deposits. A Graphic Approach to Facies Interpretation. Amsterdam, Elsevier Scientific Publ. 168 p.

Bouma, A. H. and Hollister, C. D., 1973, Deep ocean basin sedimentation, p. 79–118. *In*: Chm., Turbidites and Deep-Water Sedimentation. (G. V. Middleton and A. H. Bouma, eds.). SEPM Pacific Section, Short Course Lecture Notes, 158 p.

Bouma, A. H., 1969, Methods for the Study of Sedimentary Structures. New York, John Wiley Interscience, 446 p.

Bourgeois, J., 1980, A transgressive shelf sequence exhibiting hummocky cross-stratification: The Cape Sebastian Sandstone (Upper Cretaceous), southwestern Oregon. *J. Sediment. Petrology*, **50**:681–702.

Bourgeois, J. and Leithold, E. L., 1984, Wave-worked conglomerates: Depositional processes and criteria for recognition, p. 331–343. *In*: *Sedimentology of Gravels and Conglomerates*, (Koster, E. H. and Steel, R. J., eds.). Can. Soc. Petrol. Geologists, Memoir 10, 441 p.

Bourgeois, J., Hansen, T. A., Wilberg, P. L. and Kauffman, E. G., 1988, A tsunami deposit at the Cretaceous-Tertiary boundary in Texas. *Science*, **241**:567–570.

Brenchley, P. J., 1985, Storm influenced sandstone beds. *Modern Geology*, **9**:369–396.

Brenchley, P. J., Newell, G. and Stainstreet, I. G., 1979, A storm-wave origin for sandstone beds in the epicontinental platform sequence, Ordovician, Norway. Sediment. *Geology*, **22**:185–217.

Brenner, R. L. and Davies, D. K., 1973, Storm-generated coquinoid sandstone: genesis of high-energy marine sediments from the Upper Jurassic of Wyoming and Montana. *Geol. Soc. Am., Bull.*, **84**:1685–1698.

Brewer, R., 1964, Fabric and mineral Analysis of Soils. New York, John Wiley & Sons, 470 p.

Bridge J. S. and Leeder M. R., 1979, A simulation model of alluvial stratigraphy. *Sedimentolgy*, **29**:617–644.

Bridge, J. S., 1993, Description and interpretation of fluvial deposits: a critical perspective. *Sedimentology*, **40**:801–810.

Bridges, P. H., 1972, The significance of tool marks on a Silurian erosional furrow. *Geol. Magazine*, **109**:405–410.

Bromley, R. G. and Asgaard, U., 1979, Triassic fresh water ichnocoenoses from Carlsberg Fjord, East Greenland. *Paleogeo. Paleoclimato. Paleoecology*, **28**:39–80.

Bromley, R.G., 1975, Trace fossils at omission surfaces, p. 399–428. *In*: *The Study of Trace Fossils*, (R.W. Frey, ed.). New York, Springer-Verlag, 562 p.

Brown, M. A., Archer, A. W. and Kvale, E. P., 1990, Neap-spring tidal cyclicity in laminated carbonate channel-fill deposits and its implications: Salem Limestone (Mississippian), south-central Indiana, U.S.A. *J. Sediment. Petrology*, **60**:152–159.

Brunsden, D., Doornkamp, J. C., Green, C. P., and Jones, D. K. C., 1976, Tertiary and Cretaceous sediments in solution pipes in the Devonian Limestone of South Devon, England. *Geol. Magazine*, **113**:441–447.

Buick, R., Dunlop, J. S. R. and Groves, D. I., 1981, Stromatolite recognition in ancient rocks: An appraisal of irregularly laminated structures in an early Archaean chert-barite unit from North Pole, Western Australia. *ALCHERINGA*, **5**:161–181.

Buol, S. W., Hole, F. D. and McCracken, R. J., 1980, Soil Genesis and Classification. The Iowa State University, Ames, 360 p.

Burchette, T. P. and Wright, V. P., 1992, Carbonate ramp depositional systems. *Sediment. Geology*, **79**:3–57.

Burchette, T.P, 1988, Tectonic control on carbonate platform facies distribution and sequence development: Miocene Gulf of Suez. *Sediment. Geology*, **59**:179–204.

Burns, B. A., Heller, P. L., Marzo, M. and Paola, C., 1997, Fluvial Response in a Sequence Stratigraphic Framework: Example from the Montserrat Fan Delta, Spain. *J. Sediment. Research*, **67**:311–321.

Burri, P., Du Dresnay, R. and Wagner, C. W., 1973, Tepee structures and associated diagenetic features in intertidal carbonate sands (Lower Jurassic, Morocco). *Sediment. Geology*, **9**:221–228.

Buurman, P., 1980, Paleosols in the Reading Beds (Paleocene) of Alum Bay, Isle of Wight, U.K. *Sedimentology*, **27**:593–606.

Burst, J. F., 1965, Subaqueously formed shrinkage cracks in clay. *J. Sediment. Petrology*, **35**:348–353.

Busby, C. J. and Ingersoll, R. V., editors, 1995, Tectonics of Sedimentary Basins. Oxford, Blackwell Scientific Publications, 580 p.

Button, A. and Tyler, N., 1981, The character and economic significance of Precambrian paleoweathering and erosion surfaces in southern Africa, p. 686–709. *In*: Economic Geology, 75th Anniversary volume, (Brain Skinner, ed.).

Buxton, T. M. and Sibley, D. F., 1981, Pressure solution features in a shallow buried limestone. *J. Sediment. Petrology*, **51**:19–26.

Calvet, F. and Julia, R., 1983, Pisoids in the caliche profiles of Tarragona (N.E. Spain), p. 456–473. *In*: Coated Grains, (T. Peryt, ed.). New York, Springer-Verlag.

Campbell, C. V., 1966, Truncated wave-ripple laminae. *J. Sediment. Petrology*, **36**:825–828.

Campbell, C. V., 1967, Lamina, Lamina set, Bed and Bed set. *Sedimentology*, **8**:7–26.

Cant D. J. and Stockmal, G. S., 1993, Some controls on sedimentary sequences in foreland basin: examples from the Alberta basin, p. 49–65, IAS Spec. Publ., 20, , Oxford, Blackwell Scientific Publications.

Capo, Rosemary, C., Whipkey, Charles, E., Blache're, Jean R. and Chadwick, Oliver A., 2000, Pedogenic origin of dolomite in a basaltic weathering profile, Kohala peninsula, Hawaii. *Geology*, **28**(3): 271–274.

Catuneanu, O., Wills, A. J. and Miall, A. D., 1998, Temporal significance of sequence boundaries. *Sediment. Geology*, **121**:157–178.

Cerling, T. E, 1984, The stable isotopic composition of modern soil carbonate and its relationship to climate. *Earth and Planetary Sci. Letters*, **71**:229–240.

Cerling, T. E., 1991, Carbon dioxide in the atmosphere: evidence from Cenozoic and Mesozoic paleosols. *Am. J. Science*, **291**:377–400.

Cerling, T. E., Bowman, J. R. and O'Neil, J. R., 1988, An isotopic study of fluvial- lacustrine sequence: the Plio-pleistocene Koobi Fora sequence, East Africa. *Paleogeo. Paleoclimato. Paleoecology*, **63**:335–356.

Cerling, T. E., Quade, J., Wang, Y. and Bowman, J. R., 1989, Carbon isotopes in soils and paleosols as ecology and paleoecology indicators. *Nature*, **341**:138–139.

Chafetz, H. S. and Buczynski, C., 1992, Bacterially induced lithification of microbial mats. *Palaios*, **7**:277–293.

Chakraborty, C., 1993, Morphology, internal structure and mechanics of small longitudinal (seif) dunes in an aeolian horizon of the Proterozoic Dhandraul Quartzite, India. *Sedimentology*, **40**:79–85.

Chakraborty, C., 1995, Gutter casts from the Proterozoic Bijoygarh Shale Formation, India: their implication for storm-induced circulation in shelf settings. *Geol. Journal*, **30**:69–78.

Chakraborty, C. and Bose, P. K., 1992, Ripple/dune to upper stage plane bed transition: some observations from the ancient record. *Geol. Journal*, **27**:349–359.

Cherns, L., 1982, Paleokarst, tidal erosion surfaces and stromatolites in the Silurian Eke Formation of Scotland, Sweden. *Sedimentology*, **29**:819–833.

Choquette, P. W., and Pray, L. C., 1970, Geological nomenclature and classification of porosity in sedimentary carbonates. *Am. Assoc. Petrol. Geologists, Bull.*, **54**:207–250.

Choquette, P. W. and James, N. P., 1988, Introduction, p. 1–21, *In*: Paleokarst, (N.P. James and P.W. Choquette, eds.). New York, Springer-Verlag, 576 p.

Christie-Blick, N. and Biddle, K. T., 1985, Deformation and basin formation along strike-slip faults, p. 1–34. *In*: Strike-slip Deformation, Basin Formation, and Sedimentation. (K.T. Biddle and N. Christie-Blick, eds.). Tulsa, Oklahoma, SEPM Spec. Publ. No., 37, 386 p.

Ciari, P. A., Pierre, F. D. and Martire, L., 1995, Discontinuities in carbonate successions: identification, interpretation and classification of some Italian examples. *Sediment. Geology*, **100**:97–121.

Clifton, H. E., 1973, Pebble segregation and bed lenticularity in wave-worked versus alluvial gravel. *Sediment. Geology*, **20**:173–187.

Clifton, H. E., 1981, Progradational sequences in Miocene shoreline depsoits, southeastern Caliente Range, California. *J. Sediment. Petrology*, **51**:165–184.

Clifton, H. E., 1988, Sedimentologic relevance of convulsive geologic events, p. 1–5. *In*: Sedimentologic Consequences of Convulsive Geologic Events, (H. Edward Clifton, ed.). Special Paper 229, *Geol. Soc. Am.*, 157 p.

Cloos, H., 1939, Hebung-Spaltung-Vulkanismus: *Geologische Rundschau*, **30**: 405–525.

Cloud, P. E. and Semikhatov, M. A., 1969, Proterozoic stromatolite zonation. *Am. J. Science*, **267**:1017–1061.

Cohen, Andrew S., 1982, Paleoenvironments of root casts from the Koobi Fora Formation, Kenya. *J. Sediment. Petrology*, **52**:401–414.

Cohen, C.R., 1989, Plate tectonic model for the Oligo-Miocene evolution of the western Mediterranean. *Tectonophysics*, **68**: 283–311.

Cohen, C.R., 1989, Facies relationships and sedimentation in large rift lakes: examples from Lakes Turkana and Tanganiyka, p. 65–80, IAS Spec. Publ., 13, London, Blackwell Scientific Publications.

Coleman, M.L., 1993, Microbial processes: controls on the shape and composition of the carbonate concretions. *Marine Geology*, **113**:127–140.

Coleman, M.L. and Raiswell, R., 1995, Source of the carbonate and origin of zonation in pyritiferous carbonate concretions: evaluation of a dynamic model. *Am. Jour. Science*, **295**:282–308.

Collinson, J. D., 1978, Vertical sequence and sand body shape in alluvial sequences, p.577–586. *In*: Fluvial Sedimentology. (A. D. Miall, ed.). Calgary, Alberta, Can. Soc. Petrol. Geologists, Memoir **5**:859 p.

Collinson, J. D., and Thompson, D. B., 1982, Sedimentary Structures. London, George Allen & Unwin, 194 p.

Coward, M. P., Dewey, J. F. and Hancock, P. L., Eds., 1987, Continental Extensional Tectonics. *Geol. Soc.*, London, Spec. Publ., 28:637 p.

Cox, K. G., 1989, The role of mantle plumes in the development of continental drainage patterns. *Nature*, **342**:873–876.

Craig, D.H., 1988, Caves and other features of Permian karst in San Andres Dolomite, Yates Field reservoir, West Texas, p. 342–363, *In*: Paleokarst (James, P. Noel and P. W. Choquette, eds.). New York, Springer-Verlag, 576 p.

Crime, T. P., 1975, The stratigraphical significance of tracefossils. *In*: The Study of Trace Fossils. (R. W. Frey, ed.). Springer-Verlag, 562 p.

Currie, B. S., 1997, Sequence stratigraphy of nonmarine Jurassic-Cretaceous rocks, central Cordilleran Foreland-basin system. *Geol. Soc. Am., Bull.*, **109**:1206–1222.

Curtis, C. D., 1990, Aspects of climatic influence on the clay mineralogy and geochemistry of soils, paleosols and clastic sedimentary rocks. *J. Geol. Society*, London, **147**:351–357.

Dash, B., Sahu, K. N. and Bowes, D. R., 1987, Geochemistry and original nature of Precambrian khondalites in the Eastern Ghats, Orissa, India. Trans. Roy. Soc. Edinburgh, *Earth Sciences*, **78**:115–127.

Davies, I. C. and Walker, R. G., 1974, Transport and deposition of re-sedimented conglomerates: the Cape Enrage Formation, Cambro-Ordovician, Gaspe, Quebec. *J. Sediment. Petrology*, **44**:1200–1216.

Davies, Richard A., Knowles, Stephen C. and Bland, Michael J., 1989, Role of hurricanes in the Holocene stratigraphy of estuaries: Examples from the Gulf Coast of Florida. *J. Sediment. Petrology*, **59**:1052–1061.

Dawson, A. G., Long, D. and Smith, B. E., 1988, The Storegga Slides: Evidence from eastern Scotland for a possible tsunami. *Marine Geology*, **82**:271–276.

DeBoer, P. L., 1979, Convolute lamination in modern sands of the estuary of the Oosterschelde, the Netherlands, formed as the result of entrapped air. *Sedimentology*, **26**:283–294.

DeCelles, Peter G., 1988, Deposits of middle Tertiary convulsive geologic event, San Emigdio Range, southern California, p. 127–142, *In*: Sedimentologic Consequences of Convulsive Geologic Events, (H. E. Clifton, ed.). Geol. Soc. Am,., Special Paper 229, 197 p.

DeRaaf, J. F. M. and Boersma, J. R., 1971, Tidal deposits and their sedimentary structures. Geol. Mijnbouw. **50**:479–504.

DeRaaf, J. F. M., Boersma, J. A. and Gelder, A., 1977, Wave-generated structures and sequence from a shallow marine succession, Lower Carboniferous, County Cook, Ireland. *Sedimentology*, **24**:451–483.

Desrochers, A. and James, N. P., 1988, Early Paleozoic surface and subsurface paleokarst: Middle Ordovician carbonates, Mingan Islands, Quebec, p.183-210, *In*: Paleokarst, (N. P. James, and P. W. Choquette, eds.). New York, Springer-Verlag, 576 p.

Doglioni, C. and Prosser, G., 1997, Fold uplift versus regional subsidence and Sedimentation rate. *Marine and Petroleum Geol.*, **14**:179–190.

Doglioni, C., Agostino, N.D. and Mariotti, G., 1998, Normal faulting vs regional subsidence and sedimentation rate. Marine and Petroleum Geology. **15**:737–750.

Doglioni, C. Besellini, A. and Vail, P. R., 1990, Stratal patterns: a proposal of classification and examples from the Dolomites. *Basin Res.*, **2**:83–95.

Dott, Jr., R. H., 1963, Dynamics of subaqueous gravity depositional processes. *Am. Assoc. Petrol. Geologists, Bull.*, **47**:104–128.

Dott, Jr., R. H., 1983, Episodic sedimentation - How normal is average? How rare is rare? Does it matter? *J. Sediment. Petrology*, **53**:5–23.

Dott, Jr., R. H., 1996, Episodic event deposits versus stratigraphic sequences—shall the twain never meet? *Sediment. Geology*, **104**:243–247.

Dott, Jr., R. H. and Bourgeois, J., 1982, Hummocky stratification: significance of its variable bedding sequences. *Geol. Soc. Am., Bull.*, **93**:663–680.

Driese, S. G., Srinivasan, K., Mora, C. I. and Stapor, F. W., 1994, Paleoweathering of Mississippian Monteagle imestone preceding development of a lower Chesterian transgressive systems tract and sequence boundary, middle Tennessee and northern Alabama. *Geol. Soc. Am. Bull.*, **106**:866–878.

Driscoll, Neal W., Weissel, Jeffrey K., and Goff, John A., 2000, Potential for largescalc submarine slope failure and tsunami generation along the U.S. mid-Atlantic coast. *Geology*, 28(5): 407–410.

Duchac, K. C. and Hanor, J. S., 1987, Origin and timing of the metasomatic silicification of an Early Archaean komatite sequence, Barberton Mountain Land, South Africa. *Precambrian Res.*, **37**:125–146.

Duke, W. L., 1984, Paleohydraulic analysis of hummocky cross-stratified sands indicates equivalence with wave-formed flat bed: Pleistocene Lake Bonneville deposits, northern Utah (Abstract), *Am. Assoc. Petrol. Geologists, Bull.*, **68**:472.

Duke, W. L., 1985, Hummocky cross-stratification, tropical hurricanes, and intense winter storms. *Sedimentology*, **32**:167–194.

Dula Jr. W. F., 1991, Geometric models of listric normal faults and rollover folds. *Am. Assoc. Petrol. Geologists, Bull.*,**75**:1609–1625.

Dunbar, C. O., and Rodgers, J., 1957, Principles of Stratigraphy. New York, John Wiley & Sons, 356 p.

Dzulynski, S., 1996, Erosional and deformational structures in single sedimentary beds: A genetic commentary. *Annales Societatis Geologorum Poloniae*, **66**:101–189.

Dzulynski, S. and Slaczka, A., 1958, Directional structures and sedimentation of the Krosno beds (Carpathian flysch). *Rocz. Pol.Tow. Geol.*, **28**:205–260.

Dzulynnski, S. and Walton, E. K., 1965, Sedimentary Features of Flysch and Greywackes. Amsterdam, Elsevier Pub. Co., 274 p.

Eder, F.W., 1982, Diagenetic redistribution of carbonate-a process in forming limestone-marl alternations (Devonian-Carboniferous), Rheinisches Schiefergebirge, Germany, p. 98–112, *In*: Cyclic and Event Stratification. (G. Einsele, and A. Seilacher, eds.). Berlin-Heidelberg-New York, Springer-Verlag, 536 p.

Einsele, G., 1982, Limestone-marl cycles (periodites): Diagenesis, significance, causes—a review, p. 8–53. *In*: Cyclic and Event Stratification. (G. Einsele and A. Seilacher, eds.). Berlin-Heidelberg-New York, Springer-Verlag, 536 p.

Einsele, G., 1992, Sedimentary Basins: Evolution, Facies, and Sediment Budget. Berlin, Springer-Verlag, 628 p.

Einsele, G. and Seilacher, A., eds., 1982, Cyclic and Event Stratification. Berlin-Heidelberg-New York, Springer-Verlag, 536 p.

Einsele, G. and Seilacher, A., 1991, Distinction of tempestites and turbidites. p. 377–382, *In*: Cycles and Events in Stratigraphy. (G. Einsele, W. Ricken, and A. Seilacher, eds.). Berlin-Heidelberg, Springer-Verlag, 955 p.

Einsele, G., Ricken, W. and Seilacher, A., eds., 1991, Cycles and Events in Stratigraphy. Berlin-Heidelberg-New York Springer-Verlag, 955 p.

Ekdale, A.A., Muller, L.N. and Noval, M.T., 1984, Quantitative ichnology of modern pelagic deposits in the abyssal Atlantic. Paleogeo. *Paleoclimato. Paleoecology*, **45**:189–223.

Embry, A. F., 1995, Sequence boundaries and sequence hierarchies: problems and proposals. *In*: Sequence Stratigraphy on the Northwest European Margin. (R.J. Steel, V.L. Felt, E.P. Johannessen, C. Mathieu, eds.). Norwegian Petroleum Society Spec. Publ. 5, Elsevier, Amsterdam, 1–11.

Emery, D. and Myers, K., editors, 1996, Sequence Stratigraphy. Oxford, Blackwell Scientific. Pub., 304 p.

Emiliani, C., 1955, Pleistocene temperatures. *J. Geology*, **63**:538–578.

Emiliani, C., 1972, Quaternary paleotemperatures and the duration of high-temperature intervals. *Science*, **178**:398–401.

Emiliani, C. and Shackleton, N. J., 1974, The Brunches epoch: isotopic paleotemperatures and geochronology. *Science*, **183**:511–514.

Esteban, M., 1974, Caliche textures and Microcodium. Boll. Soc. Geol. Italy, 92, Suppl. 1973, 105–125.

Esteban, M. and Klappa, C. F., 1983, Subaerial exposure environment, p. 1–54. *In*: Carbonate Depositional Environment. (Scholle, P., Bebout, D.G. and Moore, C.H., eds.). Am. Assoc. Petrol. Geologists, Memoir 33, 708 p.

Ethridge, F. G. and Wescott, W. A., 1984, Tectonic setting, recognition and hydrocarbon reservoir potential of fan-delta deposits, p. 217–235. *In*: Sedimentology of Gravels and Conglomerates. (E.H. Koster, and R.J. Steel, eds.). Calgary, Alberta, Can. Soc. Pet. Geologists, Memoir 10, 441 p.

Fairchild, I. J., 1980, Sedimentation and origin of a late Precambrian 'Dolomite' from Scotland. *J. Sediment. Petrology*, **50**:423–446.

Fairchild, I.F., Einsele, G. and Song, T., 1997, Possible seismic origin of molar tooth structures in Neoproterozoic carbonate ramp deposits, north China. *Sedimentology*, **44**:611–636.

Farrow, G. E., 1966, Bathymetric zonation of Jurassic trace fossils from the coast of Yorkshire. England. Paleogeo. Paleoclimato. *Paleoecology*, **2**:103–151.

Fastovsky, D. E. and McSweeney, K., 1987, Paleosols spanning the Cretaceous-Paleogene transition, eastern Montana and western North Dakota. *Geol. Soc. Am., Bull.*, **99**:66–77

Feistner, K. W. A., 1989, Petrographic examination and reinterpretation of concretionary carbonate horizons from the Kimmeridge Clay, Dorset. *J. Geol. Society, London*, **146**:345–350.

Figueiredo, A. G., Sanders, J. E and Swift, D. J. P., 1982, Storm-graded layers on inner continental shelves; Examples from southern Brazil and the Atlantic coast of the central United States. *Sediment. Geology*, **31**:171–190.

FitzPatrick, E. A., 1980, Soil: their formation, classification and distribution. London-New York, Longman Group Ltd., 353 p.

Folk, R. L., 1965, Petrology of Sedimentary Rocks. Hemphills', Austin, Texas, 159 p.

Ford, D., 1988, Characteristics of dissolutional cave systems in carbonate rocks, p. 25–57, *In*: Paleokarst, (N. P. James, and P. W. Choquette, eds.). Springer-Verlag, New York, 576 p.

Frey, R. W., 1970, Environmental significance of recent marine lebensspuren near Beaufort, North Carolina. *J. Sediment. Petrology*, **44**:507–519.

Frey, R. W., 1973, Concept in the study of biogenic sedimentary structures. *J. Sediment. Petrology*, **43**:6–19.

Frey, R. W., 1975, The realm of ichnology, its strengths and limitations, p. 13–38, *In*: The Study of Trace Fossils. (R. W. Frey, ed.). New York, Springer-Verlag, 562 p.

Frey, R. W., 1978, Behavioral and ecological implications of trace fossils, p. 43–66, *In*: Trace Fossil Concepts. (Basan, P.B. ed.). Tulsa, Oklahoma, SEPM Short Course No. 5, 181 p.

Frey, R. W. and Seilacher, A., 1980, Uniformity in marine invertebrate ichnology. *Lethaia*, **13**:183–207.

Frey, R. W. and Pemberton, S. G., 1984, Trace fossil models, p. 189–207. *In*: Facies Models, 2nd edition, (R.G. Walker, ed.) Geoscience Canada, Reprint Series 1.

Frey, R. W. and Pemberton, S. G. 1985, Biogenic structures in outcrops and cores. 1.Approaches to ichnology. *Canadian Petroleum Geology, Bull.*, **33**:72–115.

Frey, R. W., Pemberton, S. G. and Saunders, T. D.A., 1990, Ichnofacies and bathymetry: A passive relationship. *J. Paleontology*, **64(1)**:155–158.

Freytet, P. and Plaziat, J. C., 1982, Continental carbonate sedimentation and pedogenesis—late Cretaceous and Early Tertiary of southern France. *In*: Contribution to Sedimentology, (B. H. Purser, ed.). **12**:1–213, Schwelzertische Verlagsbuchhandlung, Stuttgart.

Friedman, G. M. and Sanders, J. F., 1974, Positive-relief bedforms on modern tidal flat that resemble molds of flutes and grooves; Implications for geopetal criteria and for origin and classification of bedforms. *J. Sediment. Petrology*, **44**:181–189.

Friedman, G. M., Sanders, J. E. and Kopaska-Merkel, D. C., 1992, Principles of Sedimentary Deposits: Stratigraphy and Sedimentology. New York-Toronto-Oxford-Singapore-Sydney, Macmillan Publishing Company, 717 p.

Friend, P. F., Johnson, N. M. and McRae, L. E., 1989, Time level plots and accumulation patterns of sediment sequences. *Geol. Magazine*, **126**:491–498.

Frostick, L. and Reid, I., 1987, A new look at rifts. *Geology Today*, p. 122–126.

Frostick, L. E. and Read, J.F., 1989, Is structure the main control of river drainage and sedimentation in rifts? *J. Afr. Earth Sci.*, **8**:165–182.

Frostick, L.E. and Reid, I., 1990, Structural control of sedimentation patterns and implications for the economic potential of the East African Rift basins. *J. Afr. Earth Sci.*, **10**:307–318.

Frostick, L. E. and Steel, R.J., eds., 1993a,. Tectonic Controls and Signatures in Sedimentary Succession. IAS Spec. Publ., 20, Blackwell Scientific Publications, 520 p.

Frostick, L. E. and Steel, R. J., 1993b, Sedimentation in divergent plate-margin basins p. 111–128. IAS Spec. Publ., 20, Blackwell Scientific Publications, 520 p.

Frostick, L. E. and Steel, R.J., 1993c, Tectonic signatures in sedimentary basin fills: an overview 1–9. IAS Spec. Publ., 20, London, Blackwell Scientific Publications, 520 p.

Gall, Q., 1992, Precambrian paleosols in Canada, *Can. J. Earth Science*, **29**:2530-2536.

Gall, Q. and Donaldson, J.A., 1990, The sub-Thelon Formation paleosol, Northwest Territories *In*: Current Research, Part- C, Geol. Survey of Canada, Paper 90–1C, 271–277.

Gay, A. L. and Grandstaff, D. E., 1980, Chemistry and mineralogy of Precambrian paleosols at Elliot Lake, Ontario, Canada. *Precambrian Res.*, **12**:349–373.

Gebelein, C. D. and Hoffman, P., 1973, Algal origin of dolomite laminations in stromatolitic limestone. *J. Sediment. Petrology*, **43**:603–613.

Gehling, J. G., 1999, Microbial mats in terminal Proterozoic siliciclastics: Ediacaran death masks. *Palaios*, **14**:40–57.

Gerdes, G., Krumbein, W. E. and Reineck, H. -E., 1985, The depositional record of sandy, versicoloured tidal flats (Mellum Island, southern North Sea). *J. Sediment. Petrology*, **55**:265–278.

Ghosh, P., 1994, Mesozoic stratigraphy and sedimentation in and around Jabalpur, Central India. Unpub. Ph.D. thesis, University of Calcutta, 162 p.

Gibbs, A., 1987, Development of extension and mixed-mode sedimentary basins. Geol. Soc., London, Spec. Publ. No. **28**:19–33.

Gilbert, Grove Karl, 1899, Ripple marks and cross-bedding. *Geol. Soc. Am., Bull.*, **10**:135–140.

Glaessner, M. F., Preiss, W. V. and Walter, M. R., 1969, Precambrian columnar stromatolites in Australia: morphological and stratigraphic analysis. *Science*, **164**:1056–1058.

Golani, P.R., 1989, Sillimanite-corundum deposits of Sonapahar, Meghalaya. *Precambrian Res.*, **43**:175–189.

Goldring, R., 1964, Trace fossils and sedimentary surface in shallow marine sediments. p. 136–143, *In*: Deltaic and Shallow Marine deposits, Developments in Sedimentology, 1, (L.M.J.U. Van Straaten, ed.). Amsterdam-Oxford-New York, Elsevier Scientific Pub. Development Co.,

Goldbery, R., 1982a, Paleosols of the Lower Jurassic Mishhor and Ardon Formations ('Laterite Derived Facies'), Makhtesh Ramon, Israel. *Sedimentology*, **29**:669–690.

Goldbery, R., 1982b, Structural analysis of the soil microrelief in paleosols of the Lower Jurassic "Laterite Derived Facies" (Mishhor and Ardon Formations), Makhtesh Ramon, Israel. *Sediment. Geology*, **31**:119–140.

Goldring, R and Bridges, P., 1973, Sublittoral sheet sandstones. *J. Sediment. Petrology*, **43**:736–747.

Goldstein, R. H., 1988, Paleosols of Late Pennsylvanian cyclic strata, New Mexico. *Sedimentology*, **35**:777–803.

Goudie, A., 1973, Duricrusts in Tropical and Subtropical Landscapes. London, Clarendon Press, 174 p.

Grabau, A.W., 1913, Principles of Stratigraphy. New York, A. G. Seiler & Co., 1185 p.

Gray, D. I. and Benton, M. J., 1982, Multidirectional paleocurrents as indicators of shelf storm beds. p.350–353, In: Cyclic and Event Stratification, (G. Einsele, and A. Seilacher, eds.). Berlin-Heidelberg-New York, Springer-Verlag, 536 p.

Grotzinger, J. P. and Rothman, D. H., 1996, An abiotic model for stromatolite morphogenesis. *Nature*, **383**:423–425.

Hagadorn, J. W. and Bottjer, D. J., 1999, Restriction of a late Neoproterozoic biotope: Suspect microbial structures and trace fossils at the Vendian-Cambrian transition. *Palaios*, **14**:73–85.

Hamblin, A. P. and Walker, R. G., 1979, Storm-dominated shallow marine deposits the Fernie-Kootenay (Jurassic) transition, southern Rocky Mountains. *Can. Jour. Earth Sciences*, **16**:1673–1690.

Hamblin, A.P. and Rust, B. R., 1989, Tectono-sedimentary analysis of alternate-polarity half-graben basin-fill successions: Late Devonian-Early Carboniferous Horton Group, Cape Breton Island, Nova Scotia. *Basin Research* **2**:239–255.

Hampton, M. A., 1972, The role of subaqueous debris flows in generating turbidity currents. *J. Sediment. Petrology*, **42**:775–793.

Hand, Bryce M., 1997, Inverse grading resulting from coarse-sediment transport lag. *J. Sediment. Research*, **67**:124–129.

Handford, C. R., 1986, Facies and bedding sequences in shelf-storm-deposited carbonates—Fayettville Shale and Pitkin Limestone (Mississipian), Arkansas. *J. Sediment. Petrology*, **56**:123–137.

Hantzschel, W. and Frey, R. W., 1978, Bioturbation. *In*: The Encyclopedia of Sedimentology. (R. W. Fairbridge, and J. Bourgeois, eds.). Stroudsburg, Pa., Dowden,Hutchinson & Ross, 68–71.

Harbitz, C.B., 1992, Model simulations of tsunamis generated by the Storegga slides. *Marine Geology*, 105: 1–21.

Harding, S. C. and Risk, M. J., 1986, Grain orientation and electron microscope analyses of selected Phanerozoic trace fossil margins, with a possible Proterozoic example. *J. Sediment. Petrology*, **56**:684–690.

Harms, J. C., Southard, J. B., Spearing, D. R. and Walker, R. G., 1975, Depositional Environments as Interpreted from Primary Sedimentary Structures and Stratification Sequences. Tulsa, Oklahoma, SEPM Short Course No. 2, 161 p.

Harms, J. C., Southard, J. B. and Walker, R. G., 1982, Structures and Sequences in Clastic Rocks. Tulsa, Oklahoma, SEPM Short Course No. 9, (not consecutively paginated).

Harrison, R.S., 1977, Caliche profiles indicators of near-surface subaerial diagenesis, Barbados, West Indies. *Can. Petrol. Geologists, Bull.*, **25**:123–173.

Hein, F.J., 1982, The Cambro-Ordovician Cap Enrage Formation, Quebec, Canada: conglomeratic deposits of a braided submarine channel with terraces. *Sedimentology*, **29**:309–329.

Hein, F.J., 1984, Deep-sea and fluvial braided channel conglomerates: A comparison of two case studies, p. 33–49. In: Sedimentology of Gravels and Conglomerates. (Koster, E. H. and Steel, R. J., eds.), Calgary, Alberta, Can. Soc. Pet. Geologists, Memoir 10, 441 p.

Helland-Hansen, W. and Gjelberg, J.G., 1994, Conceptual basis and variability in sequence stratigraphy: a different perspective. *Sediment. Geology*, **92**:31–52.

Helland-Hansen, G. and Martinsen, O. J., 1996, Shoreline trajectories and sequences: Description of variable depositional-dip scenarios. *J. Sediment. Research.*, **66**:670–688.

Hennessy, J. and Knauth, L. P., 1985, Isotopic variations in dolomite concretions from the Monetery Formation, California. *J. Sediment. Petrology*, **55**:120–130.

Hesse, R., 1975, Turbiditic and non-turbiditic mudstones of Cretaceous flysch sections of the East Alps and other basins. *Sedimentology*, **22**:387–416.

Hiscott, R., 1982, Tidal deposits of the Lower Cambrian Random Formation, eastern Newfoundland : facies and paleoenvironments. *Can. J. Earth Science*, **19**:2028–2042.

Hobday, D. K and Reading, H. G., 1972, Fair weather versus storm processes in shallow marine bar sequences in late Precambrian of Finnmark, north Norway. *J. Sediment. Petrology*, **42**:318–321.

Hobday, D. K. and Morton, R. A., 1984, Lower Cretaceous shelf storm deposits, Northeast Texas. p. 205–213. *In*: Siliciclastic Shelf Sediments, (R. K. Tillman and C. T. Siemers, eds.). Tulsa, Oklahoma, SEPM Special Pub. 34, 268p.

Hoffman, P. E., 1967, Algal stromatolites: use in sytratigraphic correlation and paleocurrent determination. *Science*, **157**:1043–1045.

Hofmann, H. J., 1969, Attributes of stromatolites. Geol. Survey of Canada, Paper-69–39.

Hofmann, H. J., 1973, Stromatolites: characteristic and utility. *Earth-Sci. Reviews*, **9(4)**:339–373.

Hofmann, H. J., 1976, Graphic representation of fossil stromatolites: New method with improved precision In: Stromatolites: Development in Sedimentology 20, (M. R. Walter, editor), p. 15–20, Amsterdam-Oxford-New York, Elsevier Sci. Pub. Development Co., 790 p.

Hounslow, M. W., 1997, Significance of localized pore pressures to the genesis of septarian concretions. *Sedimentology*, **44**:1133–1147.

Houseknecht, D. W. and Ethridge, F. G., 1978, Depositional history of the Lamotte Sandstone of southeastern Missouri. *J. Sediment. Petrology*, **48**:575–586.

Howard, J. L., 1993, The statistics of counting clasts in rudites : a review with examples from the Upper Palaeozoic of Southern California, U.S.A. *Sedimentology*, **40(2)**:157–174.

Howard, J. D., 1975, The sedimentological significance of trace fossils. p. 131–146. *In*: The Study of Trace Fossils. (R.W. Frey, ed.). New York, Springer-Verlag 562 p.

Howard, J. D., 1978, Sedimentology and trace fossils, p. 11–42, *In*: Trace Fossil Concepts, (P. B. Basan, ed.). Tulsa, Oklahoma, SEPM Short Course No. 5, 181p.

Howard, J. D. and Frey, R. W., 1975, Estuaries in Georgia Coast, U.S.A: sedimentology and biology. II. Regional animal-sediment characteristics of Georgia Estuaries. Senckenbergiana Maritima, Vol. 7: 33–103.

Howard, James D. and Frey, Robert W., 1985, Physical and biogenic aspects of backbarrier sedimentary sequences, Georgia Coast, U.S.A. *In*: Barrier Islands. (G.F. Oertel, and S.P. Leatherman, eds.). *Marine Geology*, **63**:77–187.

Howell, J. V., 1931, Silicified shell fragments as an indication of unconformity. *Am. Assoc. Petrol. Geologists, Bull.*, **15**:1103–1104.

Huggett, J. M., 1994, Diagenesis of mudrocks and concretions from the London Clay Formation in the London Basin. Clay Mineralogy, **29**:693–707.

Hunter, R. E., 1977, Basic types of stratification in small eolian dunes. *Sedimentology*, **24**:361–387.

Hunter, R. E. and Clifton, H. E., 1982, Cyclic deposits and hummocky cross-stratification of probable storm origin in Upper Cretaceous rocks of the Cape Sebastian area, southwestern Oregon. *J. Sediment. Petrology*, **52**:127–143

Hutton, J., 1788, Theory of the Earth. *Roy. Soc. Edinburgh, Trans.*, **1**:209–304.

Ingersoll, R. V. and Busby, C. J., 1995, Tectonics of Sedimentary Basins. 1–52. *In*: Tectonics of Sedimentary Basins. (C.J. Busby, and R.V. Ingersoll, eds.). Oxford, Blackwell Scientific Publications, 1–52.

Jackson, J. A. and Leeder, M. R., 1994, Drainage systems and the development of normal faults: an example from Pleasant Valley, Nevada. *J. Struct. Geology*, **16**:1041–1059.

James, N. P., 1972, Holocene and Pleistocene calcareous crust (caliche) profiles: criteria for subaerial exposure. *J. Sediment. Petrology*, **42**:817–836.

James, N. P. and Choquette, Phil W., 1984, Diagenesis 9: limestones—the meteoric diagenetic environment. *Geoscience Canada*, 11, No. **4**:161–194.

James, N. P. and Choquette, P. W., eds.,1988, Paleokarsts. New York-Berlin-Heildelberg-London-Paris-Tokyo. Springer-Verlag, 576 p.

James, N. P and Kendall, A. C, 1992, Introduction to carbonate and evaporite facies models. *In*: Facies Models, (R.G. Walker, ed.). 3rd edition, Geoscience Canada Reprint Series 1, 317 p.

Jennings, J. N. and Sweeting, H. N., 1961, Caliche pseudo-anticlines in the Fitzroy Basin, western Australia. *Am. J. Science*, **259**:635–639.

Jervey, M. T., 1988, Quantitative geological modelling of siliciclastic rock sequences and their seismic expressions, p. 47–69. *In*: Sea-level changes: An integrated approach. (C. K. Wilgus, B. S. Hastings, C. G. St Kendall, H. W. Posamentier, C.A. Ross and J.C. Van Wagoner, eds.). Tulsa, Oklahoma, SEPM Spec. Publ., 42, 407 p.

Joeckel , R. M., 1994, Virgillian (Upper Pennsylvanian) peleosols in the Upper Lawrence Formation (Douglas Group) and in the Synderville Shale Member (Oread Formation, Shawnee Group) of the northern Midcontinent, U.S.A.: Pedologic contrast in a cyclothem sequence. *J. Sediment. Research*, **A64**:853–866.

Joffe, J. S., 1965, The ABC of Soils. Calcutta-Kharagpur-New Delhi, Oxford Book Co., 383 p.

Johnson, M. R., 1989, Paleogeographic significance of oriented calcareous concretions in the Triassic Katberg Formation, South Africa. *J. Sediment. Petrology*, **59**:1008–1010.

Jones, M. L. and Dennison, J. M., 1970, Oriented fossils as paleocurrent indicators in Paleozoic lutites of southern Appalachians. *J. Sediment. Petrology*, **40**:642–649.

Jones, M.E. and Preston, R.M.F., 1987, Deformation of sediments and sedimentary rocks. Geol. Soc. London, Spec. Pub., 29, Oxford, Blackwell Scientific Publications, 350 p.

Jordan, T. E. and Flemings, P. B., 1991, Large-scale stratigraphic architecture, eustatic variation, and unsteady tectonism: a theoretical approach. *J. Geophys. Research*, **96 (B4)**:6681–6699.

Kahle, C.F., 1977, Origin of subaerial Holocene calcareous crusts: role of algae, fungi and sparmicritization. *Sedimentology*, **24**:413–435.

Kahle, C. F., 1988, Surface and subsurface paleokarst, Silurian Lockport and Peebles Dolomites, Western Ohio, p. 229–255. *In*: Paleokarst. (N. P. James and P. W. Choquette, eds.). New York-Heidelberg-London-Paris-Tokyo, Springer-Verlag, 576 p.

Karner, G.D., Lake, S.D. and Dewey, J.F., 1987, The thermal and mechanical development of the Wessex Basin, southern England. In: Continental Extensional Tectonics. (M. P. Coward, J. F. Dewey, and P. L. Hancock, eds.). *Geol. Soc., London*, Spec. Publ., **28**:517–536.

Kelling, G. and Mullin, P. R., 1975, Graded limestones and limestone-quartzite couplets: possible storm deposits from the Moroccan Carboniferous. *Sediment*. Geology, **13**:161–190.

Kelts, K. and Arthur, M. A., 1981, Turbidites after ten years of deep-sea drilling - wringing out the mop? p. 91–127, *In*: The Deep Sea Drilling Project: A Decade of Progress (J. E. Warne, R.G. Douglas, and E.L Winterer, eds.). Tulsa, Oklahoma, SEPM Spec. Publ., **32**, 564 p.

Kerans, C. and Donaldson, J. Allan, 1988, Proterozoic paleokarst profile, Dismal Lakes Group, N.W.T., Canada, p. 167–182, In: Paleokarst, (N. P. James, and P. W. Choquette, eds.). New York-Berlin-Heidelberg-London- Paris-Tokyo Springer-Verlag, 576 p.

Kidder, D. I., 1990, Facies-controlled shrinkage-crack assemblages in Middle Proterozoic mudstones from Montana, U.S.A. *Sedimentology*, **37**:943–951.

Klappa, C. F, 1978, Biolithogenesis of Microcodium: elucidation. *Sedimentology*, **25**:489–522.

Klappa, C. F., 1979a, Lichen stromatolites: criterion for subaerial exposure and a mechanism for the formation of laminar calcretes (caliche). *J. Sediment. Petrology*, **49**:387–400.

Klappa, C. F., 1979b, Calcified filaments in Quaternary calcretes; organo-mineral interactions in the subaerial vadose environment. *J. Sediment. Petrology*, **49**:955–968.

Klappa, C. F., 1980a, Brecciation textures and tepee structures in Quaternary calcrete (caliche) profiles from eastern Spain: the plant factor in their formation. *Geol. Journal*, **15**:81–89.

Klappa, C. F., 1980b, Rhizoliths in terrestrial carbonates: classification, recognition, genesis and significance. *Sedimentology*, **27**:613–629.

Klein, G. D., 1995, Intracratonic Basins, pp. 459–478. *In*: Tectonics of Sedimentary Basins. (C. J. Busby, and R. V. Ingersoll, eds.). Oxford, Blackwell Scientific Publications.

Kocurek, G. and Dott Jr., R. H., 1981, Distinctions and uses of Stratification types in the interpretation of Eolian Sand. *J. Sediment. Petrology*, **51**:579–595.

Kocurek, G. and Fielder, G., 1980, Adhesion structures. *J. Sediment. Petrology*, **52**:1229–1241.

Kraus, M. J., 1999, Paleosols in clastic sedimentary rocks: their geologic applications. Earth Sci. Reviews, **47**:41–70.

Kreisa, R. D. and Bambach, R. K., 1982, The role of storm processes in generating shell beds in Paleozoic shelf environments. p. 200–207, *In*: Cyclic and Event Stratification. (G. Einsele, and A. Seilacher, eds.). Berlin-Heidelberg-New York, Springer-Verlag, 536 p.

Kreisa, R. D., 1981, Storm-generated sedimentary structures in subtidal marine facies with example from Middle and Upper Ordovician of southwestern Virginia. *J. Sediment. Petrology*, **51**:823–848.

Krumbein, W. C., 1941, Measurement and geologic significance of shape and roundness of sedimentary particles. *J. Sediment. Petrology*, **11**:64–72

Krumbein, W. C., 1942, Criteria for subsurface recognition of unconformities. Am. Assoc. *Petrol. Geologists, Bull.*, **26**:36–62.

Krumbein, W. C. and Sloss, L. L., 1963, Stratigraphy and Sedimentation, 2nd edition. San Francisco-London, W. H. Freeman & Co., 660 p.

Krumbein, W. E., 1983, Stromatolites—The challenge of a term in space and time. *Precambrian Res.*, **20**:493–531

Kuenen, Ph. H., 1958, Experiments in Geology. Glasgow Geol. Soc., *Trans.*, **23**:1–28.

Kuenen, P. H. and Migliorini, C. I., 1950, Turbidity currents as a cause of graded bedding. *Jour. Geology*, **58**:91–127.

Kumar, N. and Sanders, J. E., 1976, Characteristics of shoreface storm deposits: modern and ancient examples. *J. Sediment. Petrology*, **46**:145–162

Lander, R. H., Bloch, S., Mehta, S., and Atkinson, C. D., 1991, Burial diagenesis of paleosols in the giant Yacheng gas field, People's Republic of China: bearing on illite reaction pathways. *J. Sediment. Petrology*, **61**:256–268.

Leckie, D. A. and Walker, R. G., 1982, Storm- and tide-dominated shorelines in Cretaceous Moosebar-Lower Gates interval. *Am. Assoc. Petrol. Geologists, Bull.*, **66**:138–157.

Leckie, Dale, 1988, Wave-formed, coarse-grained ripples and their relationships to hummocky cross-stratification. *J. Sediment. Petrology*, **58**:607–622.

Leckie, Dale and Krystinik, L. F., 1989, Is there evidence for geostrophic currents preserved in the sedimentary record of inner- to middle shelf deposits? *J. Sediment. Petrology*, **59**:862–870.

Leeder, M. R., 1995, Continental rifts and proto-Oceanic rift troughs, p. 119–148. *In*: Tectonics of Sedimentary Basins. Busby, C. J. and Ingersoll, R. V. eds., Oxford, Blackwell Scientific Publications.

Leeder, M. R. and Alexander, J., 1987, The origin and tectonic significance of asymmetrical meander belts. *Sedimentology*, **34**:217–226.

Leeder, M. R. and Gawthorpe, R. L., 1987, Sedimentary models for extensional tilt-block/half-graben basins. *Geol. Soc., London*, Spec. Publ. No. **28**:139–152.

Leith, C.K., 1925, Silicification of erosion surfaces. Econ. *Geology*, **20**:513–523.

Leithold, A. L. and Bourgeois, J., 1984, Characteristics of coarse-grained sequences deposited in near-shore, wave-dominated environments—examples from the Miocene of southwest Oregon. *Sedimentology*, **31**:749–775.

Lohman, K. C., 1988, Geochemical patterns of meteoric diagenesis systems and their application to studies of paleokarst, p. 58–80, *In*: Paleokarst, (N. P. James, and P. W. Choquette, eds.). New York, Springer-Verlag, 576 p.

Lowe, D.R., 1982, Sediment gravity flows: II. Depositional models with special reference to deposits of high-density turbidity currents. *J. Sediment. Petrology*, **52**:279–297.

Lowe, D.R., 1983, Restricted shallow water sedimentation of Early Archaean stromatolitic and evaporitic strata of the Strelly Pool Chert, Pilbara Block, Western Australia. *Precambrian Res.*, **19**:239–283.

Lowe, D. R., Byerly, G. R., Ransom, B. L. and Nocita, B. W., 1985, Stratigraphic and sedimentological evidence bearing on structural repetition in Early Archaean rocks of the Barberton Greenstone Belt, South Africa. *Precambrian Res.*, **27**:165–186.

Lucchi, R., 1986, The Oligocene to Recent foreland basins of the northern Apennines. *In*: Foreland Basins: An Introduction. (P.A. Allen, P. Homewood, and G. D. Williams, eds.). IAS Spec. Publ., 8, Blackwell Scientific Publications, 453 p.

Lundberg, Joyce and Taggart, Bruce E., 1995, Dissolution pipes in northern Puerto Rico: An exhumed paleokarst. *Carbonates and Evaporites*, **10**:171–183.

Mack, G. H. and Seager, W. R., 1990, Tectonic control on facies distribution of the Camp Rice and Palomas Formations (Pliocene-Pleistocene) in the southern Rio Grande rift. *Geol. Soc. Am. Bull.*, **102**:45–53.

Mack, G. H. James, W. C. and Monger, H. C., 1993, Classification of paleosols. *Geol. Soc. Am. Bull.*, **105**:129–136.

Mann, P., Hempton, M. R., Bradley, D. C. and Burke, K., 1983, Development of pull-apart basins. *J. Geology*, **91**:529–554.

Marriott, S. B. and Wright, V. Paul, 1993, Paleosols as indicators of geomorphic stability in two Old Red Sandstone alluvial suites, South Wales. *J. Geol. Soc., London*, **150**:1109–1120.

Martin, C. A. L. and Turner, B. R., 1998, Origins of massive-type sandstones in braided river systems. *Earth-Sci. Reviews*, **44**:15–38.

Martinsen, O. J. and Helland-Hansen, W., 1995, Strike variability of clastic depositional systems: does it matter for sequence stratigraphic analysis? *Geology*, **23**:439–442.

Martinsen, O. J., Ryseth, A., Helland-Hansen, W., Flesche, H., Torkildsen, G. and Idil, S., 1999, Stratigraphic base level and fluvial architecture: Ericson Sandstone (Campanian), Rock Springs Uplift, SW Wyoming, USA. *Sedimentology*, **46**:235–259.

Martinsson, A., 1970, Toponomy of trace fossils 323–330. In: Trace Fossils, (T.P. Crimes, and J. C. Harper, eds.). *Geol. Jour., London, Special Issue* 3:547 p.

Massari, F., Mellere, D. and Doglioni, C., 1993, Cyclicity in non-marine foreland-basin sedimentary fill: the Messinian conglomerate-bearing succession of the Venetian Alps (Italy) 501–520. IAS Spec. Publs. **17**: Oxford, Blackwell Scientific Publications.

Mayall, M. J., 1983, An earthquake origin for synsedimentary deformatioı. in late Triassic (Rhaetian) lagoonal sequence, southwest Britain. *Geol. Magazine*, **120**:613–622.

McFarlane, M. J., 1976, Laterite and Landscape. New York, American Press.

McKenzie, Dan., 1978, Some Remarks on the Development of Sedimentary Basins. *Earth and Planetary Sci. Letters.*, **40**:25–32.

McManus, J. and Bajabaa, S., 1998, Aquifer air release features of modern and ancient wadi sediments. Sediment. *Geology*, **120**:337–343.

McNeil, D. F., Ginsburg, R. N., Chang, S. B. R. and Kirschvink, J. L., 1988, Magneto-stratigraphic dating of shallow water carbonates from San Salvador, Bahamas. *Geology*, **16**:8–12.

McPherson, J. B., 1979, Calcrete (caliche) paleosols in fluvial redbeds of the Aztec Siltstone (Upper Devonian), Southern Victoria Land, Australia. *Sediment. Geology*, **22**:267–285.

Meckel, L. D., 1967, Origin of Pottsville conglomerates (Pennsylvanian) in the Central Appalachians. *Geol. Soc. Am., Bull.*, **78**:223–258.

Meyers, W.J., 1988, Paleokarstic features on Mississippian Limestones, New Mexico, p. 306–328, In: Paleokarst, (James, N. P. and P. W. Choquette, eds.). New York, Springer-Verlag, 576 p.

Miall, A. D., 1985, Architectural-element analysis: A new method of facies analysis applied to fluvial deposits. *Earth-Sci. Reviews*, **22**:261–308.

Miall, A. D., 1987, Recent Developments in the study of fluvial facies models. *In*: Recent Developments in Fluvial Sedimentology, (F. G. Ethridge, R. M. Flores, and M. D. Harvey, eds.). Tulsa, Oklahoma, SEPM Spec. Publ., 39:1–9.

Miall, A. D., 1988, Reservoir heterogeneities in fluvial sandstones: Lessons from outcrop studies. *Am. Assoc. Petrol. Geologists, Bull.*, **72**:682–697.

Miall, A. D., 1990, Principles of Sedimentary Basin Analysis, 2nd edition. New York, Springer-Verlag, 668 p.

Miall, A. D., 1995a, Collision-related foreland basins. p. 393–424. *In*: Tectonics of Sedimentary Basins. (C.J. Busby, and R.V. Ingersoll, eds.). Oxford, Blackwell Scientific Publications.

Miall, A. D., 1995 b, Whither Stratigraphy? *Sediment. Geology*, **100**:5–20.

Miall, A. D., 1995 c. Discussion: Description and interpretation of fluvial deposits: a critical perspective. *Sedimentology*, **42**:379–389.

Middleton, G. V., 1966a, Small-scale models of turbidity currents and the criterion for auto-suspension. *J. Sediment. Petrology*, **36**:202–208.

Middleton, G. V., 1966b, Experiments on dennsity and turbidity currents. I. Motion of the head. *Can. J. Earth Sciences*, **3**:523–546.

Middleton, G.V., 1966c, Experiments on density and turbidity currents. II. Uniform flow of density currents. *Can. J. Earth Sciences*, **3**:627–637.

Middleton, G.V., 1967, Experiments on density and turbidity currents. III. Deposition of sediment. *Can. J. Earth Sciences*, **4**:475–505.

Middleton, G. V. and Hampton, M. A., 1973, Sediment gravity flows: mechanics of flow and deposition, p. 1–38. *In*: Turbidites and Deep-water Sedimentation. (G. V. Middleton, and A. H. Bouma, Chm.). SEPM, Pacific Section, Short Course, Anaheim, 158 p.

Middleton, G. V. and Southard, J. B., 1978, Mechanics of Sediment Movement. Tulsa, Oklahoma, SEPM Short Course 3.

Mills, P.C., 1983, Genesis and diagnostic value of soft-sediment deformation structures: A review. *Sediment. Geology*, **35**:83–104.

Minoura, K. and Nakaya, S., 1991, Traces of tsunami preserved in intertidal lacustrine and marsh deposits: Some examples from northeast *Japan. J. Geology*, **99**:265–287.

Minoura, K., Gusiakov, V.G., Kurbatov, A., Takeuti, S., Svendsen, J.L., Bondevik, S., and Oda, T., 1996, Tsunami sedimentation associated with 1923 Kamchatka earthquake. *Sediment. Geology*, **106**:145–154.

Mohindra, Rakesh and Bagati, T. N., 1996, Seismically induced soft-sediment deformation structures (seismites) around Sumdo in the Lower Spiti Valley (Tethys Himalayan). *Sediment. Geology*, **101**:69–83.

Molina, J. M., Ruiz-Ortiz, P. A. and Vera, J. A., 1997, Calcareous tempestites in pelagic facies (Jurassic, Beltic Cordilleras, Southern Spain). *Sediment. Geology*, **100**:95–109.

Moore, Andrew, 2000, Landward fining in onshore gravel as evidence for a late Pleistocene tsunami on Molokai, Hawaii. *Geology*, **28**(3): 247–250.

Moore, G. W. and Moore, J. G., 1988, Large-scale bedforms in boulder gravel produced by giant waves in Hawaii, p. 101–110. *In*: Sedimentologic Consequences of Convulsive Geologic Events. (H. E. Clifton, ed.). Geol. Soc. Am., Special Paper **229**, 197 p.

Morton, R. A., 1988, Near-shore responses to great storms, p. 7–21. *In*: Sedimentologic Consequences of Convulsive Geologic Events, (H. E. Clifton, ed.). Geol. Soc. Am., Special Paper **229**, 197 p.

Mount, J. F., 1984, Mixing of siliciclastic and carbonate sediments in shallow shelf environments. *Geology*, **12**:432–435

Mount, J. F. and Cohen, A. S., 1984, Petrology and geochemistry of rhizoliths from Plio-Pleistocene fluvial and marginal lacustrine deposits, East Lake Turkana, Kenya. *J. Sediment. Petrology*, **54**:263–275.

Mozley, P. S. and Burns, S. J., 1993, Oxygen and carbon isotopic composition of marine carbonate concretions: an overview. *J. Sediment. Petrology*, **63**:73–83.

Mozley, P. S., 1989, Complex compositional zonation in concretionary siderite: implications for geochemical studies. *J. Sediment. Petrology*, **59**:815–818.

Mozley, P. S., 1996, The internal structure of carbonate concretions in mudrocks: a critical evaluation of onvenntional concentric model of concretion growth. Sediment. *Geology*, **103**:85–91.

Myrow, P. M., 1992a, By-pass zone tempestite facies model and proximality trends for an ancient muddy shoreline and shelf. *J. Sediment. Petrology*, **62**:99–115.

Myrow, P. M., 1992b, Pot and gutter casts from the Chappel Island Formation, south-east Newfoundland. *J. Sediment. Petrology*, **62**:992–1007.

Myrow, P. M. and Hiscott, R. H., 1991, Shallow-water gravity- flow deposits, Chapel Island Formation, southeast Newfoundland, *Canada. Sedimentology*, **38**:935–959.

Myrow, P. M. and Southard, J. B., 1991, Combined flow model for vertical stratification sequences in shallow marine storm-deposited beds. *J. Sediment. Petrology*, **61**:202–210.

Myrow, P. M. and Southard, J. B., 1996, Tempestite deposition. *J. Sediment. Research*, **A66**:875–887.

Nelson, C. H., 1982, Modern shallow-water graded sand layers from storm surges, Bering Shelf : A mimic of Bouma sequences and turbidite systems. *J. Sediment. Petrology*, **52**:537–545.

Nemec, W., Porebski, S. J and Steel, R. J., 1980, Texture and structure of re-sedimented conglomerates; examples from Ksiaz Formation (Famennian-Tournaisian), southeastern Poland. *Sedimentology*, **27**:519–538.

Nemec, W. and Steel, R. J., 1984, Alluvial and coastal conglomerates: Their significant features and some comments on gravelly mass flow deposits, p. p.1–31. *In*: Sedimentology of Gravels and Conglomerates, (E. H. Koster and R. J. Steel, eds.). Calgary, Alberta, Can. Soc. Petrol. Geologists, Memoir **10**, 441 p.

Nemec, W, Steel, R. J., Porebski, S. J. and Spinnanger, A., 1984, Domba Conglomerate, Devonian, Norway: Process and lateral variability in a mass-flow dominated, lacustrine fan-delta. p. 295–320 *In*: Sedimentology of Gravels and Conglomerates. (E. H. Koster and R. J. Steel, eds.). Calgary, Alberta, Can. Soc. Petrol. Geologists, Memoir 10. 441p.

Nilsen, T. H. and McLaughlin, R. J., 1985, Comparison of tectonic framework and depositional patterns of the Hornelen strike-slip basin of Norway and the Ridge and Little Sulphur Creek strike-slip basins of California, p. 79–103. Tulsa, Oklahoma, SEPM. Spec. Publ., 37.

Nilsen, T.H. and Syivester, A.G., 1995 Strike-slip Basins, p. 425–458. *In*: Tectonics of Sedimentary Basins. (C.J. Busby, and R.V. Ingersoil, eds.). Oxford, Blackwell Scientific Publications.

O'Brien, N. R., Nakazawa, K., and Tokuhashi, S., 1980, Use of clay fabric to distinguish turbidite and hemipelagic siltstones and silts. *Sedimentology*, **27**:47–61.

Odin, G. S., 1985, Significance of green particles (glaucony, berthierine, chlorite) in arenites, p. 279–307. *In*: Provenance of Arenites. (G.G. Zuffa, ed.) Dordrecht, D. Reidel Pub. Co. 361 p.

Olsen, K. H. and Morgan, P., 1995, Introduction: Progress in understanding continental rifts, p. 1–26. *In*: Continental Rifts: Evolution, Structure, Tectonics. (K.H. Olsen, ed.). Elsevier Scientific Pub., 466 p.

Olsen, K. H., ed, 1995, Continental Rifts: Evolution, Structure, Tectonics. Elsevier Scientific Pub., 466 p.

Olsen, T., Steel, R, Hogseth, K., Skar, T. and Roe, S. L., 1995, Sequential Architecture in a Fluvial Succession: Sequence Stratigraphy in the Upper Cretaceous Mesaverde Group, Price Canyon, Utah. *J. Sediment Research*., **B65**:265–280.

Orton, G. J. and Reading, H. G., 1993, Variability of deltaic processes in terms of sediment supply, with particular emphasis on grain size. *Sedimentology*, **40**:475–512.

Owen, D. E., 1987, Commentary: Usage of Stratigraphic Terminology in Papers, Illustrations, and Talks. *J. Sediment. Petrology*, **57**:363–372.

Paola, C., Wielo, S. W. and Reinhardt, M. A., 1989, Upper-regime parallel lamination as the result of turbulent sediment transport and low amplitude bedforms. *Sedimentology*, **36**:47–59.

Paton, T. R., 1974, Origin and terminology for gilgai in Australia. *Geoderma*, **11**:221–242.

Pelletier, B. R., 1958, Pocono paleocurrents in Pennsylvania and Maryland. *Geol. Soc. Am., Bull.*, **69**:1033–1064.

Pettijohn, F. J., 1975, Sedimentary Rocks, 3rd edition, Harper & Row, 628 p.

Pfluger, Friedrich. 1999, Mat ground structures and redox facies. *Palaios*, **14**:25–39.

Picard, M.B., 1971, Classification of fine-grained sedimentary rocks. J. Sediment. *Petrology*, **41**:179–195.

Pimentel, N. L., Wright, V. Paul and Azevedo, T. M., 1996, Distinguishing early groundwater alteration effects from pedogenesis in ancient alluvial basins: examples from the Paleogene of southern Portugal. *Sediment. Geology*, **105**:1–10.

Piper, D. J. W, 1972, Turbidite origin of some laminated mudstones. *Geol. Magazine*, **109**:115–126.

Piper, D. J. W., 1978, Turbidite muds and silts on deep-sea fans and abyssal plains. p.163–176. *In*: Sedimentation in Submarine Canyons, Fans and Trenches. (D.J. Stanley and G. Kelling, eds.). Stroudsberg, Pa. Dowden, Hutchinson & Ross, 395 p.

Piper, D. J. W., Stow, D.A.V. and Alam, M., 1991, Fine-grained turbidites, p. 361–376. *In*: Cycles and Events in Stratigraphy. (G. Einsele, W. Ricken, and A. Seilacher, eds.). Berlin-Heidelberg, Springer-Verlag, 955 p.

Pirsson, L. V., and Schichert, C., 1920, Introductory Geology. New York, John Wiley & Sons, Inc.

Pluhar, A. and Ford, Derek, 1970, Dolomite karren of the Niagara escarpment, Ontario, Canada. *Zeitschrift fur Geomorphologie*, **14**:392–410.

Plumley, W. J., 1948, Rock Hills terrace gravels: a study in sediment transport. *J. Geology*, **56**:526–577.

Plummer, P. S., Gostin, V. A., 1981, Shrinkage cracks: desiccation or syneresis. J. Sediment. *Petrology*, **51**:1147–1156.

Posamentier, H. W., Jervey, M. T. and Vail, P. R., 1988 a, Eustatic controls on clastic deposition I Conceptual framework, p. 109–124, *In*: Sea Level Changes: An Integrated Approach. (C.

K. Wilgus, B. S. Hastings, C. G. St. C. Kendall, H. W. Posamentier, C. A. Ross and J. C. Van Wagoner, eds.). Tulsa, Oklahoma, SEPM Special Pub. 42, 407 p.

Posamentier, H. W. and Vail, P.R, 1988b. Eustatic controls on clastic deposition 2: sequences and systems tract models, p. 125–154, *In*: Sea Level Changes: An Integrated Approach. (C. K Wilgus, B. S. Hastings, C. G. St. C. Kendall, H. W. Posamentier, C. A. Ross and J. C. Van Wagoner, eds.). Tulsa, Oklahoma, SEPM Spec. Publ. 42, 407 p.

Posamentier, H. W., Allen, H. W., James, D. P and Tesson, M, 1992, Forced regression in a sequence stratigraphic framework: concepts, examples and sequence stratigraphic significance. *Am. Assoc. Petrol. Geologists, Bull.*, **76**:1687–1709.

Posamentier, H. W. and James, D. P, 1993, An overview of sequence stratigraphic concepts: Uses and abuses, p. 3–18. *In*: Sequence Stratigraphy and Facies Associations. (H. W. Posamentier, C. P. Summerhayes, B. K. Haq, and G. P. Allen, eds.). IAS Spec. Pub. 18, Oxford, Blackwell Scientific Publications.

Potter, P. E., 1955, Petrology and origin of the Lafayette gravel. *J. Geology*, **63**:1–38.

Potter, P. E., Maynard, J. B. and Pryor, W. A., 1980, Sedimentology of Shales: Study Guide and Reference Source. New York-Heidelberg-Berlin. Springer-Verlag, 141 p.

Powell, J.W., 1875 Exploration of the Colorado River of the West and its Tributaries. Government Printing Office, Washington DC.

Pratt, Brian R., 1998, Syneresis cracks: subaqueous shrinkage in argillaceous sediments caused by earthquake-induced dewatering. *Sediment. Geology*, **117**:1–10.

Preiss, W.V., 1972, The systematics of South Australian Precambrian and Cambrian stromatolites, Part 1. Trans. Roy. Soc. South Australia, 96:69–100.

Preiss, W.V., 1976, Basic field and laboratory methods for the study of stromatolites, p. 5–13, In: Stromatolites. Development in Sedimentology 20, (M.R. Walter, ed.). Amsterdam-Oxford-NewYork, Elsevier Scientific Pub. Development Co., 790 p.

Raaf, De.J.F.M., Boersma, J.R. and Gelder, A. van, 1977, Wave-generated structures and sequences from a shallow marine succession, Lower Carboniferous, County Cork, Ireland. *Sedimentology*, **24**:451–483.

Raiswell, R., 1971, The growth of Cambrian and Liassic concretions. *Sedimentology*, **17**:147–171.

Raiswell, R., 1976, The microbiological formation of carbonate concretions in the Upper Lias of NE England. *Chemical Geology*, **18**:227–244.

Raiswell, R., and Fisher, Q.J., 2000, Mudrock-hosted carbonate concretions: a review of growth mechanisms and their influence on chemical and isotopic compositions. *J. Geol. Soc., London*, **157**:239–251.

Ravnas, R. and Steel, R.J., 1998, Architecture of marine rift-basin succession. *Am. Assoc. Petrol. Geologists, Bull.*, **82**:110–146.

Read, J.F., 1976, Calcretes and their distinction from stromatolites, p. 55–71,. *In*: Stromatolites, Developments in Sedimentology 20, (M. R. Walter, ed.). Amsterdam-Oxford-New York, Elsevier Scientific Pub. Development Co., 790 p.

Read, J.F., 1985, Carbonate platform facies models. *Am. Ass. Petrol. Geologists, Bull.*, **69**:1–21.

Read, J. F. and Grover, G. A., 1977, Scalloped and planar erosion surfaces, Middle Ordovician Limestones, Virginia: Analogues for Holocene exposed karst or tidal rock platforms. *J. Sediment. Petrology*, **47**:956–972.

Reading, H. G., editor, 1996, Sedimentary Environments and Facies: Processes, Facies and Stratigraphy, 3rd edition. Oxford, Blackwell Science Pub. Co., 688 p.

Reineck, H. -E., 1977, Natural indicators of energy level in recent sediments: the application of ichnology to a coastal engineering problem. *Geol. Jour.*, London, Special Issue **9**:265–272.

Reineck, H. -E. and Singh, I. B., 1972, Genesis of laminated sand and graded rhythmites in storm sand layers of shelf mud. *Sedimentology*, **18**:123–128.

Reineck, H. -E. and Singh, I.B., 1980, Depositional Sedimentary Environments. NewYork-Heidelberg-Berlin, Springer-Verlag, 439 p.

Reineck, H. -E. and Wunderlich, F., 1968, Classification and origin of flaser and lenticular bedding. *Sedimentology*, **11**:99–104.

Retallack, G. J., 1976, Triassic paleosols in the Upper Narrabeen group of New South Wales. Part I. Features of the paleosols. *J. Geol. Soc. Australia*, **23**:383–399.

Retallack, G. J., 1986a, Reappraisal of a 2200 Ma-old paleosol from near Waterval Onder, South Africa. *Precambrian Res.*, **32**:195–252.

Retallack, G. J., 1986b, The fossil record of soils. p. 1–57. *In*: Paleosols: Their Recognition and Interpretation, (Wright, V.P., ed.). Oxford, Blackwell Scientific Publication, 315 p.

Retallack, G. J., 1988, Field recognition of paleosols, p. 1–20. *In*: Paleosols and Weathering through Geologic Time: Principles and Applications. (J. Reinhardt, and W. R. Sigleo, eds.). Geol. Soc. Am., Special Paper **216**,

Retallack, G. J., 1990, Soils of the Past. An Introduction to Paleopedology. Boston, Unwin Hymen, 520 p.

Retallack, G.J., 1992, How to find a Precambrian paleosol, p. 16–30, *In*: Early Organic Evolution. (M. Schidlowski, S. Golubic, M. M. Kimberley, D. M. Mckirdy, P. A. Trudinger, eds.). Berlin, Springer-Verlag, 555 p.

Retallack, G.J. and Mindszenty, A., 1994, Well-preserved late Precambrian paleosols from northwest Scotland. *J. Sediment. Research*, **A64**:264–281.

Rhoads, D.C., 1975, The paleoecologic and environmental significance of trace fossils, p. 147–160. *In*: The Study of Trace Fossils, (R.W. Frey, ed.). New York, Springer-Verlag, Inc., 562 p.

Riba, O., 1976, Syntectonic unconformities of the Alto Cardener, Spanish Pyrenees: A Genetic Interpretation. *Sediment. Geology*, **15**:213–233.

Ricken, W., 1986, Diagenetic Bedding: Lecture Notes in Earth Sciences, Berlin-Heidelberg-New York-London-Paris-Tokyo, Springer-Verlag, 210 p.

Ricken, W. and Hemleben, C., 1982, Origin of marl-limestone alternations (Oxford 2) in southwest Germany, p. 63–71, *In*: Cyclic and Event Stratification, (G. Einsele, and A. Seilacher, eds.). Berlin-Heidelberg-New York, Springer-Verlag, 536 p

Ricken, W. and Eder, F. W., 1991, Diagenetic modification of calcareous beds—an overview, p.430–449. *In*: Cycles and Events in Stratigraphy, (G. Einsele, Werner Ricken, and Adolf Seilacher, eds.). Berlin-Heidelberg-New York, Springer-Verlag, 955 p.

Robert, C. and Kennett, J.P., 1994, Antarctic subtropical humid episode at the Paleocene-Eocene boundary: clay mineral evidence. *Geology*, **22**:211–214.

Rosendahl, B. R., 1987, Architecture of continental rifts with special reference to East Africa. *Ann. Rev. Earth Planet. Sci.*, **15**:445–503.

Rossinsky, Jr., V., Wanless, H. R. and Swart, P. K., 1992, Penetrative calcretes and their stratigraphic implications. *Geology*, **20**:331–334.

Rubin, D. H., 1987, Cross-bedding, Bedforms and Paleocurrents. Tulsa, Oklahoma, SEPM. Concepts in Sedimentology and Paleontology, Vol. 1, 187 p.

Rubin, D. H. and Hunter, R. E., 1985, Why deposits of longitudinal dunes are rarely recognized in the geologic record. *Sedimentology*, **32**:147–157.

Rust, B. R., 1979, Coarse alluvial deposits. p. 9–21. In: Walker, R. G., ed., Facies Models. Geoscience Canada Reprint Series 1,

Sanders, J. E., 1965, Primary sedimentary structures formed by turbidity currents and related resedimentation mechanisms, p. 192–219. In: Primary Sedimentary Structures and Their Hydrodynamic Interpretation,(G. V. Middleton, ed.). Tulsa, Oklahoma, SEPM Special Pub., **12**, 265 p.

Sarg, J. F., 1988, Carbonate sequence stratigraphy. *In*: Sea-level changes: an integrated approach. (C.K. Wilgus, B.S. Hastings, C. G. St. C. Kendall, H.W. Posamentier, C.A. Ross, J.C. Van Wagoner, eds.). Tulsa, Oklahoma, SEPM Spec. Publ., **42**, 407 p.

Sarjeant, W. A. S., 1975, Plant trace fossils, p. 163–179.. In: The Study of Trcae Fossils. (R.W. Frey, ed.). New York: Springer-Verlag, 562 p.

Schieber, J., 1998, Possible indicators of microbial mat deposits in shales and sandstones: Examples from the mid-Proterozoic Belt Supergroup, Montana, U.S.A. *Sediment. Geology*, **120**:105–124.

Schieber, J., 1999, Microbial mats in terrigenous clastics: The challenge of identification in the rock record. *Palaios*, **14**:3–12.

Schlager, W., 1988, Drowning unconformities on carbonate platforms. Tulsa, Oklahoma, SEPM Special Pub. **44**:15–25.

Schlager, W., 1993, Accommodation and supply—a dual control on stratigraphic sequences. *Sediment. Geology*, **86**:111–136.

Schlee, J., 1957, Upland gravels of southern Maryland. *Geol. Soc. Am., Bull.*, **68**:1371–1410.

Schlische, R. W., 1991, Half-graben basin filling models: new constraints on continental extensional basin development. *Basin Res.* **3**:123–141.

Schumm, S. A., 1993, River response to baselevel change: implications for sequence stratigraphy. *J. Geology*, **101**:279–294.

Scotchman, I. C., 1991, The geochemistry of concretions from the Kimmeridge Clay Formation of southern and eastern England. *Sedimentology*, **38**:79–106.

Scott, B. and Price, S., 1988, Earthquake-induced structure in young sediments. *Tectonophysics*, **147**:165–170.

Seilacher, A., 1964, Biogenic sedimentary structures, p. 296–316. *In*: Approaches to Paleoecology, (Imbrie, John, and N.D. Newell, eds.). New York, John Wiley & Sons.

Seilacher, A., 1967, Bathymetry of trace fossils. *Marine Geology*, **5**:413–428.

Seilacher, A., 1969, Fault-graded beds interpreted as seismites. *Sedimentology*, **13**:155–160.

Seilacher, A., 1982, Distinctive features of sandy tempestites, p. 333-349, *In*: Cyclic and Event Stratification. (G. Einsele, and A. Seilacher, eds.). Berlin-Heidelberg-New York, Springer-Verlag, 536 p.

Seilacher, A., 1984, Sedimentary structures tentatively attributed to seismic events. *Marine Geology*, **55**:1–12.

Seilacher, A., 1999, Biomat-related lifestyles in the Precambrian. *Palaios*, **14**: 86–90.

Seilacher, A. and Meischner, D., 1964, Fazies-analyse im Palaozoikum des Oslo-Gebietes. *Geol. Rundschau*, **54**:596–619.

Seilacher, A. and Aigner, T., 1991, Storm deposition at the bed, facies, and basin scale: the geologic perspectives, p. 249–267. In: Cycles and Events in Stratigraphy. (G. Einsele, W. Ricken and A. Seilacher, eds.). Berlin, Springer-Verlag, 955 p.

Selles-Martinez, J., 1996, Concretion morphology, classification and genesis. *Earth-Sci. Reviews*, **41**:177–210.

Sengor, A.M.C., 1995, Sedimentation and tectonics of fossil rifts, p. 53–118. *In*: Tectonics of Sedimentary Basins. (C.J. Busby, and R. V. Ingersoll, eds.). IAS Special Publ. **17**, Oxford, Blackwell Scientific Publications.

Shanley, K. W. and McCabe, P. J., 1993, Alluvial architecture in a sequence stratigraphic framework: a case history from the Upper Cretaceous of southern Utah, U.S.A, p. 21–55. IAS Spec. Publ., **15**, Oxford, Blackwell Scientific Publications.

Shanley, K. W. and McCabe, P. J., 1994, Perspective on sequence stratigraphy of continental strata. *Am. Assoc. Pet. Geologists, Bull.*, **78**:544–568.

Shanmugam, G., 1988, Origin, recognition, and importance of erosional unconformities in sedimentary basins, p. 83–108. *In*: New Perspective in Basin Analysis. (K. L. Kleinsphen, and C. Paola, eds.). New York, Springer-Verlag.

Shanmugam, G., 1997, The Bouma sequence and the turbidite mindset. *Earth-Sci. Reviews*, **42(4)**:201–229.

Shanmugam, G., and Moiola, R. J., 1988, Submarine fans: characteristics, models, classification and reservoir potential. *Earth-Sci. Reviews*, **24**:383–428.

Sharma, R. P., 1979, Origin of the pyrophyllite-diaspore deposits of the Bundelkhand Complex, Central India. *Mineralium Deposita*, **14**:343–352.

Sharp, R. P. 1940, Ep-Archean and Ep-Algonkian erosion surfaces, Grand Canyon, Arizona. *Geol. Soc. Am., Bull.*, **51**:1235–1270.

Simonson, B. M and Carney, Karen E, 1999, Roll-up structures: Evidence of in situ microbial mats in late Archaean deep shelf environments. *Palaios*, **14**:13–24.

Simpson, John, 1985, Stylolite-controlled layering in an homogeneous limestone: pseudo-bedding produced by burial diagenesis. *Sedimentology*, **32**:495–505.

Singer, A., 1980, The paleoclimatic interpretation of clay minerals in soils and weathering profiles. *Earth-Sci. Reviews*, **15**:303–326.

Singer, A., 1984, The paleoclimatic interpretation of clay minerals in sediments: A review. *Earth-Sci. Reviews*, **21**:251–293.

Sleep, N. H., Nunn, J. A. and Chou, L., 1980, Platform Basins. *Ann. Rev. Earth Planet. Sciences*, **8**:17–34.

Sloss, L. L., 1962, Stratigraphic models in exploration. *Am. Assoc. Petrol. Geologists Bull.*, **46**:1050–1057.

Smith, D. B., 1974, Origin of tepees in the Upper Permian shelf carbonate rocks of Guadalupe Mountains, New Mexico. *Am. Assoc. Petrol. Geologists, Bull.*, **58**: 63–70.

Soil Survey Staff, 1975, Soil Taxonomy, U. S. Department of Agriculture Handbook 436, Washington, D.C., Government Printing Office, 754 p.

Soreghan, G. S., Elmore, R. D., Katz, B., Cogoini, M., Banerjee, S., 1997, Pedogenically enhanced magnetic susceptibility variations preserved in Paleozoic loessite. *Geology*, **25**:1003–1006.

Specht, R. W. and Brenner, R. L., 1979, Storm-wave genesis of bioclastic carbonates in Upper Jurassic epicontinental mudstones, East-Central Wyoming. *J. Sediment. Petrology*, **49**:1307–1322.

Stear, W. M., 1983, Morphological characteristics of ephemeral stream channel and overbank splay sandstone bodies in the Permian Lower Beaufort Group, Karoo Basin, South Africa. IAS Spec. Publ., No. **6**, Oxford, Blackwell Scientific Publications, 408 p.,

Stewart, A. D., 1982, Late Proterozoic rifting in NW Scotland: the genesis of the "Torridinian". *Jour. Geol. Soc., London*, **139**:413–420.

Stow, D. A. V., Alam, M., and Piper, D. J. W., 1984, Sedimentology of the Halifax Formation, Nova Scotia: Lower Paleozoic fine-grained turbidites. *Geol. Soc. London, Spec. Publ.*, **16**:127–144.

Strasser, A., 1984, Black pebble occurrence and genesis in Holocene carbonate sedements (Florida keys, Bahamas and Tunisia). *J. Sediment. Petrology*, **54**:1097–1123.

Summerfield, M. A., 1983, Petrography and diagenesis of silcrete from the Kalahari Basin and Cape coastal zone, southern Africa. *J. Sediment. Petrology*, **53**:895–909.

Sweeting, M. M., 1972, Karst Landforms. London, Macmillan Press, p. 362.

Swift, D. J. P., Figueiredo, A. G., Jr., Freeland, G. L., and Oertel, G. F., 1983, Hummocky cross-stratification and megaripples: a geological double standard? *J. Sediment. Petrology*, **53**:1295–1317.

Tandon, S. K., Good, A., Andrews, J. E., Dennis, P. F., 1995, Paleoenvironments of the dinosaur-bearing Lameta Beds (Maastrichtian), Narmada Valley, Central India. Paleogeo. Paleoclimat. *Paleoecology*, **117**:153–184.

Tandon, S. K. and Gibling, M. R., 1997, Calcretes at sequence boundaries in Upper Carboniferous cyclothems of the Sydney Basin, Atlantic Canada. *Sediment. Geology*, **112**:43–67.

Tankard, A. J., 1986, On the depositional response to thrusting and lithospheric flexure: examples from the Appalachian and Rocky Mountain basins, p. 369–394. *In*: Foreland Basins. (P. A. Allen, and P. Homewood, eds.). IAS Spec. Publ., **8**, Oxford, Blackwell Scientific Publications.

Tanner, P. W. G., 1998, Interstratal dewatering origin for polygonal patterns of sand-filled cracks: a case study from late Proterozoic metasediments of islay, Scotland. *Sedimentology*, **45**:71–89.

Taylor, A. M., and Goldring, R., 1993, Description and analysis of bioturbation and ichnofabric. *Jour. Geol. Soc., London*, **150**:141–148.

Taylor, J. M. C. and Illing, L.V., 1969, Holocene intertidal carbonate cementation, Qatar Persian Gulf. *Sedimentology*, **12**:69–107.

Thrailkill, John, 1976, Speleothems, p. 55–71. *In*: Stromatolites, Developments in Sedimentology 20, (M. R. Walter, editor), Amsterdam-Oxford-New York, Elsevier Scientific Pub. Development Co., 790 p.

Tillmann, R. W., Swift, D. J. P. and Walker, R. G., Eds., 1985, Shelf Sands and Sandstone Reservoirs. Tulsa, Oklahoma, SEPM Short Course Notes 13, 708 p.

Tourtelot, H. A., 1960, Origin and the use of the word 'shale'. *Am. Jour. Science*, Bradley volume, **258-A**:335–343.

Tucker, Maurice, 1982a, The Field Description of Sedimentary Rocks. Geol. Soc., London, Handbook Series, Open University Press & Halsted Press,

Tucker, M., 1982b, Storm-surge sandstones and the deposition of interbedded limestone: Late Precambrian, southern Norway. p. 363–370. *In*: Cyclic and Event Stratification. (G. Einsele. and A. Seilacher, eds.). Berlin-Heidelberg-New York, Springer-Verlag, 536 p.

Tucker, M., 1991, Sedimentary petrology, an introduction to the origin of sedimentary rocks. Second Edition, Oxford, Blackwell Scientific Publications. 260p.

Tuckhole, B. E. and Embley, R. W., 1984, Cenozoic regional erosion of the abyssal seafloor off South Africa, p. 145–164. *In*: Interregional Unconformities and Hydrocarbon Accumulation. (J. S. Schlee, ed.). Am. Assoc. Petrol. Geologists, Memoir **36**.

Twenhofel, W. H., 1932, Treatise on Sedimentation. Baltimore, Williams & Williams Co., 926 p.

Vail, P. R., Mitchum, R. M., Jr., Todd, R. G., Widmier, J. M., Thompson, S., III, Sangree, J. B., Bubb, J. N. and Hatlelid, W. G., 1977, Seismic stratigraphy and global changes of sea-level, p. 49–212. *In*: Seismic Stratigraphy—Application to Hydrocarbon Exploration. (C. E. Payton, ed.). Am. Assoc. Petrol. Geologists, Memoir **26**.

Valentine, K. W. G. and Dalrymple, J. B, 1976, Quaternary buried paleosols: a critical review. *Quaternary Research*, **6**:209–222.

Valeton, I., 1972, Bauxite—Developments in Soil Science I. Amsterdam, Elsevier Scientific Pub., 226 p.

Van Wagoner, J. C., Mitchum, R. M. Campion, K. M. and Rahmanian, V. D., 1990, Siliciclastic sequence stratigraphy in well logs, cores and outcrops. *Am. Assoc. Pet. Geologists*, Methods in Exploration Series 7, 55 p.

Van Wagoner, J. C., Posamentier, H. W., Mitchum, R. M., Vail, P. R., Sarg, J. F., Loutit, T. S. and Hardenbol, J., 1988, An overview of the fundamentals of sequence stratigraphy and key definitions, p. 39-45. *In*: Sea-level Changes: An Integrated Approach. (C. K. Wilgus, B. S. Hastings, C .G. St. C. Kendall, H. W. Posamentier, C. A. Ross, J. C. Van Wagoner, eds.). Tulsa, Oklahoma, SEPM Spec Publ. 42, 407 p.

Walkden, G. M. and Davies, J., 1983, Polyphase subaerial erosion surfaces in the Late Dinantian of Anglesey, North Wales. *Sedimentology*, **30**:861–878.

Walkden, G. M., 1974, Paleokarstic surfaces in Upper Visean (Carboniferous) limestones of the Derbyshire block, England. *J. Sediment. Petrology*, **44**: 1232–1247.

Walker, R. G., 1965, The origin and significance of internal sedimentary structures of turbidites. *Yorkshire Geol. Soc.*, **35**:1–29.

Walker, R. G., 1967, Turbidite sedimentary structures and their relationship to proximal and distal depositional environments. J. Sediment. Petrology, **37**: 25–43.

Walker, R. G., 1973, Mopping up of the turbidite mess, p. 1–37. *In*: Evolving Concepts in Sedimentology. (R. N. Ginsburg, ed.). Baltimore-London, The Johns Hopkins Univ. Press, 191 p.

Walker, R. G., 1992, Facies, facies models and modern stratigraphic concepts. p. 1–14. *In*: Facies Models: Response to Sea Level Changes. (R.G. Walker, and N.P. James, eds.). Geol. Assoc. Canada.

Walker, R. G. and Mutti, E., 1973, Turbidite facies and facies associations, p. 119–157. *In*: Turbidites and Deep-water Sedimentation. (G. V. Middleton and A. H. Bouma Chm.). Tulsa, Oklahoma, SEPM, Pacific Section, Short Course Lecture Notes, 157 p.

Walker, R. G., Duke, W. L. and Leckie, D. A., 1983, Hummocky stratification: significance of its variable bedding sequences: Discussion and reply and discussion. *Geol. Soc. Am., Bull.*, **94**:1245–1249.

Walker, R. G. and James, N. P., 1992, Facies Models—Response to Sea-Level Change. *Geoscience Canada*, Reprint Series **1**, 317 p.

Walter, M. R., 1972, Stromatolites and the biostratigraphy of the Australian Precambrian and Cambrian. Palaeontol. Assoc., Special Publ., Vol. 11, 191 p.

Walter, M. R., 1976, Geyserites of Yellowstone National Park: An example of abiogenic 'stromatolites', p. 87–112, *In*: Stromatolites, Developments in Sedimentology 20, (M. R. Walter, ed.). Amsterdam-Oxford-New York, Elsevier Sci. Pub. Development Co., 790 p.

Wanless, H. R., 1979, Limestone response to stress, pressure solution, and dolomitization. *J. Sediment. Petrology*, **49**:437–462.

Ward, W. C., Folk, R. L., and Wilson, J. L., 1970, Blackening of eolianite and caliche adjacent to saline lakes, Isla Mujeres, Quintana Roo, Mexico. *J. Sediment. Petrology*, **40**:548–555.

Watkins, N. D., and Kennett, J. P., 1972, Regional sedimentary disconformities and Upper Cenozoic changes in bottom water velocities between Australasia and Antarctica. *Antarctic Research Series*, **19**:273–293.

Watts, N. L., 1977, Pseudoanticlines and other structures in some calcretes of Botswana and South Africa. *Earth Surface Processes*, **2**:63–74.

Weaver, J. D., 1976, Seismically induced load structures in the Basal Coal Measures, South Wales. *Geol. Magazine*, **113**:535–543.

Webb, G. E., 1994, Paleokarst, paleosol and rocky-shore deposits at the Mississippian-Pennsylvanian unconformity, northwestern Arkansas. *Geol. Soc. Am., Bull.*, **106**:634–648.

Wernicke, B., Axen, G. J., 1988, On the role of isostasy in the evolution of normal fault systems. *Geology*, **16**:848–851.

Wheeler, J. E., 1964, Base level, lithostratigraphic surface and time stratigraphy. *Geol. Soc. Am., Bull.*, **75**:599–610.

Whitaker, J. H. McD., 1973, "Gutter casts", a new name for scour-and-fill structures: with examples from the Llandoverian of Ringerike and Malmoya, southern Norway. *Norsk Geologisk Tidsskrift*, **53**:403–417

White, N. and McKenzie, D., 1988, Formation of the "steer's head" geometry of sedimentary basins by differential stretching of the crust and mantle. *Geology*, **16**:250–253.

Whittaker, A, C., J. C. W, Cowie, J. W., Gibbons, W. H., E. A., House, M. R., Jenkins, D. G., Rawson, P. F, Rushton, A. W. A., Smith, D. G., Thomas A. T. and Wimbledon W. A., 1991, A Guide to Stratigraphical Procedure. *J. Geol. Soc., London*, **148**:813–824.

Williams, G. E., 1986, Precambrian permafrost horizons as indicators of paleoclimate. *Precambrian Res.*, **32**:233–242.

Wray, R. A. L., 1997, A global review of solutional weathering forms on quartz sandstones. *Earth-Science Rev.*, **42**:137–160.

Wright, M. E. and Walker, R. G., 1981, Cardium Formation (U. Cretaceous) at Seebe, Alberta—Storm transported sandstones and conglomerates in shallow marine depositional environments below fair-weather wave base. *Canadian J. Earth Sciences*, **18**:795–809.

Wright, V. P., 1982, The recognition and interpretation of paleokarsts: two examples from the Lower Carboniferous of South Wales. *J. Sediment. Petrology*, **52**:83–94.

Wright, V. P., 1984, The significance of needle fibre calcite in a Lower Carboniferous paleosol. *Geol. J.*, **19**:23–32.

Wright, V. P., 1992, A guide to early diagenesis in terrestrial setting. *In*: Diagenesis III, (K. H. Wolf, and G. V. Chilingarian, eds.). Elsevier Science-NL, 598 p.

Wright, V. P., 1994a, Losses and gains in weathering profiles and duripans, p. 95–123,. *In*: Quantitative Diagenesis: Recent Developments and Application to Reservoir Geology, (A. Parker, and B.W. Selwood, eds.). Netherlands. Kluwer Academic Publishers,

Wright, V. P., 1994b, Paleosols in shallow marine carbonate sequences. *Earth-Sci. Reviews.*, **35**:367–395.

Wright, V. P., and Vanstone, S. D., 1991a, Assessing the carbon dioxide content of ancient atmosphere using paleocalcretes: theoretical and empirical constraints. *Geol. Soc., London*, **148**:945–947.

Wright, V. P., Vanstone, V. D. and Robinson, D, 1991b, Ferrolysis in Arundian alluvial paleosols: evidence of a shift in the early Carboniferous monsoonal system. *J. Geol. Soc., London*, **148**:9–12.

Wright, V. P. and Smart, P.L., 1994, Paleokarst (dissolution diagenesis): its occurrence and hydrocarbon exploration significance. *In*: Diagenesis, IV, Development in Sedimentology, 51, (K. H. Wolf, and Chilingarian, G. V. eds.). Amsterdam, Elsevier Science B.V.

Wright, V. P., Platt, N. H., and Wimbledon, W. A., 1988 Biogenic laminar calcretes : evidence of root-mat horizons in paleosols. *Sedimentology*, **35**: 603–620.

Wright, V. P. and Marriott, S.B., 1993, The sequence stratigraphy of fluvial depositional systems: the role of floodplain sediment storage. *Sediment. Geology*, **86**:203–210.

Wright, V. Paul and Platt, N. H., 1995, Seasonal wetland carbonate sequences and dynamic catenas: a re-appraisal of palustrine limestones. *Sediment. Geology*, **99**:65–71.

Wright, V.P., Platt, N. H., Marriott, S.B. and Beck, V. H., 1995, A classification of rhizogenic (root-formed) calcretes, with examples from the Upper Jurassic-Lower Cretaceous of Spain and Upper Cretaceous of southern France. *Sediment. Geology,* **100**:143–158.

Wunderlich, F., 1979, Die Insel Mellum (sudliche Nordsec). Dynamische Prozesse und Sedimentgefuge. I. Sudwatt, ubergangszme und Hochflache. *Senckenbergiana Marit.*, **11**:59–113.

Wurster, P., 1958. Geometrie und geologie von kreuzschichtungskorpern. *Geol. Rundschau*, **47**:322–359.

Yaalon, D. H., 1960, Some implications of fundamental concepts of pedology in soil classification. 7th Internat. Cong. Soil Science, Madison, Wisconsin, U.S.A., **16**:119–123.

Yaalon, D. H., 1966, Chart for quantitative estimation of mottling and of nodules in soil profiles. *Soil Sci.*, **102,(3)**: 212–213.

Yaalon, D. H., 1988, Calcic horizon and calcrete in Aridic soils and paleosols: progress in the last twentytwo years. *Soil Sci. Soc. Am.*, Agron., (Abstract).

Yeakel, L. S., 1959, Tuscarora, Juniata, and Bald Eagle paleocurrents and paleogeography in the central Appalachians. *Geol. Soc. America, Bull.*, **73**:1515–1540.

Zbinden, E. A., Holland, H. D. and Feakes, C. R., 1988, The Sturgeon Falls paleosols and the composition of the atmosphere 1.1 Ga B.P. *Precambrian Res.*, **42**:141–163.

APPENDIX

North American Stratigraphic Code[1]

NORTH AMERICAN COMMISSION ON STRATIGRAPHIC NOMENCLATURE

Foreword

This code of recommended procedures for classifying and naming stratigraphic and related units has been prepared during a four-year period, by and for North American earth scientists, under the auspices of the North American Commission on Stratigraphic Nomenclature. It represents the thought and work of scores of persons, and thousands of hours of writing and editing. Opportunities to participate in and review the work have been provided throughout its development, as cited in the Preamble, to a degree unprecedented during preparation of earlier codes.

Publication of the International Stratigraphic Guide in 1976 made evident some insufficiencies of the American Stratigraphic Codes of 1961 and 1970. The Commission considered whether to discard our codes, patch them over, or rewrite them fully, and chose the last. We believe it desirable to sponsor a code of stratigraphic practice for use in North America, for we can adapt to new methods and points of view more rapidly than a worldwide body. A timely example was the recognized need to develop modes of establishing formal nonstratiform (igneous and high-grade metamorphic) rock units, an objective which is met in this Code, but not yet in the Guide.

The ways in which this Code differs from earlier American codes are evident from the Contents. Some categories have disappeared and others are new, but this Code has evolved from earlier codes and from the International Stratigraphic Guide. Some new units have not yet stood the test of long practice, and conceivably may not, but they are introduced toward meeting recognized and defined needs of the profession. Take this Code, use it, but do not condemn it because it contains something new or not of direct interest to you. Innovations that prove unacceptable to the profession will expire without damage to other concepts and procedures, just as did the geologic-climate units of the 1961 Code.

[1]Reprinted by permission from American Association of Petroleum Geologists Bulletin, v. 67, no. 5 (May 1983), p. 841–875.

This code is necessarily somewhat innovative because of: (1) the decision to write a new code, rather than to revise the old; (2) the open invitation to members of the geologic profession to offer suggestions and ideas, both in writing and orally; and (3) the progress in the earth sciences since completion of previous codes. This report strives to incorporate the strength and acceptance of established practice, with suggestions for meeting future needs perceived by our colleagues; its authors have attempted to bring together the goods from the past, the lessons of the Guide, and carefully reasoned provisions for the immediate future.

Participants in preparation of this Code are listed in Appendix I, but many others helped with their suggestions and comments. Major contributions were made by the members, and especially the chairmen, of the named subcommittees and advisory groups under the guidance of the Code Committee, chaired by Steven S. Oriel, who also served as principal, but not sole, editor. Amidst the noteworthy contributions by many, those of James D. Aitken have been outstanding. The work was performed for and supported by the Commission, chaired by Malcolm P. Weiss from 1978 to 1982.

This Code is the product of a truly North American effort. Many former and current commissioners representing not only the ten organizational members of the North American Commission on Stratigraphic Nomenclature (Appendix II), but other institutions as well, generated the produce. Endorsement by constituent organizations is anticipated, and scientific communication will be fostered if Canadian, United States, and Mexican scientists, editors, and administrators consult Code recommendations for guidance in scientific reports. The Commission will appreciate reports of formal adoption or endorsement of the Code, and asks that they be transmitted to the Chairman of the Commission (c/o American Association of Petroleum Geologists, Box 979, Tulsa, Oklahoma 74101, U.S.A.).

Any code necessarily represents but a stage in the evolution of scientific communication. Suggestions for future changes of, or additions to, the North American Stratigraphic Code are welcome. Suggested and adopted modifications will be announced to the profession, as in the past, by serial Notes and Reports published in the *Bulletin* of the American Association of Petroleum Geologists. Suggestions may be made to representatives of your association or agency who are current commissioners, or directly to the Commission itself. The Commission meets annually during the national meetings of the Geological Society of American.

1982 North American Commission on Stratigraphic Nomenclature

Contents

PART I. PREAMBLE

Background

Perspective

Codes of Stratigraphic Nomenclature prepared by the American Commission on Stratigraphic Nomenclature (ACSN, 1961) and its predecessor (Committee on Stratigraphic Nomenclature, 1933) have been used widely as a basis for stratigraphic terminology. Their formulation was a response to needs recognized during the past century by government surveys (both national and local) and by editors of scientific journals for uniform standards and common procedures in defining and classifying formal rock bodies, their fossils, and the time spans represented by them. The most recent Code (ACSN, 1970) is a slightly revised version of that published in 1961, incorporating some minor amendments adopted by the Commission between 1962 and 1969. The Codes have served the profession admirably and have been drawn upon heavily for codes and guides prepared in other parts of the world (ISSC, 1976, p. 104-106). The principles embodied by any code, however, reflect the state of knowledge at the time of its preparation, and even the most recent code is now in need of revision.

New concepts and techniques developed during the past two decades have revolutionized the earth sciences. Moreover, increasingly evident have been the limitations of previous codes in meeting some needs of Precambrian and Quaternary geology and in classification of plutonic, high-grade metamorphic, volcanic, and intensely deformed rock assemblages. In addition, the important contributions of numerous international stratigraphic organizations associated with both the International Union of Geological Sciences (IUGS) and UNESCO, including working groups of the International Geological Correlation Program (IGCP), merit recognition and incorporation into a North American code.

For these and other reasons, revision of the American Code has been undertaken by committees appointed by the North American Commission on Stratigraphic Nomenclature (NACSN). The Commission, founded as the American Commission on Stratigraphic Nomenclature in 1946 (ACSN, 1947), was renamed the NACSN in 1978 (Weiss, 1979b) to emphasize that delegates from ten organizations in Canada, the United States, and Mexico represent the geological profession throughout North America (Appendix II).

Although many past and current members of the Commission helped prepare this revision of the Code, the participation of all interested geologists has been sought (for example, Weiss, 1979a). Open forums were held at the national meetings of both the Geological Society of America at San Diego in November, 1979, and the American Association of Petroleum Geologists at Denver in June 1980, at which comments and suggestions were offered by more than 150 geologists. The resulting draft of this report was printed, through the courtesy of the Canadian Society of Petroleum Geologists, on October 1, 1981, and additional comments were invited from the profession for a period of one year before submittal of this report to the Commission for adoption. More than 50 responses were received with sufficient suggestions for improvement to prompt moderate revision of the printed draft (NACSN, 1981). We are particularly indebted to Hollis D. Hedberg and Amos Salvador for their exhaustive and perceptive reviews of early drafts of this Code, as well as to those who responded to the request for comments. Participants in the preparation and revisions of this report, and conferees, are listed in Appendix 1.

Some of the expenses incurred in the course of this work were defrayed by National Science Foundation Grant EAR 7919845, for which we express appreciation. Institutions represented by the participants have been especially generous in their support.

Scope

The North American Stratigraphic Code seeks to describe explicit practices for classifying and naming all formally defined geologic units. *Stratigraphic procedures* and principles, although developed initially to bring order to strata and the events recorded therein, are applicable to all earth materials, not solely to strata. They promote systematic and rigorous study of the composition, geometry, sequence, history and genesis of rocks and unconsolidated materials. They provide the framework within which time and space relations among rock bodies that constitute the Earth are ordered systematically. Stratigraphic procedures are used not only to reconstruct the history of the Earth and of extra-terrestrial bodies, but also to define the distribution and geometry of some commodities needed by society. *Stratigraphic classification* systematically arranges and partitions bodies of rock or unconsolidated materials of the Earth's crust into units based on their inherent properties or attributes.

A *stratigraphic code* or guide is a formulation of current views on stratigraphic principles and procedures designed to promote standardized classification and formal nomenclature of rock materials. It provides the basis for formalization of the language used to denote rock units and their spatial and temporal relations. To be effective, a code must be widely accepted and used; geologic organizations and journals may adopt its recommendations for nomenclatural procedure. Because any code embodies only current concepts and principles, it should have the flexibility to provide for both changes and additions to improve its relevance to new scientific problems.

Any system of nomenclature must be sufficiently explicit to enable users to distinguish objects that are embraced in a class from those that are not. This stratigraphic code makes no attempt to systematize structural, petrographic, paleontologic, or physiographic terms. Terms from these other fields that are used as part of formal stratigraphic names should be sufficiently general as to be unaffected by revisions of precise petrographic or other classifications.

The objective of a system of classification is to promote unambiguous communication in a manner not so restrictive as to inhibit scientific progress. To minimize ambiguity, a code must promote recognition of the distinction between observable features (reproducible data) and inferences or interpretations. Moreover, it should be sufficiently adaptable and flexible to promote the further development of science.

Stratigraphic classification promotes understanding of the *geometry* and *sequence* of rock bodies. The development of stratigraphy as a science required formulation of the Law of Superposition to explain sequential stratal relations. Although superposition is not applicable to many igneous, metamorphic, and tectonic rock assemblages, other criteria (such as cross-cutting relations and isotopic dating) can be used to determine sequential arrangements among rock bodies.

The term *stratigraphic unit* may be defined in several ways. Etymological emphasis requires that it be a stratum or assemblage of adjacent strata distinguished by any or several of the many properties that rocks may possess (ISSC, 1976, p. 13). The scope of stratigraphic classification and procedures, however, suggests a broader definition: a naturally occurring body of rock or rock material distinguished from adjoining rock on the basis of some stated property or properties. Commonly used properties include composition, texture, included fossils, magnetic signature, radioactivity, seismic velocity, and age. Sufficient care is required in defining the boundaries of a unit to enable others to distinguish the material body from those adjoining it. Units based on one property commonly do not coincide with those based on another and, therefore, distinctive terms are needed to identify the property used in defining each unit.

The adjective *stratigraphic* is used in two ways in the remainder of this report. In discussions of lithic (used here as synonymous with "lithologic") units, a conscious attempt is made to restrict the term to lithostratigraphic or layered rocks and sequences that obey the Law of Superposition. For nonstratiform rocks (of plutonic or tectonic origin, for example), the term *lithodemic* (see Article 27) is used. The adjective *stratigraphic* is also used in a broader sense to refer to those procedures derived from stratigraphy which are now applied to all classes of earth materials.

An assumption made in the material that follows is that the reader has some degree of familiarity with basic principles of stratigraphy as outlined, for example, by Dunbar and Rodgers (1957), Weller (1960), Shaw (1964), Matthews (1974), or the International Stratigraphic Guide (ISSC, 1976).

Relation of Codes of International Guide

Publication of the International Stratigraphic Guide by the International Subcommission on the Stratigraphic Classification (ISSC, 1976), which is being endorsed and adopted throughout the world, played a part in prompting examination of the American Stratigraphic Code and the decision to revise it.

The International Guide embodies principles and procedures that had been adopted by several national and regional stratigraphic committees and commissions. More than two decades of effort by H.D. Hedberg and other members of the Subcommission (ISSC, 1976, p. VI, I, 3) developed the consensus required for preparation of the Guide. Although the Guide attempts to cover, all kinds of rocks and the diverse ways of investigating them, it is necessarily incomplete. Mechanisms are needed to stimulate individual innovations toward promulgating new concepts, principles, and practices which subsequently may be found worthy of inclusion in later editions of the Guide. The flexibility of national and regional committees or commissions enables them to perform this function more readily than an international subcommission, even while they adopt the Guide as the international standard of stratigraphic classification.

A guiding principle in preparing this Code has been to make it as consistent as possible with the International Guide, which was endorsed by the ACSN in 1976, and at the same time to foster further innovations to meet the expanding and changing needs of earth scientists on the North American continent.

Overview

Categories Recognized

An attempt is made in this Code to strike a balance between serving the needs of those in evolving specialties and resisting the proliferation of categories of units. Consequently, more formal categories are recognized here than in previous codes or in the International Guide (ISSC, 1976). On the other hand, no special provision is made for formalizing certain kinds of units (deep oceanic, for example) which may be accommodated by available categories.

Four principal categories of units have previously been used widely in traditional stratigraphic work; these have been termed lithostratigraphic, biostratigraphic, chronostratigraphic, and geochronologic and are distinguished as follows:

1. A *lithostratigraphic unit* is a stratum or body of strata, generally but not invariably layered, generally but not invariably tabular, which conforms to the Law of Superposition and is distinguished and delimited on the basis of lithic characteristics and stratigraphic position. Example: Navajo Sandstone.

2. A *biostratigraphic unit* is a body of rock defined and characterized by its fossil content. Example: *Discoaster multiradiatus* Interval Zone.

3. A *chronostratigraphic unit* is body of rock established to serve as the material reference for all rocks formed during the same span of time. Example: Devonian System. Each boundary of a chronostratigraphic unit is synchronous. Chronostratigraphy provides a means of organizing strata into units based on their age relations. A chronostratigraphic body also serves as the basis for defining the specific interval of geologic time, or geochronologic unit, represented by the referent.

4. A *geochronologic unit* is a division of time distinguished on the basis of the rock records preserved in a chronostratigraphic unit. Example: Devonian Period.

The first two categories are comparable in that they consist of material units defined on the basis of content. The third category differs from the first two in that it serves primarily as the standard for recognizing and isolating materials of a specific age. The fourth, in contrast, is not a material, but rather a conceptual, unit; it is a division of time. Although a geochronologic unit is not a stratigraphic body, it is so intimately tied to chronostratigraphy that the two are discussed properly together.

Properties and procedures that may be used in distinguishing geologic units are both diverse and numerous (ISSC, 1976, p. 1, 96; Harland, 1977, p. 230), but all may be assigned to the following principal classes of categories used in stratigraphic classification (Table 1), which are discussed below:

I. Material categories based on content, inherent attributes, or physical limits,

II. Categories distinguished by geologic age,

A. Material categories used to define temporal spans, and

B. Temporal categories.

Material Categories Based on Content or Physical Limits

The basic building blocks for most geologic work are rock bodies defined on the basis of composition and related lithic characteristics, or on their physical, chemical, or biologic content or properties. Emphasis is placed on the relative objectivity and reproducibility of data used in defining units within each category.

Table 1. Categories of Units Defined*

MATERIAL CATEGORIES BASED ON CONTENT OR PHYSICAL LIMITS

- *Lithostratigraphic* (22)
- *Lithodemic* (31)**
- *Magnetopolarity* (44)
- Biostratigraphic (48)
- *Pedostratigraphic* (55)
- *Allostratigraphic* (58)

CATEGORIES EXPRESSING OR RELATED TO GEOLOGIC AGE

- Material Categories Used to Define Temporal Spans
  - Chronostratigraphic (66)
  - *Polarity-Chronostratigraphic* (83)
- Temporal (Non-Material) Categories
  - Geochronologic (80)
  - *Polarity-Chronologic* (88)
  - *Diachronic* (91)
  - *Geochronometric* (96)

*Numbers in parentheses are the numbers of the Articles where units are defined.
**Italicized categories are those introduced or developed since publication of the previous code (ACSN, 1970).

Foremost properties of rocks are composition, texture, fabric, structure, and color, which together are designated *lithic characteristics*. These serve as the basis for distinguishing and defining the most fundamental of all formal units. Such units based primarily on composition are divided into two categories (Henderson and others, 1980): lithostratigraphic (Article 22) and lithodemic (defined here in Article 31). A lithostratigraphic unit obeys the Law of Superposition, whereas a lithodemic unit does not. A *lithodemic unit* is a defined body of predominantly intrusive, highly metamorphosed, or intensely deformed rock that, because it is intrusive or has lost primary structure through metamorphism or tectonism, generally does not conform to the Law of Superposition.

Recognition during the past several decades that remanent magnetism in rocks records the Earth's past magnetic characteristics (Cox, Doell, and Dalrymple, 1963) provides a powerful new tool encompassed by magnetostratigraphy (McDougall, 1977; McElhinny, 1978). Magnetostratigraphy (Article 43) is the study of remanent magnetism in rocks; it is the record of the Earth's magnetic polarity (or field reversals), dipole-field-pole position (including apparent polar wander), the non-dipole component (secular variation), and field intensity. Polarity is of particular utility and is used to define a *magnetopolarity unit* (Article 44) as a body of rock identified by its remanent magnetic polarity (ACSN, 1976; ISSC, 1979). Empirical demonstration of uniform polarity does not necessarily have direct temporal connotations because the remanent magnetism need not be related to rock deposition or crystallization. Nevertheless, polarity is a physical attribute that may characterize a body of rock.

Biologic remains contained in, or forming, strata are uniquely important in stratigraphic practice. First, they provide the means of defining and recognizing material units based on fossil content (biostratigraphic units). Second, the irreversibility of organic evolution makes it possible to partition enclosing strata temporally. Third, biologic remains provide important data for the reconstruction of ancient environments of deposition.

Composition also is important in distinguishing pedostratigraphic units. A *pedostratigraphic unit* is a body of rock that consists of one or more pedologic horizons developed in one or more lithic units now buried by a formally defined lithostratigraphic or allostratigraphic unit or units. A pedostratigraphic unit is the part of a buried soil characterized by one or more clearly defined soil horizons containing pedogenically formed minerals and organic compounds. Pedostratigraphic terminology is discussed below and in Article 55.

Many upper Cenozoic, especially Quaternary, deposits are distinguished and delineated on the basis of content, for which lithostratigraphic classification is appropriate. However, others are delineated on the basis of criteria other than content. To facilitate the reconstruction of geologic history, some compositionally similar deposits in vertical sequence merit distinction as separate stratigraphic units because they are the products of different processes; others merit distinction because they are of demonstrably different ages. Lithostratigraphic classification of these units is impractical and a new approach, allostratigraphic classification, is introduced here and may prove applicable to older deposits as well. An *allostratigraphic unit* is a mappable stratiform body of sedimentary rock defined and identified on the basis of bounding discontinuities (Article 58 and related Remarks).

Geologic-Climate units, defined in the previous Code (ACSN, 1970, p. 31), are abandoned here because they proved to be of dubious utility. Inferences regarding climate are subjective and too tenuous a basis for the definition of formal geologic units. Such inferences commonly are based on deposits assigned more appropriately to lithostratigraphic or allostratigraphic units and may be expressed in terms of diachronic units (defined below).

Categories Expressing or Related to Geologic Age

Time is a single, irreversible continuum. Nevertheless, various categories of units are used to define intervals of geologic time, just as terms having different bases, such as Paleolithic, Renaissance, and Elizabethan, are used to designate, specific periods of human history. Different temporal categories are established to express intervals of time distinguished in different ways.

Major objectives of stratigraphic classification are to provide a basis for systematic ordering of the time and space relations of rock bodies and to establish a time framework for the discussion of geologic history. For such purposes, units of geologic time traditionally have been named to represent the span of time during which a well-described sequence of rock, or a

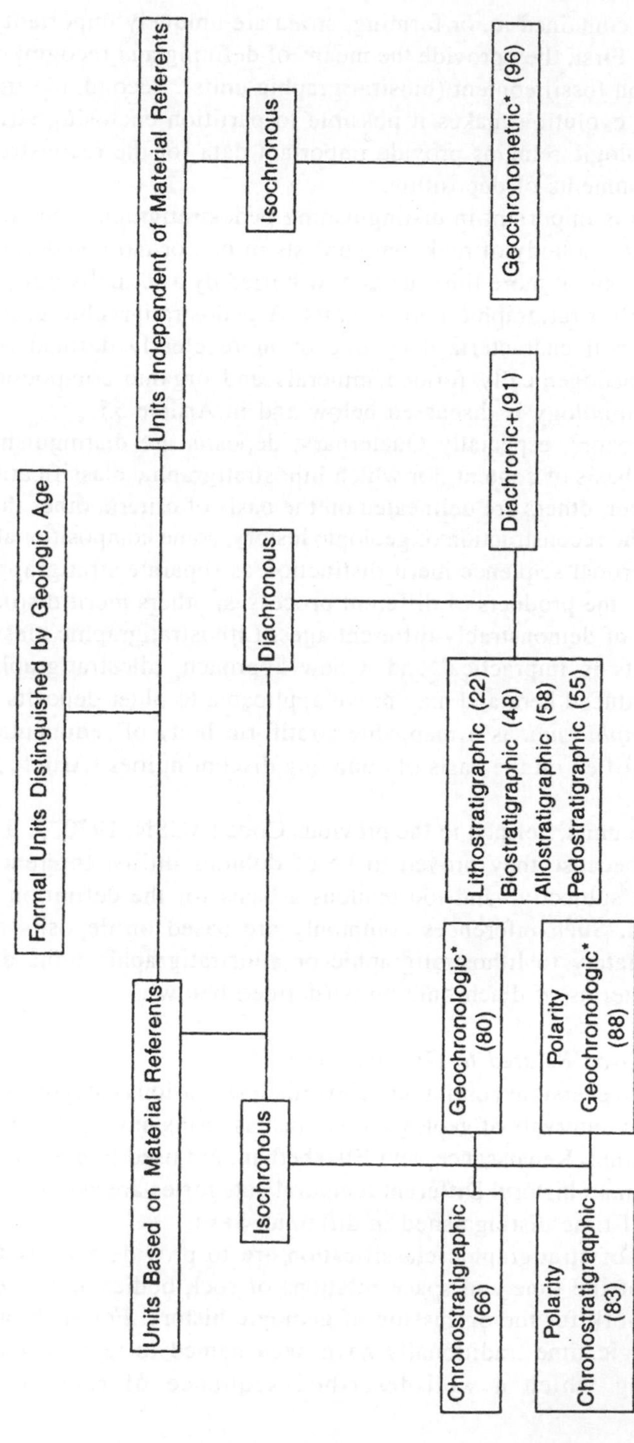

Fig. 1: Relation of geologic time units to the kinds of rock-unit referents on which most are based.

chronostratigraphic unit, was deposited ("time units based on material referents," Fig. 1). This procedure continues, to the exclusion of other possible approaches, to be standard practice in studies of Phanerozoic rocks. Despite admonitions in previous American codes and the International Stratigraphic Guide (ISSC, 1976, p. 81) that similar procedures should be applied to the Precambrian, no comparable chronostratigraphic units, or geochronologic units derived therefrom, proposed for the Precambrian have yet been accepted worldwide. Instead, the IUGS Subcommission on Precambrian Stratigraphy (since 1979) and its Working Groups (Harrison and Peterman, 1980) recommend division of Precambrian time into *geochronometric units* having no material referents.

A distinction is made throughout this report between *isochronous* and *synchronous*, as urged by Cumming, Fuller, and Porter (1959, p. 730), although the terms have been used synonymously by many. *Isochronous* means of equal duration; *synchronous* means simultaneous, or occurring at the same time. Although two rock bodies of very different ages may be formed during equal durations of time, the term *isochronous* is not applied to them in the earth sciences. Rather, isochronous bodies are those bounded by synchronous surfaces and formed during the same span of time. *Isochron,* in contrast, is used for a line connecting points of equal age on a graph representing physical or chemical phenomena; the line represents the same or equal time. The adjective *diachronous* is applied either to a rock unit with one or two bounding surfaces which are not synchronous, or to a boundary which is not synchronous (which "transgresses time").

Two classes of time units based on material referents, or stratotypes, are recognized (Fig. 1). The first is that of the traditional and conceptually isochronous units, and includes *geochronologic units,* which are based on *chronostratigraphic units*, and *polarity-geochronologic units*. These isochronous units have worldwide applicability and may be used even in areas lacking a material record of the named span of time. The second class of time units, newly defined in this Code, consists of *diachronic units* (Article 91), which are based on rock bodies known to be diachronous. In contrast to isochronous units, a diachronic term is used only where a material referent is present; a diachronic unit is coextensive with the material body or bodies on which it is based.

A *chronostratigraphic unit*, as defined above and in Article 66, is a body of rock established to serve as the material reference for all rocks formed during the same span of time; its boundaries are synchronous. It is the referent for a *geochronologic unit,* as defined above and in Article 80. Internationally accepted and traditional chronostratigraphic units were based initially on the time spans of lithostratigraphic units, biostratigraphic units, or other features of the rock record that have specific durations. In sum, they form the Standard Global Chronostratigraphic Scale (ISSC, 1976, p. 76-81; Harland, 1978), consisting of established systems and series.

A *polarity-chronostratigraphic unit* is a body of rock that contains a primary magnetopolarity record imposed when the rock was deposited or crystallized (Article 83). It serves as a material standard or referent for a part of geologic time during which the Earth's magnetic field had a characteristic polarity or sequence of polarities; that is, for a *polarity-chronologic unit* (Article 88).

A *diachronic unit* comprises the unequal spans of time represented by one or more specific diachronous rock bodies (Article 91). Such bodies may be lithostratigraphic, biostratigraphic, pedostratigraphic, allostratigraphic, or an assemblage of such units. A diachronic unit is applicable only where its material referent is present.

A *geochronometric* (or chronometric) *unit* is an isochronous direct division of geologic time expressed in years (Article 96). It has no material referent.

Pedostratigraphic Terms

The definition and nomenclature for pedostratigraphic[2] units in this Code differ from those for soil-stratigraphic units in the previous Code (ACSN, 1970, Article 18), by being more specific with regard to content, boundaries, and the basis for determining stratigraphic position.

The term "soil" has different meanings to the geologist, the soil scientists, the engineer, and the layman, and commonly has no stratigraphic significance. The term *paleosol* is currently used in North America for any soil that formed on a landscape of the past; it may be a buried soil, a relict soil, or an exhumed soil (Ruhe, 1965; Valentine and Dalrymple, 1976).

A *pedologic soil* is composed of one or more soil horizons[3]. A *soil horizon* is a layer within a pedologic soil that (1) is approximately parallel to the soil surface, (2) has distinctive physical, chemical, biological, and morphological properties that differ from those of adjacent, genetically related, soil horizons, and (3) is distinguished from other soil horizons by objective compositional properties that can be observed or measured in the field. The physical boundaries of buried pedologic horizons are objective traceable boundaries with stratigraphic significance. A buried pedologic soil provides the material basis for definition of a stratigraphic unit in pedostratigraphic classification (Article 55), but a buried pedologic soil may be somewhat more inclusive than a pedostratigraphic unit. A pedologic soil may contain both an 0-horizon and the entire C-horizon (Fig. 6), whereas the former is excluded and the latter need not be included in a pedostratigraphic unit.

The definition and nomenclature for pedostratigraphic units in this Code differ from those of soil stratigraphic units proposed by the International Union for Quaternary Research and International Society of Soil Science (Parsons,

---

[2]From Greek, *pedon,* ground or soil.

[3]As used in a geological sense, a *horizon* is a surface or line. In pedology, however, it is a body of material, and such usage is continued here.

1981). The pedostratigraphic unit, geosol, also differs from the proposed INQUA-ISSS soil-stratigraphic unit, pedoderm, in several ways, the most important of which are: (1) a geosol may be in any part of the geologic column, whereas a pedoderm is a surficial soil; (2) a geosol is a buried soil, whereas a pedoderm may be a buried, relict, or exhumed soil; (3) the boundaries and stratigraphic position of a geosol are defined and delineated by criteria that differ from those for a pedoderm; and (4) a geosol may be either all or only a part of a buried soil, whereas a pedoderm is the entire soil.

The term *geosol*, as defined by Morrison (1967, p. 3), is a laterally traceable, mappable, geologic weathering profile that has a consistent stratigraphic position. The term is adopted and redefined here as the fundamental and only unit in formal pedostratigraphic classification (Article 56).

Formal and Informal Units

Although the emphasis in this Code is necessarily on formal categories of geologic units, informal nomenclature is highly useful in stratigraphic work.

Formally named units are those that are named in accordance with an established scheme of classification; the fact of formality is conveyed by capitalization of the initial letter of the *rank* or *unit* term (for example, Morrison Formation). Informal units, whose unit terms are ordinary nouns, are not protected by the stability provided by proper formalization and recommended classification procedures. Informal terms are devised for both economic and scientific reasons. Formalization is appropriate for those units requiring stability of nomenclature, particularly those likely to be extended far beyond the locality in which they were first recognized. Informal terms are appropriate for casually mentioned, innovative, and most economic units, those defined by unconventional criteria, and those that may be too thin to map at usual scales.

Casually mentioned geologic units not defined in accordance with this Code are informal. For many of these, there may be insufficient need or information, or perhaps an appropriate basis, for formal designations. Informal designations as beds or lithozones (the pebbly beds, the shaly zone, third coal) are appropriate for many such units.

Most economic units, such as aquifers, oil sands, coal beds, querry layers, and ore-bearing "reefs," are informal, even though they may be named. Some such units, however, are so significant scientifically and economically that they merit formal recognition as beds, members, or formations.

Innovative approaches in regional stratigraphic studies have resulted in the recognition and definition of units best left as informal, at least for the time being. Units bounded by major regional unconformities on the North American craton were designated "sequences" (example: Sauk sequence) by Sloss (1963).

Major unconformity-bounded units also were designated "synthems' by Chang (1975), who recommended that they be treated formally. Marker-defined units that are continuous from one lithofacies to another were designated "formats" by Forgotson (1957). The term "chronosome" was proposed by Schultz (1982) for rocks of diverse facies corresponding to geographic variations in sedimentation during an interval of deposition identified on the basis of bounding stratigraphic markers. Successions of faunal zones containing evolutionally related forms, but bounded by non-evolutionary biotic discontinuities, were termed "biomeres" (Palmer, 1965). The foregoing are only a few selected examples to demonstrate how informality provides a continuing avenue for innovation.

The terms *magnafacies* and *parvafacies*, coined by Caster (1934) to emphasize the distinction between lithostratigraphic and chronostratigraphic units in sequences displaying marked facies variation, have remained informal despite their impact on clarifying the concepts involved.

Tephrochronologic studies provide examples of informal units too thin to map at conventional scales but yet invaluable for dating important geologic events. Although some such units are named for physiographic features and places where first recognized (e.g., Guaje pumice bed, where it is not mapped as the Guaje Member of the Bandelier Tuff), others bear the same name as the volcanic vent (e.g., Huckleberry Ridge ash bed of Izett and Wilcox, 1981).

Informal geologic units are designated by ordinary nouns, adjectives or geographic terms and lithic or unit-terms that are not capitalized (chalky formation or beds, St. Francis coal).

No geologic unit should be established and defined, whether formally or informally, unless its recognition serves a clear purpose.

Correlation

Correlation is a procedure for demonstrating correspondence between geographically separated parts of a geologic unit. The term is a general one having diverse meanings in different disciplines. Demonstration of temporal correspondence is one of the most important objectives of stratigraphy. The term "correlation" frequently is misused to express the idea that a unit has been identified or recognized.

Correlation is used in this Code as the demonstration of correspondence between two geologic units in both some defined property and relative stratigraphic position. Because correspondence may be based on various properties, three kinds of correlation are best distinguished by more specific terms. *Lithocorrelation* links units of similar lithology and stratigraphic position (or sequential or geometric relation, for lithodemic units). *Biocorrrelation* expresses similarity of fossil content and biostratigraphic position. *Chronocorrelation* expresses correspondence in age and in chronostratigraphic position.

Other terms that have been used for the similarity of content and stratal succession are homotaxy and chronotaxy. *Homotaxy* is the similarity in separate regions of the serial arrangement or succession of strata of comparable compositions or of included fossils. The term is derived from *homotaxis*, proposed by Huxley (1862, p. xlvi) to emphasize that similarity in succession does not prove age equivalence of comparable units. The term *chronotaxy* has been applied to similar stratigraphic sequences composed of units which are of equivalent age (Henbest, 1952, p. 310).

Criteria used for ascertaining temporal and other types of correspondence are diverse (ISSC, 1976, p. 86-93) and new criteria will emerge in the future. Evolving statistical tests, as well as isotopic and paleomagnetic techniques, complement the traditional paleontologic and lithologic procedures. Boundaries defined by one set of criteria need not correspond to those defined by others.

Part II Articles

Introduction

*Article 1.—**Purpose***

This Code describes explicit stratigraphic procedures for classifying and naming geologic units accorded formal status. Such procedures, if widely adopted, assure consistent and uniform usage in classification and terminology and therefore promote unambiguous communication.

*Article 2.—**Categories***

Categories of formal stratigraphic units, though diverse, are of three classes (Table 1). The first class is of rock-material categories based on inherent attributes or content and stratigraphic position, and includes lithostratigraphic, lithodemic, magnetopolarity, biostratigraphic, pedostratigraphic, and allostratigraphic units. The second class is of material categories used as standards for defining spans of geologic time, and includes chronostratigraphic and polarity-chronostratigraphic units. The third class is of non-material temporal categories, and includes geochronologic, polarity-chronologic, geochronometric, and diachronic units.

General Procedures

Definition of Formal Units

*Article 3.-**Requirements for Formally Named Geologic Units***

Naming, establishing, revising, redefining, and abandoning formal geologic units require publication in a recognized scientific medium of a comprehensive statement which includes: (i) intent to designate or modify a formal unit; (ii) designation of category and rank of unit; (iii) selection and derivation of name; (iv) specification of stratotype (where applicable); (v) description of unit; (vi) definition of boundaries; (vii) historical background; (viii) dimensions, shape, and other regional aspects; (ix) geologic age; (x) correlations; and possibly (xi) genesis (where applicable). These requirements apply to subsurface and offshore, as well as exposed, units.

*Article 4.—**Publication***[4]

"Publication in a recognized scientific medium" in conformance with this Code means that a work, when first issued, must (1) be reproduced in ink on paper or by some method that assures numerous identical copies and wide distribution; (2) be issued for the purpose of scientific, public, permanent record; and (3) be readily obtainable by purchase or free distribution.

Remarks: (a) ***Inadequate publication***—The following do not constitute publication within the meaning of the Code: (1) distribution of microfilms, microcards, or matter reproduced by similar methods; (2) distribution to colleagues or students of a note, even if printed, in explanation of an accompanying illustration; (3) distribution of proof sheets; (4) open-file release; (5) theses, dissertations, and dissertation abstracts; (6) mention at a scientific or other meeting; (7) mention in an abstract, map explanation, or figure caption; (8) labeling of a rock specimen in a collection; (9) mere deposit of a document in a library; (10) anonymous publication; or (11) mention in the popular press or in a legal document.

(b) ***Guidebooks.***—A guidebook with distribution limited to participants of a field excursion does not meet the test of availability. Some organizations publish and distribute widely large editions of serial guidebooks that include referred regional papers; although these do meet the tests of scientific purpose and availability, and therefore constitute valid publication, other media are preferable.

*Article 5.—**Intent and Utility***

To be valid, a new unit must serve a clear purpose and be duly proposed and duly described, and the intent to establish it must be specified. Casual mention of a unit, such as "the granite exposed near the Middleville schoolhouse," does not establish a new formal unit, nor does mere use in a table, columnar section, or map.

Remark: (a) ***Demonstration of purpose served.***—The initial definition or revision of a named geologic unit constitutes, in essence, a proposal. As such, it lacks status until use by others demonstrates that a clear purpose has been served. A unit becomes established through repeated demonstration of its utility. The decision not to use a newly proposed or a newly revised term requires a full discussion of its unsuitability.

*Article 6.—**Category and Rank***

The category and rank of a new or revised unit must be specified.

Remark: (a) ***Need for specification.***—Many stratigraphic controversies have arisen from confusion or misinterpretation of the category of a unit (for exam-

---

[4]This article is modified slightly from a statement by the International Commission of Zoological Nomenclature (1964, p. 7–9).

Table 2. **Categories and Ranks of Units Defined in This Code***

A. Material Units

Lithostratigraphic	Lithodemic	Magnetopolarity	Biostratigraphic	Pedostratigraphic	Allostratigraphic
Supergroup	Supersuite (Complex)				
Group	Suite (Complex)	Polarity Superzone			Allogroup
Formation	*Lithodeme*	*Polarity sons*	*Biosons* (Interval Assemblage or Abundance)	*Geosol*	*Alloformation*
Member (or Lens, or Tongue)		Polarity Subzone	Subbiozone		*Allomember*
Bed(s) or Flow(s)					

B. Temporal and Related Chronostratigraphic Units

Chronostratigraphic	Geochronologic Geochronometric	Polarity Chrono-Stratigraphic	Polarity Chronologic	Diachronic
Eonothem	Eon	Polarity Superchronozone	Polarity Superchron	
Erathem (Supersystem)	Era (Superperiod)			
System *(Subsystem)*	*Period* *(Subperiod)*	*Polarity Chronosone*	*Polarity Chron*	*Episode* (Diachron)
Series	Epoch			Phase (Diachron)
Stage (Substage)	Age (Subage)	Polarity Subchronozone	Polarity Subchron	Span (Diachron)
Chronozone	Chron			Cline

ple, lithostratigraphic vs. chronostratigraphic). Specification and unambiguous description of the category is of paramount importance. Selection and designation of an appropriate rank from the distinctive terminology developed for each category help serve this function (Table 2).

Article 7.—Name

The name of a formal geologic unit is compound. For most categories, the name of a unit should consist of a geographic name combined with an appropriate rank (Wasatch Formation) or descriptive term (Viola Limestone). Biostratigraphic units are designated by appropriate biologic forms (*Exus albus* Assemblage Biozone). Worldwide chronostratigraphic units bear long established and generally accepted names of diverse origins (Triassic System). The first letters of all words used in the names of formal geologic units are capitalized (except for the trivial species and subspecies terms in the name of a biostratigraphic unit).

Remarks: (a) ***Appropriate geographic terms.***—Geographic names derived from permanent natural or artificial features at or near which the unit is present are preferable to those derived from impermanent features such as farms, schools, stores, churches, crossroads, and small communities. Appropriate names may be selected from those shown on topographic, state, provincial, county, forest service, hydrographic, or comparable maps, particularly those showing names approved by a national board for geographic names. The generic part of a geographic name, e.g., river, lake, village, should be omitted from new terms, unless required to distinguish between two otherwise identical names (e.g., Redstone Formation and Redstone River Formation). Two names should not be derived from the same geographic feature. A unit should not be named for the source of its components; for example, a deposit inferred to have been derived from the Keewatin glaciation center should not be designated the "Keewatin Till".

(b) ***Duplication of names.***—Responsibility for avoiding duplication, either in use of the same name for different units (homonymy) or in use of different names for the same unit (synonomy), rests with the proposer. Although the same geographic term has been applied to different categories of units (example: the lithostratigraphic Word Formation and the chronostratigraphic Wordian Stage) now entrenched in the literature, the practice is undesirable. The extensive geologic nomenclature of North America, including not only names but also nomenclatural history of formal units, is recorded in compendia maintained by the Committee on Stratigraphic Nomenclature of the Geological Survey of Canada, Ottawa, Ontario; by the Geologic Names Committee of the United States Geological Survey, Reston, Virginia; by the Instituto de Geologia, Ciudad Universitaria, Mexico, D.F.; and by many state and provincial geological surveys. These organizations respond to inquiries regarding the availability of names, and some are prepared to reserve names for units likely to be defined in the next year or two.

(c) ***Priority and preservation of established names.***—Stability of nomenclature is maintained by use of the rule of priority and by preservation of well-established names. Names should not be modified without explaining the need. Priority in publication is to be respected, but priority alone does not justify displacing a well-established name by one neither well-known nor commonly used; nor should an inadequately established name be preserved merely on the basis of priority. Redefinitions in precise terms are preferable to abandonment of the names of well-established units which may have been defined imprecisely but nonetheless in conformance with older and less stringent standards.

(d) ***Differences of spelling and changes in name.***—The geographic component of a well-established stratigraphic name is not changed due to differences in spelling or changes in the name of a geographic feature. The name Bennett Shale, for example, used for more than half a century, need not be altered because the town is named Bennet. Nor should the Mauch Chunk Formation be changed because the town has been renamed Jim Thorpe. Disappearance of an impermanent geographic feature, such as a town, does not affect the name of an established geologic unit.

(e) ***Names in different countries and different languages.***—For geologic units that cross local and international boundaries, a single name for each is preferable to several. Spelling of a geographic name commonly conforms to the usage of the country and linguistic group involved. Although geographic names are not translated (Cuchillo is not translated to Knife), lithologic or rank terms are (Edwards Limestone, Caliza Edwards; Formation La Casita, La Casita Formation).

*Article 8.—**Stratotypes***

The designation of a unit of boundary stratotype (type section or type locality) is essential in the definition of most formal geologic units. Many kinds of units are best defined by reference to an accessible and specific sequence of rock that may be examined and studied by others. A stratotype is the standard (original or subsequently designated) for a named geologic unit or boundary and constitutes the basis for definition or recognition of that unit or boundary; therefore, it must be illustrative and representative of the concept of the unit or boundary being defined.

Remarks: (a) ***Unit stratotype.***—A unit stratotype is the type section for a stratiform deposit or the type area for a nonstratiform body that serves as the standard for definition and recognition of a geologic unit. The upper and lower limits of a unit stratotype are designated points in a specific sequence or locality and serve as the standards for definition and recognition of a stratigraphic unit's boundaries.

(b) ***Boundary stratotype.***—A boundary stratotype is the type locality for the boundary reference point for a stratigraphic unit. Both boundary stratotypes for any unit need not be in the same section or region. Each boundary stratotype serves as the standard for definition and recognition of the base of a stratigraphic unit. The top of a unit may be defined by the boundary stratotype of the next higher stratigraphic unit.

(c) ***Type locality.***—A type locality is the specified geographic locality where the stratotype of a formal unit or unit boundary was originally defined and named. A type area is the geographic territory encompassing the type locality. Before the concept of a stratotype was developed, only type localities and areas were designated for many geologic units which are now long and well-established. Stratotypes, though now mandatory in defining most stratiform units, are impractical in definitions of many large nonstratiform rock bodies whose diverse major components may be best displayed at several reference localities.

(d) ***Composite-stratotype.***—A composite stratotype consists of several reference sections (which may include a type section) required to demonstrate the range or totality of a stratigraphic unit.

(e) ***Reference sections.***—Reference sections may serve as invaluable standards in definitions or revisions of formal geologic units. For those well-established stratigraphic units for which a type section never was specified, a principal reference section (lectostratotype of ISSC, 1976, p. 26) may be designated. A principal reference section (neostratotype of ISSC, 1976, p. 26) also may be designated for those units or boundaries whose stratotypes have been destroyed, covered, or otherwise made inaccessible. Supplementary reference sections often are designated to illustrate the diversity or heterogeneity of a defined unit or some critical feature not evident or exposed in the stratotype. Once a unit or boundary stratotype section is designated, it is never abandoned or changed, however, if a stratotype proves inadequate, it may be supplemented by a principal reference section or by several reference sections that may constitute a composite stratotype.

(f) ***Stratotype descriptions.***—Stratotypes should be described both geographically and geologically. Sufficient geographic detail must be included to enable others to find the stratotype in the field, and may consist of maps and/or aerial photographs showing location and access, as well as appropriate coordinates or bearings. Geologic information should include thickness, descriptive criteria appropriate to the recognition of the unit and its boundaries, and discussion of the relation of the unit to other geologic units of the area. A carefully measured and described section provides the best foundation for definition of stratiform units. Graphic profiles, columnar sections, structure-sections, and photographs are useful supplements to a description; a geologic map of the area including the type locality is essential.

Article 9.— ***Unit Description***

A unit proposed for formal status should be described and defined so clearly that any subsequent investigator can recognize that unit unequivocally. Distinguishing features that characterize a unit may include any or several of the following: composition, texture, primary structures, structural attitudes, biologic remains, readily apparent mineral composition (e.g., calcite vs. dolomite), geochemistry, geophysical properties (including magnetic signatures), geomorphic expression, uncomfortable or cross-cutting relations, and age. Although all distinguishing features pertinent to the unit category should be described sufficiently to characterize the unit, those not pertinent to the category (such as age and inferred genesis for lithostratigraphic units, or lithology for biostratigraphic units) should not be made part of the definition.

Article 10.— ***Boundaries***

The criteria specified for the recognition of boundaries between adjoining geologic units are of paramount importance because they provide the basis for scientific reproducibility of results. Care is required in describing the criteria, which must be appropriate to the category of unit involved.

Remarks: (a) ***Boundaries between intergradational units.***—Contacts between rocks of markedly contrasting composition are appropriate boundaries of lithic units, but some rocks grade into, or intertongue with, others of different lithology. Consequently, some boundaries are necessarily arbitrary as, for example, the top of the uppermost limestone in a sequence of interbedded limestone and shale. Such arbitrary boundaries commonly are diachronous.

(b) ***Overlaps and gaps.***—The problem of overlaps and gaps, between long-established adjacent chronostratigraphic units is being addressed by international IUGS and IGCP working groups appointed to deal with various parts of the geologic column. The procedure recommended by the Geological Society of London (George and others, 1969; Holland and others, 1978), of defining only the basal boundaries of chronostratigraphic units, has been widely adopted (e.g., McLaren, 1977) to resolve the problem. Such boundaries are defined by a carefully selected and agreed-upon boundary-stratotype (marker-point type section or "golden spike") which becomes the standard for the base of a chronostratigraphic unit. The concept of the mutual-boundary stratotype (ISSC, 1976, p. 84-86), based on the assumption of continuous deposition in selected sequences, also has been used to define chronostratigraphic units.

Although international chronostratigraphic units of series and higher rank are being redefined by IUGS and IGCP working groups, there may be a continuing need for some provincial series. Adoption of the basal boundary-stratotype concept is urged.

Article 11.—***Historical Background***

A proposal for a new name must include a nomenclatorial history of rocks assigned to the proposed unit, describing how they were treated previously and by whom (references), as well as such matters as priorities, possible synonymy, and other pertinent considerations. Consideration of the historical background of an older unit commonly provides the basis for justifying definition of a new unit.

*Article 12.—**Dimensions and Regional Relations***

A perspective on the magnitude of a unit should be provided by such information as may be available on the geographic extent of a unit; observed ranges in thickness, composition, and geomorphic expression; relations to other kinds and ranks of stratigraphic units; correlations with other nearby sequences; and the bases for recognizing and extending the unit beyond the type locality. If the unit is not known anywhere but in an area of limited extent, informal designation is recommended.

*Article 13.—**Age***

For most formal material geologic units, other than chronostratigraphic and polarity-chronostratigraphic, inferences regarding geologic age play no proper role in their definition. Nevertheless, the age, as well as the basis for its assignment, are important features of the unit and should be stated. For many lithodemic units, the age of the protolith should be distinguished from that of the metamorphism or deformation. If the basis for assigning an age is tenuous, a doubt should be expressed.

Remarks: (a) ***Dating.***—The geochronologic ordering of the rock record, whether in terms of radioactive-decay rates or other processes, is generally called "dating". However, the use of the noun "date" to mean "isotopic age" is not recommended. Similarly, the term "absolute age" should be suppressed in favor of "isotopic age" for an age determined on the basis of isotopic ratios. The more inclusive term 'numerical age" is recommended for all ages determined from isotopic ratios, fission tracks, and other quantifiable age-related phenomena.

(b) ***Calibration.***—The dating of chronostratigraphic boundaries in terms of numerical ages is a special form of dating for which the word "calibration" should be used. The geochronological time-scale now in use has been developed mainly through such calibration of chronostratigraphic sequences.

(c) ***Convention and abbreviations.***—The age of a stratigraphic unit or the time of a geologic event, as commonly determined by numerical dating or by reference to a calibrated time-scale, may be expressed in years before the present. The unit of time is the modern year as presently recognized worldwide. Recommended (but not mandatory) abbreviations for such ages are SI (International System of Units) multipliers coupled with "a" for annum: ka, Ma, and Ga[5] for kilo-annum (10^3 years), Mega-annum (10^6 years), and Giga-annum (10^9 years), respectively. Use of these terms after the age value follows the convention established in the field of C-14 dating. The "present" refers to 1950 AD, and such qualifiers as "ago" or "before the present" refers to 1950 AD, and such qualifiers as "ago" or "before the present" are omitted after the value because

---

[5]Note that the initial letters of Mega-and Giga-are capitalized, but that of kilo-is not, by SI convention.

measurement of the duration from the present to the past is implicit in the designation. In contrast, the duration of a remote interval of geologic time, as a number of years, should not be expressed by the same symbols. Abbreviations for numbers of years, without reference to the present, are informal (e.g., y or yr for years; my, m.y., or m.yr. for millions of years; and so forth, as preference dictates). For example, boundaries of the Late Cretaceous Epoch currently are calibrated at 63 Ma and. 96 Ma, but the interval of time represented by this epoch is 33 m.y.

(d) ***Expression of "age" of lithodemic units.***—The adjectives "early", "middle", and "late" should be used with the appropriate geochronologic term to designate the age of lithodemic units. For example, a granite dated isotopically at 510 Ma should be referred to using the geochronologic term "Late Cambrian granite" rather than either the chronostratigraphic term "Upper Cambrian granite" or the more cumbersome designation "granite of Late Cambrian age."

Article 14.— ***Correlation***

Information regarding spatial and temporal counterparts of a newly defined unit beyond the type area provides readers with an enlarged perspective. Discussions of criteria used in correlating a unit with those in other areas should make clear the distinction between data and inferences.

Article 15.— ***Genesis***

Objective data are used to define and classify geologic units and to express their spatial and temporal relations. Although many of the categories defined in this Code (e.g., lithostratigraphic group, plutonic suite) have genetic connotations, inferences regarding geologic history or specific environments of formation may play no proper role in the definition of a unit. However, observations, as well as inferences, that bear on genesis are of great interest to readers and should be discussed.

Article 16.— ***Subsurface and Subsea Units***

The foregoing procedures for establishing formal geologic units apply also to subsurface and offshore or subsea units. Complete lithologic and paleontologic descriptions or logs of the samples or cores are required in written or graphic form, or both. Boundaries and divisions, if any, of the unit should be indicated clearly with their depths from an established datum.

Remarks: (a) ***Naming subsurface units.***—A subsurface unit may be named for the borehole (Eagle Mills Formation), oil field (Smackover Limestone), or mine which is intended to serve as the stratotype, or for a nearby geographic feature. The hole or mine should be located precisely, both with map and exact geographic coordinates, and identified fully (operator or company, farm or lease block, dates drilled or mined, surface elevation and total depth, etc.).

(b) ***Additional recommendations.***—Inclusion of appropriate borehole geophysical logs is urged. Moreover, rock and fossil samples and cores and all pertinent accompanying materials should be stored, and available for examination, at appropriate federal, state, provincial, university, or museum depositories. For offshore or subsea units (Clipperton Formation of Tracey and others, 1971, p. 22; Argo Salt of McIver, 1972, p. 57), the names of the project and vessel, depth of sea floor, and pertinent regional sampling and geophysical data should be added.

(c) ***Seismostratigraphic units.***—High-resolution seismic methods now can delineate stratal geometry and continuity at a level of confidence not previously attainable. Accordingly, seismic surveys have come to be the principal adjunct of the drill in subsurface exploration. On the other hand, the method identifies rock types only broadly and by inference. Thus, formalization of units known only from seismic profiles is inappropriate. Once the stratigraphy is calibrated by drilling, the seismic method may provide objective well-to-well correlations.

Revision and Abandonment of Formal Units

*Article 17.—**Requirements for Major Changes***

Formally defined and named geologic units may be redefined, revised, or abandoned, but revision and abandonment require as much justification as establishment of a new unit.

Remark: (a) ***Distinction between redefinition and revision.***—Redefinition of a unit involves changing the view or emphasis on the content of the unit without changing the boundaries or rank, and differs only slightly from redescription. Neither redefinition nor redescription is considered revision. A redescription corrects an inadequate or inaccurate description, whereas a redefinition may change a descriptive (for example, lithologic) designation. Revision involves either minor changes in the definition of one or both boundaries or in the rank of a unit (normally, elevation to a higher rank). Correction of a misidentification of a unit outside its type area is neither redefinition nor revision.

*Article 18.—**Redefinition***

A correction or change in the descriptive term applied to a stratigraphic or lithodemic unit is a redefinition which does not require a new geographic term.

Remarks: (a) ***Change in lithic designation.***—Priority should not prevent more exact lithic designation if the original designation is not everywhere applicable; for example, the Niobrara Chalk changes gradually westward to a unit in which shale is prominent, for which the designation "Niobrara Shale" or "Formation" is more appropriate. Many carbonate formations originally designated "limestone" or "dolomite" are found to be geographically inconsistent as to prevailing rock type. The appropriate lithic term or "formation" is again preferable for such units.

(b) ***Original lithic designation inappropriate.***—Restudy of some long established lithostratigraphic units has shown that the original lithic designation was incorrect according to modern criteria; for example, some "shales" have the chemical and mineralogical composition of limestone, and some rocks described as felsic lavas now are understood to be welded tuffs. Such new knowledge is recognized by changing the lithic designation of the unit, while retaining the original geographic term. Similarly, changes in the classification of igneous rocks have resulted in recognition that rocks originally described as quartz monzonite now are more appropriately termed granite. Such lithic designations may be modernized when the new classification is widely adopted. If heterogeneous bodies of plutonic rock have been misleadingly identified with a single compositional term, such as "gabbro", the adoption of a neutral term, such as "intrusion" or "pluton", may be advisable.

*Article 19.—**Revision***

Revision involves either minor changes in the definition of one or both boundaries of a unit, or in the unit's rank.

Remarks: (a) ***Boundary change.***—Revision is justifiable if a minor change in boundary or content will make a unit more natural and useful. If revision modifies only a minor part of the content of a previously established unit, the original name may be retained.

(b) ***Change in rank.***—Change in rank of a stratigraphic or temporal unit requires neither redefinition of its boundaries nor alteration of the geographic part of its name. A member may become a formation or vice versa, a formation may become a group or vice versa, and a lithodeme may become a suite or vice versa.

(c) ***Examples of changes from area to area.***—The Conasauga Shale is recognized as a formation in Georgia and as a group in eastern Tennessee; the Osgood Formation, Laurel Limestone, and Waldron Shale in Indiana are classed as members of the Wayne Formation in a part of Tennessee; the Virgelle Sandstone is a formation in western Montana and a member of the Eagle Sandstone in central Montana; the Skull Creek Shale and the Newcastle Sandstone in North Dakota are members of the Ashville Formation in Manitoba.

(d) ***Example of change in single area.***—The rank of a unit may be changed without changing its content. For example, the Madison Limestone of early work in Montana later became the Madison Group, containing several formations.

(e) ***Retention of type section.***—When the rank of a geologic unit is changed, the original type section or type locality is retained for the newly ranked unit (see Article 22c).

(f) ***Different geographic name for a unit and its parts.***—In changing the rank of a unit, the same name may not be applied both to the unit as a whole and to a part of it. For example, the Astoria Group should not contain an

Astoria Sandstone, nor the Washington Formation, a Washington Sandstone Member.

(g) ***Undesirable restriction.***—When a unit is divided into two or more of the same rank as the original, the original name should not be used for any of the divisions. Retention of the old name for one of the units precludes use of the name in a term of higher rank. Furthermore, in order to understand an author's meaning, a later reader would have to know about the modification and its date, and whether the author is following the original or the modified usage. For these reasons, the normal practice is to raise the rank of an established unit when units of the same rank are recognized and mapped within it.

*Article 20.—**Abandonment***

An improperly defined or obsolete stratigraphic, lithodemic, or temporal unit may be formally abandoned, provided that (a) sufficient justification is presented to demonstrate a concern for nomenclatural stability, and (b) recommendations are made for the classification and nomenclature to be used in its place.

Remarks: (a) ***Reasons for abandonment.***—A formally defined unit may be abandoned by the demonstration of synonymy or homonymy, of assignment to an improper category (for example, definition of a lithostratigraphic unit in a chronostratigraphic sense), or of other direct violations of a stratigraphic code or procedures prevailing at the time of the original definition. Disuse, or the lack of need or useful purpose for a unit, may be a basis for abandonment; so, too, may widespread misuse in diverse ways which compound confusion. A unit also may be abandoned if it proves impracticable, neither recognizable nor mappable elsewhere.

(b) ***Abandoned names.***—A name for a lithostratigraphic or lithodemic unit, once applied and then abandoned, is available for some other unit only if the name was introduced casually, or if it has been published only once in the last several decades and is not in current usage, and if its reintroduction will cause no confusion. An explanation of the history of the name and of the new usage should be a part of the designation.

(c) ***Obsolete names.***—Authors may refer to national and provincial records of stratigraphic names to determine whether the name is obsolete (see Article 7b).

(d) ***Reference to abandoned names.***—When it is useful to refer to an obsolete or abandoned formal name, its status is made clear by some such term as "abandoned" or "obsolete", and by using a phrase such as "La Plata Sandstone of Cross (1898)". (The same phrase also is used to convey that a named unit has not yet been adopted for usage by the organization involved.)

(e) ***Reinstatement.***—A name abandoned for reasons that seem valid at the time, but which subsequently are found to be erroneous, may be reinstated. Example: the Washakie Formation, defined in 1869, was abandoned in 1918 and reinstated in 1973.

Code Amendment

*Article 21.—**Procedure for Amendment***

Additions to, or changes of, this Code may be proposed in writing to the Commission by any geoscientist at any time. If accepted for consideration by a majority vote of the Commission, they may be adopted by a two-thirds vote of the Commission at an annual meeting not less than a year after publication of the proposal.

Formal Units Distinguished by Content, Properties, or Physical Limits

Lithostratigraphic Units

Nature and Boundaries

*Article 22.—**Nature of Lithostratigraphic Units***

A lithostratigraphic unit is a defined body of sedimentary, extrusive igneous, metasedimentary, or metavolcanic strata which is distinguished and delimited on the basis of lithic characteristics and stratigraphic position. A lithostratigraphic unit generally conforms to the Law of Superposition and commonly is stratified and tabular in form.

Remarks: (a) ***Basic units.***—Lithostratigraphic units are the basic units of general geologic work and serve as the foundation for delineating strata, local and regional structure, economic resources, and geologic history in regions of stratified rocks. They are recognized and defined by observable rock characteristics; boundaries may be placed at clearly distinguished contacts or drawn arbitrarily within a zone of gradation. Lithification or cementation is not a necessary property; clay, gravel, till, and other unconsolidated deposits may constitute valid lithostratigraphic units.

(b) ***Type section and locality.***—The definition of a lithostratigraphic unit should be based, if possible, on a stratotype consisting of readily accessible rocks in place, e.g., in outcrops, excavations, and mines, or of rocks accessible only to remote sampling devices, such as those in drill holes and underwater. Even where remote methods are used, definitions must be based on lithic criteria and not on the geophysical characteristics of the rocks, nor the implied age of their contained fossils. Definitions must be based on descriptions of actual rock material. Regional validity must be demonstrated for all such units. In regions where the stratigraphy has been established through studies of surface exposures, the naming of new units in the subsurface is justified only where the subsurface section differs materially from the surface section, or where there is doubt as to the equivalence of a subsurface and a surface unit. The establishment of subsurface reference sections for units originally defined in outcrop is encouraged.

(c) ***Type section never changed.***—The definition and name of a lithostratigraphic unit are established at a type section (or locality) that, once specified, must not be changed. If the type section is poorly designated or delimited, it may be redefined subsequently. If the originally specified stratotype is incomplete, poorly exposed, structurally complicated, or unrepresentative of the unit, a principal reference section or several reference sections may be designated to supplement, but not to supplant, the type section (Article 8c).

(d) ***Independence from inferred geologic history.***—Inferred geologic history, depositional environment, and biological sequence have no place in the definition of a lithostratigraphic unit, which must be based on composition and other lithic characteristics; nevertheless, considerations of well-documented geologic history properly may influence the choice of vertical and lateral boundaries of a new unit. Fossils may be valuable during mapping in distinguishing between two lithologically similar, non-contiguous lithostratigraphic units. The fossil content of a lithostratigraphic unit is a legitimate lithic characteristics; for example, oyster-rich sandstone, coquina, coral reef, or graptolitic shale. Moreover, otherwise similar units, such as the Formation Mendez and Formation Velasco mudstones, may be distinguished on the basis of coarseness of contained fossils (foraminifera).

(c) ***Independence from time concepts.***—The boundaries of most lithostratigraphic units may transgress time horizons, but some may be approximately synchronous. Inferred time-spans, however measured, play no part in differentiating or determining the boundaries of any lithostratigraphic unit. Either relatively short or relatively long intervals of time may be represented by a single unit. The accumulation of material assigned to a particular unit may have begun or ended earlier in some localities than in others; also, removal of rock by erosion, either within the time-span of deposition of the unit or later, may reduce the time-span represented by the unit locally. The body in some places may be entirely younger than in other places. On the other hand, the establishment of formal units that straddle known, identifiable, regional disconformities is to be avoided, if at all possible. Although concepts of time or age play no part in defining lithostratigraphic units nor in determining their boundaries, evidence of age may aid recognition of similar lithostratigraphic units at localities far removed from the type sections or areas.

(f) ***Surface form.***—Erosional morphology or secondary surface form may be a factor in the recognition of a lithostratigraphic unit, but properly should play a minor part at most in the definition of such units. Because the surface expression of lithostratigraphic units is an important aid in mapping, it is commonly advisable, where other factors do not countervail, to define lithostratigraphic boundaries so as to coincide with lithic changes that are expressed in topography.

(g) ***Economically exploited units.***—Aquifers, oil sands, coal beds, and quarry layers are, in general, informal units even though named. Some such units, however, may be recognized formally as beds, members, or formations because they are important in the elucidation of regional stratigraphy.

(h) ***Instrumentally defined units.***—In subsurface investigations, certain bodies or rock and their boundaries are widely recognized on bore-hole geophysical logs showing their electrical resistivity, radioactivity, density, or other physical properties. Such bodies and their boundaries may or may not correspond to formal lithostratigraphic units and their boundaries. Where other considerations do not countervail, the boundaries of subsurface units should be defined so as to correspond to useful geophysical markers; nevertheless, units defined exclusively on the basis of remotely sensed physical properties, although commonly useful in stratigraphic analysis, stand completely apart from the hierarchy of formal lithostratigraphic units and are considered informal.

(i) ***Zone.***—As applied to the designation of lithostratigraphic units, the term "zone" is informal. Examples are "producing zone", "mineralized zone", "metamorphic zone", and "heavy-mineral zone". A zone may include all or parts of a bed, a member, a formation, or even a group.

(j) ***Cyclothemes.***—Cyclic or rhythmic sequences of sedimentary rocks, whose repetitive divisions have been named cylothemes, have been recognized in sedimentary basins around the world. Some cyclothems have been identified by geographic names, but such names are considered informal. A clear distinction must be maintained between the division of a stratigraphic column into cyclothems and its division into groups, formations, and members. Where a cyclothem is identified by a geographic name, the word *cyclothem* should be part of the name, and the geographic term should not be the same as that of any formal unit embraced by the cyclothem.

(k) ***Soils and paleosols.***—Soils and paleosols are layers composed of the in-situ products of weathering of older rocks which may be of diverse composition and age. Soils and paleosols differ in several respects from lithostratigraphic units, and should not be treated as such (see "Pedostratigraphic Units," Article 55 et seq.)

(l) ***Depositional facies.***—Depositional facies are informal units, whether objective (conglomeratic, black shale, graptolitic) or genetic and environmental (platform, turbiditic, fluvial), even when a geographic term has been applied, e.g., Lantz Mills facies. Descriptive designations convey more information than geographic terms and are preferable.

Article 23.—**Boundaries**

Boundaries of lithostratigraphic units are placed at positions of lithic change. Boundaries are placed at distinct contacts or may be fixed arbitrarily within zones of gradation (Fig. 2a). Both vertical and lateral boundaries are based on the lithic criteria that provide the greatest unity and utility.

Remarks: (a) ***Boundary in a vertically gradational sequence.***—A named lithostratigraphic unit is preferably bounded by a single lower and a single upper surface so that the name does not recur in a normal stratigraphic succession (see Remark b). Where a rock unit passes vertically into another by intergrading or interfingering of two or more kinds of rock, unless the gradational strata are sufficiently thick to warrant designation of a third, independent unit, the boundary is necessarily arbitrary and should be selected on the basis of practicality (Fig. 2b). For example, where a shale unit overlies a unit of interbedded limestone and shale, the boundary commonly is placed at the top of the highest readily traceable limestone bed. Where a sandstone unit grades upward into shale, the boundary may be so gradational as to be difficult to place even arbitrarily; ideally it should be drawn at the level where the rock is composed of one-half of each component. Because of creep in outcrops and caving in boreholes, it is generally best to define such arbitrary boundaries by the highest occurrence of a particular rock type, rather than the lowest.

(b) ***Boundaries in lateral lithologic change.***—Where a unit changes laterally through abrupt gradation into, or intertongues with, a markedly different kind of rock, a new unit should be proposed for the different rock type. An arbitrary lateral boundary may be placed between the two equivalent units. Where the area of lateral intergradation or intertonguing is sufficiently extensive, a transitional interval of interbedded rocks may constitute a third independent unit (Fig. 2c). Where tongues (Article 25b) of formations are mapped separately or otherwise set apart without being formally named, the unmodified formation name should not be repeated in a normal stratigraphic sequence, although the modified name may be repeated in such phrases as "lower tongue of Mancos Shale" and "upper tongue of Mancos Shale". To show the order of superposition on maps and cross sections, the unnamed tongues may be distinguished informally (Fig. 2d) by number, letter, or other means. Such relationships may also be dealt with informally through the recognition of depositional facies (Article 22-I).

(c) ***Key beds used for boundaries.***—Key beds (Article 26b) may be used as boundaries for a formal lithostratigraphic unit where the internal lithic characteristics of the unit remain relatively constant. Even though bounding key beds may be traceable beyond the area of the diagnostic overall rock type, geographic extension of the lithostratigraphic unit bounded thereby is not necessarily justified. Where the rock between key beds becomes drastically different from that of the type locality, a new name should be applied (Fig. 2e), even though the key beds are continuous (Article 26b). Stratigraphic and sedimentologic studies of stratigraphic units (usually informal) bounded by key beds may be very informative and useful, especially in subsurface work where the key beds may be recognized by their geophysical signatures. Such units, however, may be a kind of chronostratigraphic, rather than lithostratigraphic, unit (Article 75, 75c), although others are diachronous because one, or both, of the key beds are also diachronous.

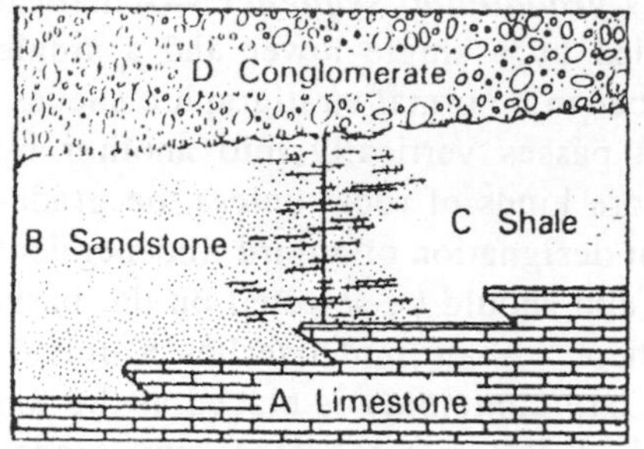

A. Boundaries at sharp lithologic contacts and in laterally gradational sequence.

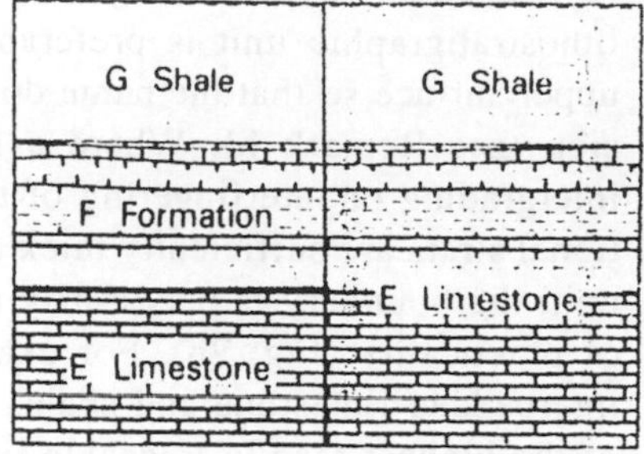

B. Alternative boundaries in a vertically gradational or interlayered sequence.

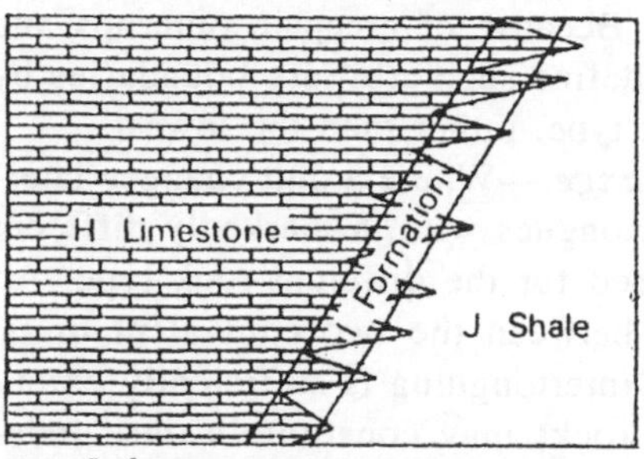

C. Possible boundaries for a laterally intertonguing sequence.

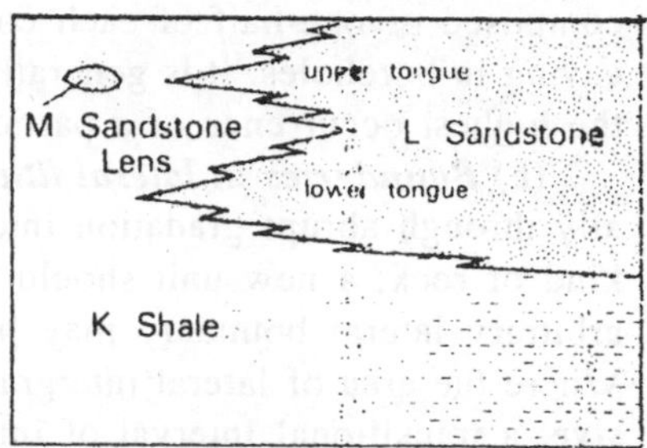

D. Possible classification of parts of an intertonguing sequence.

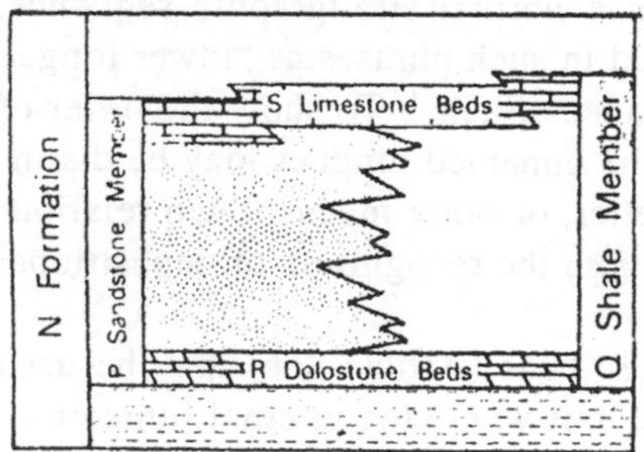

E. Key beds, here designated the R Dolostone beds and the S Limestone Beds, are used as boundaries to distinguish the Q Shale Member from the other parts of the N Formation. A lateral change in composition between the key beds requires that another name, P Sandstone member, be applied. The key beds are part of each member.

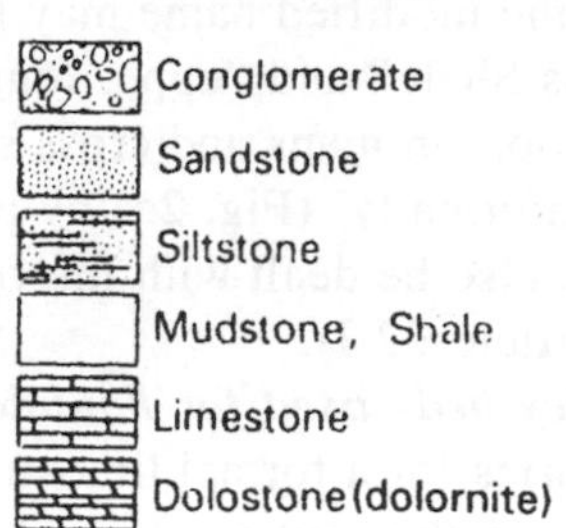

Fig. 2: Diagrammatic examples of lithostratigraphic boundaries and classification.

(d) ***Unconformities as boundaries.***—Unconformities, where recognizable objectively on lithic criteria, are ideal boundaries for lithostratigraphic units. However, a sequence of similar rocks may include an obscure unconformity so that separation into two units may be desirable but impracticable. If no lithic distinction adequate to define a widely recognizable boundary can be made, only one unit should be recognized, even though it may include rock that accumulated in different epochs, periods, or eras.

(c) ***Correspondence with genetic units.***—The boundaries of lithostratigraphic units should be chosen on the basis of lithic changes and, where feasible, to correspond with the boundaries of genetic units, so that subsequent studies of genesis will not have to deal with units that straddle formal boundaries.

Ranks of Lithostratigraphic Units

Article 24. ***Formation***

The formation is the fundamental unit in lithostratigraphic classification. A formation is a body of rock identified by lithic characteristics and stratigraphic position; it is prevailingly but not necessarily tabular and is mappable at the Earth's surface or traceable in the subsurface.

Remarks: (a) ***Fundamental unit.***—Formations are the basic lithostratigraphic units used in describing and interpreting the geology of a region. The limits of a formation normally are those surfaces of lithic change that give it the greatest practicable unity of constitution. A formation may represent a long or short time interval, may be composed of materials from one or several sources, and may include breaks in deposition (see Article 23d).

(b) ***Content.***—A formation should possess some degree of internal lithic homogeneity or distinctive lithic features. It may contain between its upper and lower limits (i) rock of one lithic type, (ii) repetitions of two or more lithic types, or (iii) extreme lithic heterogeneity which in itself may constitute a form of unity when compared to the adjacent rock units.

(c) ***Lithic characteristics.***—Distinctive lithic characteristics include chemical and mineralogical composition, texture, and such supplementary features as color, primary sedimentary or volcanic structures, fossils (viewed as rock-forming particles), or other organic content (coal, oil-shale). A unit distinguishable only by the taxonomy of its fossils is not a lithostratigraphic but a biostratigraphic unit (Article 48). Rock type may be distinctively represented by electrical, radioactive, seismic, or other properties (Article 22h), but these properties by themselves do not describe adequately the lithic character of the unit.

(d) ***Mappability and thickness.***—The proposal of a new formation must be based on tested mappability. Well-established formations commonly are divisible into several widely recognizable lithostratigraphic units; where formal recognition of these smaller units serves a useful purpose, they may be established as members and beds, for which the requirement of mappability is not mandatory. A unit formally recognized as a formation in one area may be treated elsewhere as a group, or as a member of another formation, without change of name. Example: the Niobrara is mapped at different places as a member of the Mancos Shale, of the Cody Shale, or of the Colorado Shale, and also as the Niobrara Formation, as the Niobrara Limestone, and as the Niobrara Shale.

Thickness is not a determining parameter in dividing a rock succession into formations; the thickness of a formation may range from a feather edge at its depositional or erosional limit to thousands of meters elsewhere. No formation is considered valid that cannot be delineated at the scale of geologic mapping practiced in the region when the formation is proposed. Although representation of a formation on maps and cross sections by a labeled line may be justified, proliferation of such exceptionally thin units is undesirable. The methods of subsurface mapping permit delineation of units much thinner than those usually practicable for surface studies; before such thin units are formalized, consideration should be given to the effect on subsequent surface and subsurface studies.

(e) ***Organic reefs and carbonate mounds.***—Organic reefs and carbonate mounds ("buildups") may be distinguished formally, if desirable, as formations distinct from their surrounding, thinner, temporal equivalents. For the requirements of formalization, see Article 30f.

(f) ***Interbedded volcanic and sedimentary rock.***—Sedimentary rock and volcanic rock that are interbedded may be assembled into a formation under one name which should indicate the predominant or distinguishing lithology, such as Mindego Basalt.

(g) ***Volcanic rock.***—Mappable distinguishable sequences of stratified volcanic rock should be treated as formations or lithostratigraphic units of higher or lower rank. A small intrusive component of a dominantly stratiform volcanic assemblage may be treated informally.

(h) ***Metamorphic rock.***—Formations composed of low-grade metamorphic rock (defined for this purpose as rock in which primary structures are clearly recognizable) are, like sedimentary formations, distinguished mainly by lithic characteristics. The mineral facies may differ from place to place, but these variations do not require definition of a new formation. High-grade metamorphic rocks whose relation to established formations is uncertain are treated as lithodemic units (see Articles 31 et seq).

Article 25. ***Member***

A member is the formal lithostratigraphic unit next in rank below a formation and is always a part of some formation. It is recognized as a named entity within a formation because it possesses characteristics distinguishing it from adjacent parts of the formation. A formation need not be divided into members unless a useful purpose is served by doing so. Some formations may be divided completely into members; others may have only certain parts designated as members; still others may have no members. A member may extend laterally from one formation to another.

Remarks: (a) ***Mapping of members.***—A member is established when it is advantageous to recognize a particular part of a heterogeneous formation. A member, whether formally or informally designated, need not be mappable at the scale required for formations. Even if all members of a formation are locally mappable, it does not follow that they should be raised to formational rank, because proliferation of formation names may obscure rather than clarify relations with other areas.

(b) ***Lens and tongue.***—A geographically restricted member that terminates on all sides within a formation may be called a lens (lentil). A wedging member that extends outward beyond a formation or wedges ("pinches") out within another formation may be called a tongue.

(c) ***Organic reefs and carbonate mounds.***—Organic reefs and carbonate mounds may be distinguished formally, if desirable, as members within a formation. For the requirements of formalization, see Article 30f.

(d) ***Division of members.***—A formally or informally recognized division of a member is called a bed or beds, except for volcanic flow-rocks, for which the smallest formal unit is a flow. Members may contain beds or flows, but may never contain other members.

(e) ***Laterally equivalent members.***—Although members normally are in vertical sequence, laterally equivalent parts of a formation that differ recognizable may also be considered members.

*Article 26.—**Bed(s)***

A bed, or beds, is the smallest formal lithostratigraphic unit of sedimentary rocks.

Remarks. (a) ***Limitations.***—The designation of a bed or a unit of beds as a formally named lithostratigraphic unit generally should be limited to certain distinctive beds whose recognition is particularly useful. Coal beds, oil sands, and other beds of economic importance commonly are named, but such units and their names usually are not a part of formal stratigraphic nomenclature (Article 22g and 30g).

(b) ***Key or marker beds.***—A key or marker bed is a thin bed of distinctive rock that is widely distributed. Such beds may be named, but usually are considered informal units. Individual key beds may be traced beyond the lateral limits of a particular formal unit (Article 23c).

*Article 27—**Flow***

A flow is the smallest formal lithostratigraphic unit of volcanic flow rocks. A flow is a discrete, extrusive, volcanic body distinguishable by texture, composition, order of superposition, paleomagnetism, or other objective criteria. It is part of a member and thus is equivalent in rank to a bed or beds of sedimentary-rock classification. Many flows are informal units. The designation and naming of flows as formal rock-stratigraphic units should be limited to those that are distinctive and widespread.

*Article 28.—**Group***

A group is the lithostratigraphic unit next higher in rank to formation; a group may consist entirely of named formations, or alternatively, need not be composed entirely of named formations.

Remarks: (a) ***Use and content.***—Groups are defined to express the natural relationships of associated formations. They are useful in small-scale mapping and regional stratigraphic analysis. In some reconnaissance work, the term "group" has been applied to lithostratigraphic units that appear to be divisible into formations, but have not yet been so divided. In such cases, formations may be erected subsequently for one or all of the practical divisions of the group.

(b) ***Change in component formations.***—The formations making up a group need not necessarily be everywhere the same. The Rundle Group, for example, is widespread in western Canada and undergoes several changes in formational content. In southwestern Alberta, it comprises the Livingstone, Mount Head, and Etherington Formations in the Front Ranges, whereas in the foothills and subsurface of the adjacent plains, it comprises the Pekisko, Shunda, Turner Valley, and Mount Head Formations. However, a formation or its parts may not be assigned to two vertically adjacent groups.

(c) ***Change in rank.***—The wedge-out of a component formation or formations may justify the reduction of a group to formation rank, retaining the same name. When a group is extended laterally beyond where it is divided into formations, it becomes in effect a formation, even if it is still called a group. When a previously established formation is divided into two or more component units that are given formal formation rank, the old formation, with its old geographic name, should be raised to group status. Raising the rank of the unit is preferable to restricting the old name to a part of its former content, because a change in rank leaves the sense of a well-established unit unchanged (Articles 19b, 19g).

*Article 29.—**Supergroup***

A supergroup is a formal assemblage of related or superposed groups, or of groups and formations. Such units have proved useful in regional and provincial syntheses. Supergroups should be named only where their recognition serves a clear purpose.

Remark: (a) ***Misuse of "series" for group or supergroup.***—Although "series" is a useful general term, it is applied formally only to a chronostratigraphic unit and should not be used for a lithostratigraphic unit. The term "series" should no longer be employed for an assemblage of formations or an assemblage of formations and groups, as it has been, especially in studies of the Precambrian. These assemblages are groups or supergroups.

Lithostratigraphic Nomenclature

*Article 30.—**Compound Character***

The formal name of a lithostratigraphic unit is compound. It consists of a geographic name combined with a discriptive lithic term or with the appropriate rank term, or both. Initial letters of all words used in forming the names of formal rock-stratigraphic units are capitalized.

Remarks: (a) ***Omission of part of a name.***—Where frequent repetition would be cumbersome, the geographic name, the lithic term, or the rank term may be used alone, once the full name has been introduced; as "the Burlington," "the limestone," or "the formation," for the Burlington Limestone.

(b) ***Use of simple lithic terms.***—The lithic part of the name should indicate the predominant or diagnostic lithology, even if subordinate lithologies are included. Where a lithic term is used in the name of a lithostratigraphic unit, the simplest generally acceptable term is recommended (for example, limestone, sandstone, shale, tuff, quartzite). Compound terms (for example, clay shale) and terms that are not in common usage (for example, calcirudite, orthoquartzite) should be avoided. Combined terms, such as "sand and clay," should not be used for the lithic part of the names of lithostratigraphic units, nor should an adjective be used between the geographic and the lithic terms, as "Chattanooga Black Shale" and "Biwabik Iron-Bearing Formation."

(c) ***Group names.***—A group name combines a geographic name with the term "group", and no lithic designation is included; for example, San Rafael Group.

(d) ***Formation names.***—A formation name consists of a geographic name followed by a lithic designation or by the word "formation". Examples: Dakota Sandstone, Mitchell Mesa Rhyolite, Monmouth Formation, Halton Till.

(e) ***Member names.***—All member names include a geographic term and the word "member"; some have an intervening lithic designation, if useful; for example, Wedington Sandstone Member of the Fayetteville Shale. Members designated solely by lithic character (for example, siliceous shale member), by position (upper, lower), or by letter or number, are informal.

(f) ***Names of reefs.***— Organic reefs identified as formations or members are formal units only where the name combines a geographic name with the appropriate rank term, e.g., Leduc Formation (a name applied to the several reefs enveloped by the Ireton Formation), Rainbow Reef Member.

(g) ***Bed and flow names.***—The names of bed or flows combine a geographic term, a lithic term, and the term "bed" or "flow" for example, Knee Hills Tuff Bed, Ardmore Bentonite Beds, Negus Variolitic Flows.

(h) ***Informal units.***—When geographic names are applied to such informal units as oil sands, coal beds, mineralized zones, and informal members (see Articles 22g and 26a), the unit term should not be capitalized. A name is not necessarily formal because it is capitalized, nor does failure to capitalize a name render it informal. Geographic names should be combined with the terms "formation" or "group" only in formal nomenclature.

(i) ***Informal usage of identical geographic names.***—The application of identical geographic names to several minor units in one vertical sequence is considered informal nomenclature (lower Mount Savage coal, Mount Savage fireclay, upper Mount Savage coal, Mount Savage rider coal, and Mount Savage sandstone). The application of identical geographic names to the several lithologic units constituting a cyclothem likewise is considered informal.

(j) ***Metamorphic rock.***—Metamorphic rock recognized as a normal stratified sequence, commonly low-grade metavolcanic or metasedimentary rocks, should be assigned to named groups, formations and members, such as the Deception Rhyolite, a formation of the Ash Creek Group, or the Bonner Quartzite, a formation of the Missoula Group. High-grade metamorphic and metasomatic rocks are treated as lithodemes and suites (see Articles 31, 33, 35).

(k) ***Misuse of well-known name.***—A name that suggests some well-known locality, region, or political division should not be applied to a unit typically developed in another less well-known locality of the same name. For example, it would be inadvisable to use the name "Chicago Formation" for a unit in California.

Lithodemic Units

Nature and Boundaries

*Article 31.—**Nature of Lithodemic Units***

A lithodemic[6] unit is a defined body of predominantly intrusive, highly deformed, and/or highly metamorphosed rock, distinguished and delimited on the basis of rock characteristics. In contrast to lithostratigraphic units, a lithodemic unit generally does not conform to the Law of Superposition. Its contacts with other rock units may be sedimentary, extrusive, intrusive, tectonic, or metamorphic (Fig. 3).

Remarks: (a) ***Recognition and definition.***—Lithodemic units are defined and recognized by observable rock characteristics. They are the practical units of general geological work in terranes in which rocks generally lack primary stratification; in such terranes they serve as the foundation for studying, describing, and delineating lithology, local and regional structure, economic resources, and geologic history.

---

[6]From the Greek demas, os: "living body, frame".

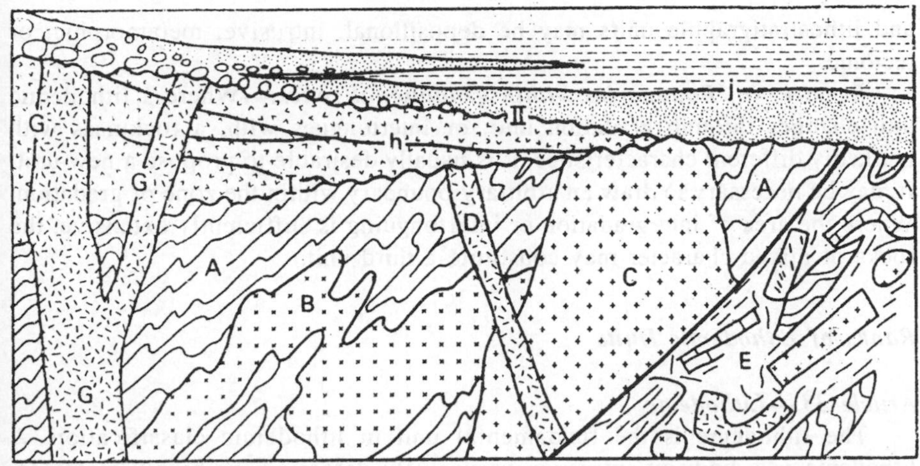

Fig. 3: Lithodemic (upper case) and lithostratigraphic (lower case) units. A *lithodeme* of *gneiss* (A) contains an *intrusion* of diorite (B) that was deformed with the gnesis. A and B may be treated jointly as a *complex*. A younger *granite* (C) is cut by a dike of *syenite* (D), that is cut in turn by unconformity I. All the foregoing are in fault contact with a *structural complex* (E). A *volcanic complex* (G) is built upon unconformity I, and its feeder dikes cut the unconformity. Laterally equivalent volcanic strata in orderly, mappable succession (h) are treated as lithostratigraphic units. A *gabbro* feeder (G'), to the volcanic complex, where surrounded by gneiss is readily distinguished as a separate lithodeme and named as a *gabbro* or an *intrusion*. All the foregoing are overlain, at unconformity II, by sedimentary rocks (j) divided into formations and members.

(b) ***Type and reference localities.***—The definition of a lithodemic unit should be based on as full a knowledge as possible of its lateral and vertical variations and its contact relationships. For purposes of nomenclatural stability, a type locality and, wherever appropriate, reference localities should be designated.

(c) ***Independence from inferred geologic history.***—Concepts based on inferred geologic history properly play no part in the definition of a lithodemic unit. Nevertheless, where two rock masses are lithically similar but display objective structural relations that preclude the possibility of their being even broadly of the same age, they should be assigned to different lithodemic units.

(d) ***Use of "zone".***—As applied to the designation of lithodemic units, the term "zone" is informal. Examples are "mineralized zone," "contact zone," and "pegmatitic zone".

*Article 32.—**Boundaries***

Boundaries of lithodemic units are placed at positions of lithic change. They may be placed at clearly distinguished contacts or within zones of gradation. Boundaries, both vertical and lateral, are based on the lithic criteria that provide the greatest unity and practical utility. Contacts with other lithodemic and lithostratigraphic units may be depositional, intrusive, metamorphic, or tectonic.

Remark: (a) ***Boundaries within gradational zones.***—Where a lithodemic unit changes through gradation into, or intertongues with, a rockmass with markedly different characteristics, it is usually desirable to propose a new unit. It may be necessary to draw an arbitrary boundary within the zone of gradation. Where the area of intergradation or intertonguing is sufficiently extensive, the rocks of mixed character may constitute a third unit.

Ranks of Lithodemic Units

*Article 33.—**Lithodeme***

The lithodeme is the fundamental unit in lithodemic classification. A lithodeme is a body of intrusive, pervasively deformed, or highly metamorphosed rock, generally non-tabular and lacking primary depositional structures, and characterized by lithic homogeneity. It is mappable at the Earth's surface and traceable in the subsurface. For cartographic and hierarchical purposes, it is comparable to a formation (see Table 2).

Remarks: (a) ***Content.***—A lithodeme should possess distinctive lithic features and some degree of internal lithic homogeneity. It may consist of (i) rock of one type, (ii) a mixture of rocks of two or more types, or (iii) extreme heterogeneity of composition, which may constitute in itself a form of unity when compared to adjoining rock-masses (see also "complex," Article 37).

(b) ***Lithic characteristics.***—Distinctive lithic characteristics may include mineralogy, textural features such as grain size, and structural features such as schistose or gneissic structure. A unit distinguishable from its neighbors only by means of chemical analysis is informal.

(c) ***Mappability.***—Practicability of surface or subsurface mapping is an essential characteristic of a lithodeme (see Article 24d).

*Article 34.—**Division of Lithodemes***

Units below the rank of lithodeme are informal.

*Article 35.—**Suite***

A *suite* (metamorphic suite, intrusive suite, plutonic suite) is the lithodemic unit next higher in rank to lithodeme. It comprises two or more associated lithodemes of the same class (e.g., plutonic, metamorphic). For cartographic and hierarchical purposes, suite is comparable to group (see Table 2).

Remarks: (a) ***Purpose.***—Suites are recognized for the purpose of expressing the natural relations of associated lithodemes having significant lithic features in common, and of depicting geology at compilation scales too small to allow delineation of individual lithodemes. Ideally, a suite consists entirely of named lithodemes, but may contain both named and unnamed units.

(b) ***Change in component units.***—The named and unnamed units constituting a suite may change from place to place, so long as the original sense of natural relations and of common lithic features is not violated.

(c) ***Change in rank.***—Traced laterally, a suite may lose all of its formally named divisions but remain a recognizable, mappable entity. Under such circumstances, it may be treated as a lithodeme but retain the same name. Conversely, when a previously established lithodeme is divided into two or more mappable divisions, it may be desirable to raise its rank to suite, retaining the original geographic component of the name. To avoid confusion, the original name should not be retained for one of the divisions of the original unit (see Article 19g).

*Article 36.—**Supersuite***

A supersuite is the unit next higher in rank to a suite. It comprises two or more suites or complexes having a degree of natural relationship to one another, either in the vertical or the lateral sense. For cartographic and hierarchical purposes, supersuite is similar in rank to supergroup.

*Article 37.—**Complex***

An assemblage or mixture of rocks of *two or more genetic classes,* i.e., igneous, sedimentary, or metamorphic, with or without highly complicated structure, may be named a *complex.* The term "complex" takes the place of the lithic or rank term (for example, Boil Mountain Complex, Franciscan Complex) and, although unranked, commonly is comparable to suite or supersuite and is named in the same manner (Articles 41, 42).

Remarks (a) Use of "complex."—Identification of an assemblage of diverse rocks as a complex is useful where the mapping of each separate lithic component is impractical at ordinary mapping scales. "Complex" is unranked but commonly comparable to suite or supersuite; therefore, the term may be retained if subsequent, detailed mapping distinguishes some or all of the component lithodemes or lithostratigraphic units.

(b) ***Volcanic complex.***—Sites of persistent volcanic activity commonly are characterized by a diverse assemblage of extrusive volcanic rocks, related intrusions, and their weathering products. Such an assemblage may be designated a *volcanic complex.*

(c) ***Structural complex.***—In some terranes, tectonic processes (e.g., shearing, faulting) have produced heterogeneous mixtures or disrupted bodies of rock in which some individual components are too small to be mapped. *Where there is no doubt that the mixing or disruption is due to tectonic processes,* such a mixture may be designated as a structural complex, whether it consists of two or more classes of rock, or a single class only. A simpler solution for some mapping purposes is to indicate intense deformation by an overprinted pattern.

(d) ***Misuse of "complex".***—Where the rock assemblage to be united under a single, formal name consists of diverse types of a *single class* of rock, as in many terranes that expose a variety of either intrusive igneous or high-grade metamorphic rocks, the term "intrusive suite," plutonic suite," or "metamorphic suite" should be used, rather than the unmodified term "complex". Exceptions to this rule are the terms *structural complex* and *volcanic complex* (see Remarks c and b, above).

Article 38.—***Misuse of "Series" for Suite, Complex, or Supersuite***

The term "series" has been employed for an assemblage of lithodemes or an assemblage of lithodemes and suites, especially in studies of the Precambrian. This practice now is regarded as improper; these assemblages are suites, complexes, or supersuites. The term "series" also has been applied to a sequence of rocks resulting from a succession of eruptions or intrusions. In these cases a different term should be used; "group" should replace "series" for volcanic and low-grade metamorphic rocks, and "intrusive suite" or "plutonic suite" should replace "series" for intrusive rocks of group rank.

Lithodemic Nomenclature

Article 39.—***General Provisions***

The formal name of a lithodemic unit is compound. It consists of a geographic name combined with a descriptive or appropriate rank term. The principles for the selection of the geographic term, concerning suitability, availability, priority, etc, follow those established in Article 7, where the rules for capitalization are also specified.

Article 40.—***Lithodeme Names***

The name of a Lithodeme combines a geographic term with a lithic or descriptive term, e.g., Killarney Granite, Adamant Pluton, Manhattan Schist, Skaergaard Intrusion, Duluth Gabbro. The term *formation* should not be used.

Remarks. (a) ***Lithic term.***—The lithic term should be a common and familiar term, such as schist, gneiss, gabbro. Specialized terms and terms not widely used, such as websterite and jacupirangite, and compound terms, such as graphitic schist and augen gneiss, should be avoided.

(b) ***Intrusive and plutonic rocks.***—Because many bodies of intrusive rock range in composition from place to place and are difficult to characterize with a single lithic term, and because many bodies of plutonic rock are considered not to be intrusions, latitude is allowed in the choice of a lithic or descriptive term. Thus, the descriptive term should preferably be compositional (e.g., gabbro, granodiorite), but may, if necessary, denote form (e.g., dike, sill), or be neutral (e.g., intrusion, pluton[7]). In any event, specialized compositional terms not widely used are to be avoided, as are form terms that are not widely used, such as bysmalith and chonolith. Terms implying genesis should be avoided as much as possible, because interpretations of genesis may change.

*Article 41.—**Suite Names***

The name of a suite combines a geographic term, the term "suite", and an adjective denoting the fundamental character of the suite, for example, Idaho Springs Metamorphic Suite, Tuolumne Intrusive Suite, Cassiar Plutonic Suite. The geographic name of a suite may not be the same as that of a component lithodeme (see Article 19f). Intrusive assemblages, however, may share the same geographic name if an intrusive lithodeme is representative of the suite.

*Article 42.—**Supersuite Names***

The name of a supersuite combines a geographic term with the term "supersuite."

Magnetostratigraphic Units

Nature and Boundaries

*Article 43.—**Nature of Magnetostratigraphic Units***

A magnetostratigraphic unit is a body of rock unified by specified remanent-magnetic properties and is distinct from underlying and overlying magnetostratigraphic units having different magnetic properties.

Remarks: (a) ***Definition.***—Magnetostratigraphy is defined here as all aspects of stratigraphy based on remanent magnetism (paleomagnetic signatures). Four basic paleomagnetic phenomena can be determined or inferred from remanent magnetism: polarity, dipole-field-pole position (including apparent polar wander), the non-dipole component (secular variation), and field intensity.

(b) ***Contemporaneity of rock and remanent magnetism.***—Many paleomagnetic signatures reflect earth magnetism at the time the rock formed. Nevertheless, some rocks have been subjected subsequently to physical and/or chemical processes which altered the magnetic properties. For example, a body of rock may be heated above the blocking temperature or Curie point for one or more minerals, or a ferromagnetic mineral may be produced by low-temperature alteration long after the enclosing rock formed, thus acquiring a component of remanent magnetism reflecting the field at the time of alteration, rather than the time of original rock deposition or crystallization.

(c) ***Designations and scope.***—The prefix *magneto* is used with an appropriate term to designate the aspect of remanent magnetism used to define a

[7]Pluton-a mappable body of plutonic rock.

unit. The terms "magnetointensity" or "magnetosecular-variation" are possible examples. This Code considers only polarity reversals, which now are recognized widely as a stratigraphic tool. However, apparent-polar-wander paths offer increasing promise for correlations within Precambrian rocks.

*Article 44.—**Definition of Magnetopolarity Unit***

A magnetopolarity unit is a body of rock unified by its remanent magnetic polarity and distinguished from adjacent rock that has different polarity.

*Remarks: (a) **Nature.***—Magnetopolarity is the record in rocks of the polarity history of the Earth's magnetic-dipole field. Frequent past reversals of the polarity of the Earth's magnetic field provide a basis for magnetopolarity stratigraphy.

(b) ***Stratotype.***—A stratotype for a magnetopolarity unit should be designated and the boundaries defined in terms of recognized lithostratigraphic and/or biostratigraphic units in the stratotype. The formal definition of a magnetopolarity unit should meet the applicable specific requirements of Articles 3 to 16.

(c) ***Independence from inferred history.***—Definition of a magnetopolarity unit does not require knowledge of the time at which the unit acquired its remanent magnetism; its magnetism may be primary or secondary. Nevertheless, the unit's present polarity is a property that may be ascertained and confirmed by others.

(d) ***Relation to the lithostratigraphic and biostratigraphic units.***—Magnetopolarity units resemble lithostratigraphic and biostratigraphic units in that they are defined on the basis of an objective recognizable property, but differ fundamentally in that most magnetopolarity unit boundaries are thought not to be time transgressive. Their boundaries may coincide with those of lithostratigraphic or biostratigraphic units, or be parallel to but displaced from those of such units, or be crossed by them.

(e) ***Relation of magnetopolarity units to chronostratigraphic units.***—Although transitions between polarity reversals are of global extent, a magnetopolarity unit does not contain within itself evidence that the polarity is primary, or criteria that permit its unequivocal recognition in chronocorrelative strata of other areas. Other criteria, such as paleontologic or numerical age, are required for both correlation and dating. Although polarity reversals are useful in recognizing chronostratigraphic units, magnetopolarity alone is insufficient for their definition.

*Article 45.—**Boundaries***

The upper and lower limits of a magnetopolarity unit are defined by boundaries marking a change of polarity. Such boundaries may represent either a depositional discontinuity or a magnetic-field transition. The boundaries are either polarity-reversal horizons or polarity transition-zones, respectively.

Remark: (a) ***Polarity-reversal horizons and transition-zones.***—A polarity-reversal horizon is either a single, clearly definable surface or a thin body of strata constituting a transitional interval across which a change in magnetic polarity is recorded. Polarity-reversal horizons describe transitional intervals of 1 m or less; where the change in polarity takes place over a stratigraphic interval greater than 1 m, the term "polarity transition-zone" should be used. Polarity-reversal horizons and polarity transition-zones provide the boundaries for polarity zones, although they may also be contained within a polarity zone where they mark an internal change subsidiary in rank to those at its boundaries.

Ranks of Magnetopolarity Units

Article 46.—Fundamental Unit

A polarity zone is the fundamental unit of magnetopolarity classification. A polarity zone is a unit of rock characterized by the polarity of its magnetic signature. Magnetopolarity zone, rather than polarity zone, should be used where there is risk of confusion with other kinds of polarity.

Remarks: (a) ***Content.***—A polarity zone should possess some degree of internal homogeneity. It may contain rocks of (1) entirely or predominantly one polarity, or (2) mixed polarity.

(b) ***Thickness and duration.***—The thickness of rock of a polarity zone or the amount of time represented should play no part in the definition of the zone. The polarity signature is the essential property for definition.

(c) ***Ranks.***—When continued work at the stratotype for a polarity zone, or new work in correlative rocks elsewhere, reveals smaller polarity units, these may be recognized formally as polarity subzones. If it should prove necessary or desirable to group polarity zones, these should be termed polarity superzones. The rank of a polarity unit may be changed when deemed appropriate.

Magnetopolarity Nomenclature

Article 47.—Compound Name

The formal name of a magnetopolarity zone should consist of a geographic name and the term *Polarity Zone.* The term may be modified by *Normal, Reversed,* or *Mixed* (example: Deer Park Reversed Polarity Zone). In naming or revising magnetopolarity units, appropriate parts of Articles 7 and 19 apply. The use of informal designations, e.g., numbers or letters, is not precluded.

Biostratigraphic Units

Nature and Boundaries

*Article 48.—**Nature of Biostratigraphic Units***

A biostratigraphic unit is a body of rock defined or characterized by its fossil content. The basic unit in biostratigraphic classification is the biozone, of which there are several kinds.

Remarks: (a) ***Enclosing strata.***—Fossils that define or characterize a biostratigraphic unit commonly are contemporaneous with the body of rock that contains them. Some biostratigraphic units, however, may be represented only by their fossils, preserved in normal stratigraphic succession (e.g., on hardgrounds, in lag deposits, in certain types of remanié accumulations), which alone represent the rock of the biostratigraphic unit. In addition, some strata contain fossils derived from older or younger rocks or from essentially coeval materials of different facies; such fossils should not be used to define a biostratigraphic unit.

(b) ***Independence from lithostratigraphic units.***—Biostratigraphic units are based on criteria which differ fundamentally from those for lithostratigraphic units. Their boundaries may or may not coincide with the boundaries of lithostratigraphic units, but they bear no inherent relation to them.

(c) ***Independence from chronostratigraphic units.***—The boundaries of most biostratigraphic units, unlike the boundaries of chronostratigraphic units, are both characteristically and conceptually diachronous. An exception is an abundance biozone boundary that reflects a mass-mortality event. The vertical and lateral limits of the rock body that constitutes the biostratigraphic unit represent the limits in distribution of the defining biotic elements. The lateral limits never represent, and the vertical limits rarely represent, regionally synchronous events. Nevertheless, biostratigraphic units are effective for interpreting chronostratigraphic relations.

*Article 49.—**Kinds of Biostratigraphic Units***

Three principal kinds of biostratigraphic units are recognized: *interval, assemblage,* and *abundance* biozones.

Remark: (a) ***Boundary definitions.***—Boundaries of interval zones are defined by lowest and/or highest occurrences of single taxa; boundaries of some kinds of assemblage zones (Oppel or concurrent range zones) are defined by lowest and/or highest occurrences of more than one taxon; and boundaries of abundance zones are defined by marked changes in relative abundances of preserved taxa.

*Article 50.—**Definition of Interval Zone***

An interval zone (or subzone) is the body of strata between two specified, documented lowest and/or highest occurrences of single taxa.

Remarks: (a) ***Interval zone types.***—Three basic types of interval zones are recognized (Fig. 4). These include the range zones and interval zones of the International Stratigraphic Guide (ISSC, 1976, p. 53, 60) and are:

1. The interval between the documented lowest and highest occurrences of a single taxon (Fig. 4A). This is the *taxon range zone* of ISSC (1976, p. 53).

2. The interval included between the documented lowest occurrence of one taxon and the documented highest occurrence of another taxon (Fig. 4B). When such occurrences result in stratigraphic overlap of the taxa (Fig. 4B-1), the interval zone is the *concurrent range zone* of ISSC (1976, p. 55), that involves only two taxa. When such occurrences do not result in stratigraphic overlap (Fig. 4B-2), but are used to partition the range of a third taxon, the interval is the *partial range zone* of George and others (1969).

3. The interval between documented successive lowest occurrences or successive highest occurrences of two taxa (Fig. 4C). When the interval is between successive documented lowest occurrences within an evolutionary lineage (Fig. 4C-1), it is the *lineage zone* of ISSC (1976, p. 58). When the interval is between successive lowest occurrences of unrelated taxa or between successive highest occurrences of either related or unrelated taxa (Fig. 4C-2), it is a kind of *interval zone* of ISSC (1976, p. 60).

(b) ***Unfossiliferous intervals.***—Unfossiliferous intervals between or within biozones are the *barren interzones* and *intrazones* of ISSC (1976 p. 49).

*Article 51.—**Definition of Assemblage Zone***

An assemblage zone is a biozone characterized by the association of three or more taxa. It may be based on all kinds of fossils present, or restricted to only certain kinds of fossils.

Remarks: (a) ***Assemblage zone contents.***—An assemblage zone may consist of a geographically or stratigraphically restricted assemblage, or may incorporate two or more contemporaneous assemblages with shared characterizing taxa (*composite assemblage zones* of Kauffman, 1969) (Fig. 5c).

(b) ***Assemblage zone types.***—In practice, two assemblage zone concepts are used:

1. The *assemblage zone* (or cenozone) of ISSC (1976, p. 50), which is characterized by taxa without regard to their range limits (Fig. 5a). Recognition of this type of assemblage zone can be aided by using techniques of multivariate analysis. Careful designation of the characterizing taxa is especially important.

2. The *Oppel zone,* or the *concurrent range zone* of ISSC (1976, p. 55, 57) a type of zone characterized by more than two taxa and having boundaries based on two or more documented first and/or last occurrences of the included characterizing taxa (Fig. 5b).

*Article 52.—**Definition of Abundance Zone***

An abundance zones is a biozone characterized by quantitatively distinctive maxima of relative abundance of one or more taxa. This is the *acme zone* of

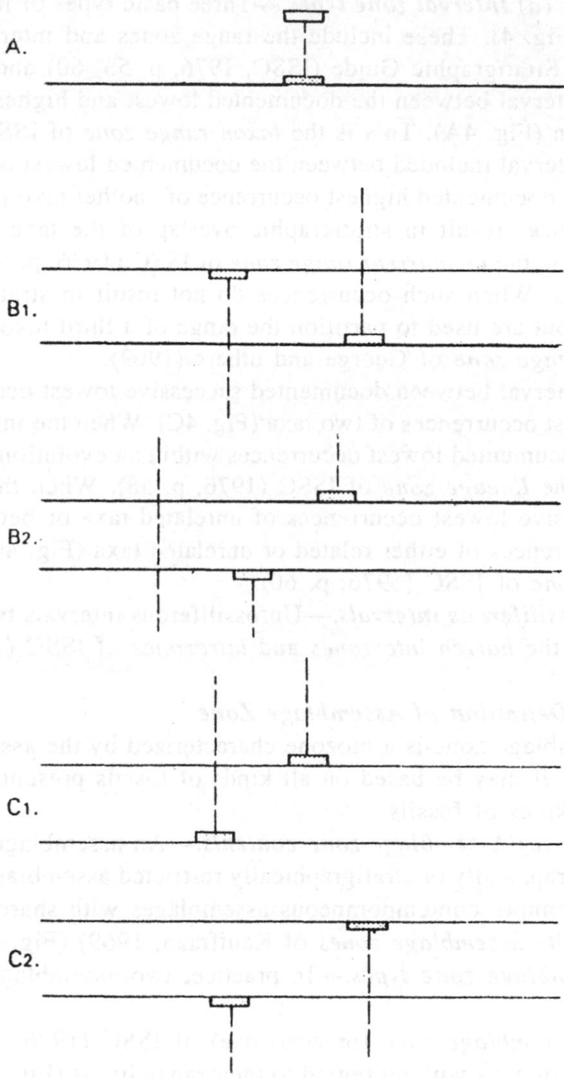

Fig. 4: Examples of biostratigraphic interval zones. Vertical broken lines indicate ranges of taxa; bars indicate lowest or highest documented occurrences

ISSC (1976, p. 59).

Remark: (a) Ecologic controls.—The distribution of biotic assemblages used to characterize some assemblage and abundance biozones may reflect strong local ecological control. Biozones based on such assemblages are included within the concepts of ecozones (Vella, 1964) and are informal.

Ranks of Biostratigraphic Units

*Article 53.—**Fundamental Unit***

The fundamental unit of biostratigraphic classification is a biozone.

Remarks: (a) ***Scope.***—A single body of rock may be divided into various kinds and scales of biozones or subzones, as discussed in the International Stratigraphic Guide (ISSC, 1976, p. 62). Such usage is recommended if it will promote clarity, but only the unmodified term *biozone* is accorded formal status.

(b) ***Divisions.***—A biozone may be completely or partly divided into formally designated sub-biozones (subzones), if such divisions serve a useful purpose.

Biostratigraphic Nomenclature

*Article 54.—**Establishing Formal Units***

Formal establishment of a biozone or subzone must meet the requirements of Article 3 and requires a unique name, a description of its content and its boundaries, reference to a stratigraphic sequence in which the zone is characteristically developed, and a discussion of its spatial extent.

Remarks: (a) ***Name.***—The name which is compound and designates the kind of biozone, may be based on:

1. One or two characteristic and common taxa that are restricted to the biozone, reach peak relative abundance within the biozone, or have their total stratigraphic overlap within the biozone. These names most commonly are those of genera or subgenera, binomial designations of species, or trinomial designations of subspecies. If names of the nominate taxa change, names of the zones should be changed accordingly. Generic or subgeneric names may be abbreviated. Trivial species or subspecies names should not be used alone because they may not be unique.

2. Combinations of letters derived from taxa which characterize the biozone. However, alpha-numeric code designations (e.g., N1, N2, N3...) are informal and not recommended because they do not lend themselves readily to subsequent insertions, combinations, or eliminations. Biozonal systems based *only* on simple progressions of letters or numbers (e.g., A, B, C, or 1, 2, 3) are also not recommended.

(b) ***Revision.***—Biozones and subzones are established empirically and may be modified on the basis of new evidence. Positions of established biozone or subzone boundaries may be stratigraphically refined, new characterizing taxa may be recognized, or original characterizing taxa may be superseded. If the concept of a particular biozone or subzone is substantially modified, a new unique designation is required to avoid ambiguity in subsequent citations.

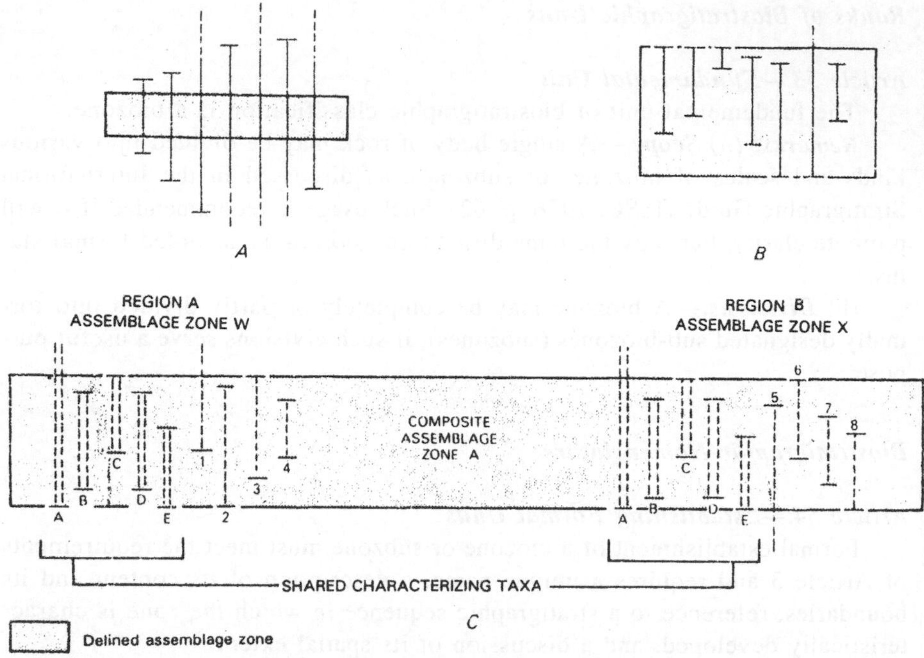

Fig. 5: Examples of assemblage zone concepts.

(c) ***Specifying kind of zone.***—Initial designation of a formally proposed biozone or subzone as an abundance zone, or as one of the types of interval zones, or assemblage zones (Article 49-52), is strongly recommended. Once the type of biozone is clearly identified, the designation may be dropped in the remainder of a text (e.g., *Exus albus* taxon range zone to *Exus albus* biozone).

(d) ***Defining taxa.***—Initial description or subsequent emendation of a biozone or subzone requires designation of the defining and characteristic taxa, and/or the documented first and last occurrences which mark the biozone or subzone boundaries.

(e) ***Stratotype.***—The geographic and stratigraphic position and boundaries of a formally proposed biozone or subzone should be defined precisely or characterized in one or more designated reference sections. Designation of a stratotype for each new biostratigraphic unit and of reference sections for emended biostratigraphic units is required.

Pedostratigraphic Units

Nature and Boundaries

*Article 55.—**Nature of Pedostratigraphic Units***

A pedostratigraphic unit is a body of rock that consists of one or more pedologic horizons developed in one or more lithostratigraphic, allostratigraphic, or lithodemic units (Fig. 6) and is overlain by one or more formally defined lithostratigraphic or allostratigraphic units.

Remarks: (a) ***Definition.***—A pedostratigraphic[8] unit is a buried, traceable, three-dimensional body of rock that consists of one or more differentiated pedologic horizons.

(b) ***Recognition.***—The distinguishing property of a pedostratigraphic unit is the presence of one or more distinct, differentiated, pedologic horizons. Pedologic horizons are products of soil development (pedogenesis) which occurred subsequent to formation of the lithostratigraphic, allostratigraphic, or lithodemic unit or units on which the buried soil was formed; these units are the parent materials in which pedogenesis occurred. Pedologic horizons are recognized in the field by diagnostic features such as color, soil structure, organic-matter accumulation, texture, clay coatings, stains, or concretions. Micromorphology, particle size, clay mineralogy, and other properties determined in the laboratory also may be used to identify and distinguish pedostratigraphic units.

(c) ***Boundaries and stratigraphic position.***—The upper boundary of a pedostratigraphic unit is the top of the uppermost pedologic horizon formed by pedogenesis in a buried soil profile. The lower boundary of a pedostratigraphic unit is the lowest *definite* physical boundary of a pedologic horizon within a buried soil profile. The stratigraphic position of a pedostratigraphic unit is determined by its relation to overlying and underlying stratigraphic units (see Remark d).

(d) ***Traceability.***—Practicability of subsurface tracing of the upper boundary of a buried soil is essential in establishing a pedostratigraphic unit because (1) few buried soils are exposed continuously for great distances, (2) the physical and chemical properties of a specific pedostratigraphic unit may vary greatly, both vertically and laterally, from place to place, and (3) pedostratigraphic units of different stratigraphic significance in the same region generally do not have unique identifying physical and chemical characteristics. Consequently, extension of a pedostratigraphic unit is accomplished by lateral tracing of the contact between a buried soil and an overlying, formally defined lithostratigraphic or allostratigraphic unit, or between a soil and two or more demonstrably correlative stratigraphic units.

---

[8]Terminology related to pedostratigraphic classification is summarized on page 850.

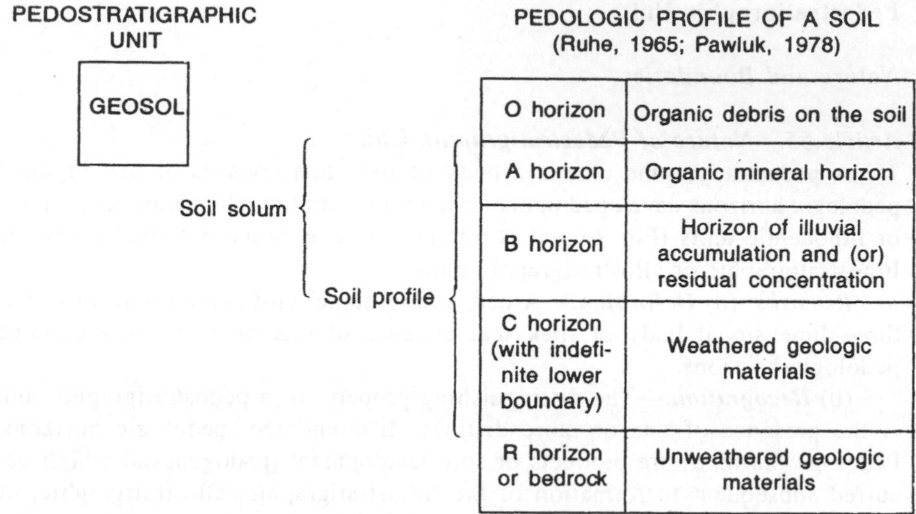

PEDOSTRATIGRAPHIC UNIT
GEOSOL
Soil solum
Soil profile
PEDOLOGIC PROFILE OF A SOIL
(Ruhe, 1965; Pawluk, 1978)
O horizon
Organic debris on the soil
A horizon
Organic mineral horizon
B horizon
Horizon of illuvial accumulation and (or) residual concentration
C horizon (with indefinite lower boundary)
Weathered geologic materials
R horizon or bedrock
Unweathered geologic materials

Fig. 6: Relationship between pedostratigraphic units and pedologic profiles. The base of a geosol is the lowest clearly defined physical boundary of a pedologic horizon in a buried soil profile. In this example it is the lower boundary of the B horizon because the base of the C horizon is not a clearly defined physical boundary. In other profiles the base may be the lower boundary of a C horizon.

(c) ***Distinction from pedologic soils.***—Pedologic soils may include organic deposits (e.g., litter zones, peat deposits, or swamp deposits) that overlie or grade laterally into differentiated buried soils. The organic deposits are not products of pedogenesis, and O horizons are not included in a pedostratigraphic unit (Fig. 6); they may be classified as biostratigraphic or lithostratigraphic units. Pedologic soils also include the entire C horizon of a soil. The C horizon in pedology is not rigidly defined; it is merely that part of a soil profile that underlies the B horizon. The base of the C horizon in many soil profiles is gradational or unidentifiable; commonly it is placed arbitrarily. The need for clearly defined and easily recognized physical boundaries for a stratigraphic unit requires that the lower boundary of a pedostratigraphic unit be defined as the lower *definite* physical boundary of a pedologic horizon in a buried soil profile, and part or all of the C horizon may be excluded from a pedostratigraphic unit.

(f) ***Relation to saprolite and other weathered materials.***—A material derived by in situ weathering of lithostratigraphic, allostratigraphic, and (or) lithodemic units (e.g., saprolite, bauxite, residuum) may be the parent material in which pedologic horizons form, but is not a pedologic soil. A pedostratigraphic unit may be based on the pedologic horizons of a buried soil developed in the product of in-situ weathering, such as saprolite. The parents of such a pedostratigraphic unit are both the saprolite and, indirectly, the rock from which it formed.

(g) ***Distinction from other stratigraphic units.***—A pedostratigraphic unit differs from other stratigraphic units in that (1) it is a product of surface alteration of one or more older material units by specific processes (pedogenesis), (2) its lithology and other properties differ markedly from those of the parent material(s), and (3) a single pedostratigraphic unit may be formed in-situ in parent material units of diverse compositions and ages.

(h) ***Independence from time concepts.***—The boundaries of a pedostratigraphic unit are time-transgressive. Concepts of time spans, however measured, play no part in defining the boundaries of a pedostratigraphic unit. Nonetheless, evidence of age, whether based on fossils, numerical ages, or geometrical or other relationships, may play an important role in distinguishing and identifying non-contiguous pedostratigraphic units at localities away from the type areas. The name of a pedostratigraphic unit should be chosen from a geographic feature in the type area, and not from a time span.

Pedostratigraphic Nomenclature and Unit

*Article 56.—**Fundamental Unit***

The fundamental and only unit in pedostratigraphic classification is a geosol.

*Article 57.—**Nomenclature***

The formal name of a pedostratigraphic unit consists of a geographic name combined with the term "geosol." Capitalization of the initial letter in each word serves to identify formal usage. The geographic name should be selected in accordance with recommendations in Article 7 and should not duplicate the name of another formal geologic unit. Names based on subjacent and superjacent rock units, for example the super-Wilcox-sub-Claiborne soil, are informal, as are those with time connotations (post-Wilcox-pre-Claiborne soil).

Remarks. (a) ***Composite geosols.***—Where the horizons of two or more merged or "welded" buried soils can be distinguished, formal names of pedostratigraphic units based on the horizon boundaries can be retained. Where the horizon boundaries of the respective merged or "welded" soils cannot be distinguished, formal pedostratigraphic classification is abandoned and a combined name such as Hallettville-Jamesville geosol may be used informally.

(b) ***Characterization.***—The physical and chemical properties of a pedostratigraphic unit commonly vary vertically and laterally throughout the geographic extent of the unit. A pedostratigraphic unit is characterized by the *range* of physical and chemical properties of the unit in the type area, rather than by "typical" properties exhibited in a type section. Consequently, a pedostratigraphic unit is characterized on the basis of a composite stratotype (Article 8d).

---

[9]From the Greek allo: "other, different."

(c) ***Procedures for establishing formal pedostratigraphic units.***—A formal pedostratigraphic unit may be established in accordance with the applicable requirements of Article 3, and additionally by describing major soil horizons in each soil facies.

Allostratigraphic Units

Nature and Boundaries

*Article 58.—**Nature of Allostratigraphic Units***

An allostratigraphic[9] unit is a mappable stratiform body of sedimentary rock that is defined and identified on the basis of its bounding discontinuities.

Remarks: (a) ***Purpose.***—Formal allostratigraphic units may be defined to distinguish between different (1) superposed discontinuity-bounded deposits of similar lithology (Figs. 7, 9) (2) contiguous discontinuity-bounded deposits of similar lithology (Fig. 8), or (3) geographically separated discontinuity-bounded units of similar lithology (Fig. 9),or to distinguish as single units discontinuity-bounded deposits characterized by lithic heterogeneity (Fig. 8).

(b) ***Internal characteristics.***—Internal characteristics (physical, chemical, and paleontological) may vary laterally and vertically through out the unit.

(c) ***Boundaries.***—Boundaries of allostratigraphic units are laterally traceable discontinuities (Fig. 7, 8 and 9).

(d) ***Mappability.***—A formal allostratigraphic unit must be mappable at the scale practiced in the region where the unit is defined.

(e) ***Type locality and extent.***—A type locality and type area must be designated; a composite stratotype or a type section and several reference sections are desirable. An allostratigraphic unit may be laterally contiguous with a formally defined lithostratigraphic unit; a vertical cut-off between such units is placed where the units meet.

(f) ***Relation to genesis.***—Genetic interpretation is an inappropriate basis for defining an allostratigraphic unit. However, genetic interpretation may influence the choice of its boundaries.

(g) ***Relation to geomorphic surfaces.***—A geomorphic surface may be used as a boundary of an allostratigraphic unit, but the unit should not be given the geographic name of the surface.

(h) ***Relation to soils and paleosols.***—Soils and paleosols are composed of products of weathering and pedogenesis and differ in many respects from allostratigraphic units, which are depositional units (see "Pedostratigraphic Units," Article 55). The upper boundary of a surface or buried soil may be used as a boundary of an allostratigraphic unit.

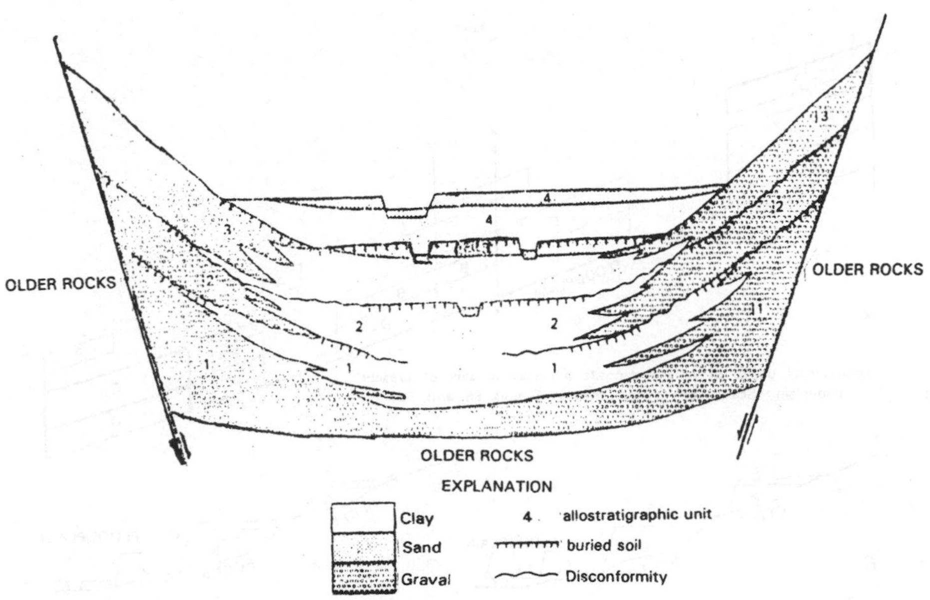

Fig. 7: Example of allostratigraphic classification of alluvial and lacustrine deposits in a graben. The alluvial and lacustrine deposits may be included in a single formation, or may be separated laterally into formations distinguished on the basis of contrasting texture (gravel, clay). Textural changes are abrupt and sharp, both vertically and laterally. The gravel deposits and clay deposits, respectively, are lithologically similar and thus cannot be distinguished as members of a formation. Four allostratigraphic units, each including two or three textural facies, may be defined on the basis of laterally traceable discontinuities (buried soils and disconformities).

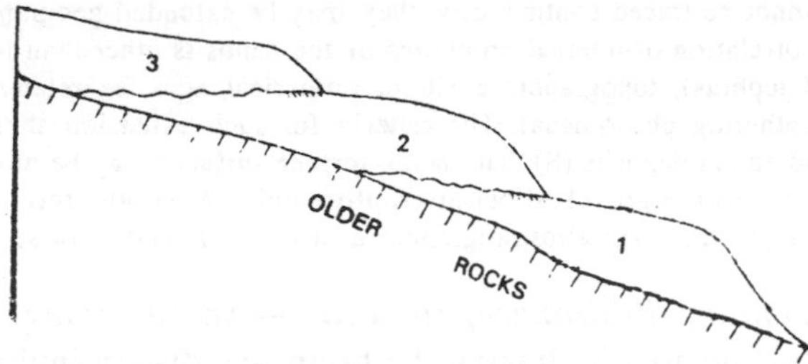

Fig. 8: Example of allostratigraphic calssification of contiguous deposits of similar lithology. Allostratigraphic units 1, 2, and 3 are physical records of three glaciations. They are lithologically similar, reflecting derivation from the same bedrock, and constitute a single lithostratigraphic unit.

(i) ***Relation to inferred geologic history.***—Inferred geologic history is not used to define an allostratigraphic unit. However, well-documented geologic history may influence the choice of the unit's boundaries.

(j) ***Relation to time concepts.***—Inferred time spans, however measured, are not used to define an allostratigraphic unit. However, age relationships may influence the choice of the unit's boundaries.

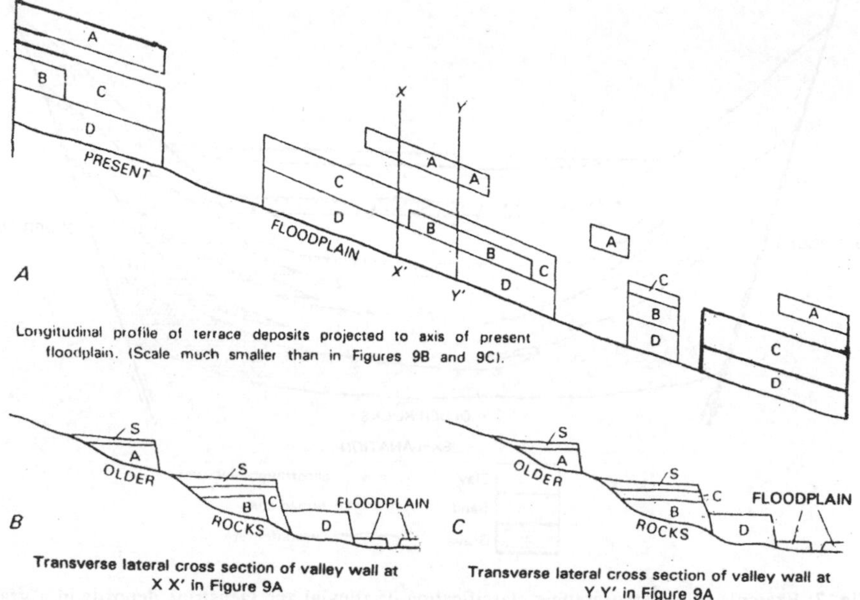

Fig. 9: Example of allostratigraphic calssification of lithologically similar discontinuous terrace deposits.

A, B, C, and D are terrace gravel units of similar lithology at different topographic positions on a valley wall. The deposits may be defined as separate formal allostratigraphic units if such units are useful and if bounding discontinuities can be traced laterally. Terrace gravels of the same age commonly are separated geographically by exposures of older rocks. Where the bounding discontinuities cannot be traced continuously, they may be extended geographically on the basis of the objective correlation of internal properties of the deposits other than lithology (e.g., fossil content, included tephras), topographic position, numerical ages, or relative-age criteria (e.g., soils or other weathering phenomena). The criteria for such extension should be documented. Slope deposits and eolian deposits (S) that mantle terrace surfaces may be of diverse ages and are not included in a terrace-gravel allostratigraphic unit. A single terrace surface may be underlain by more than one allostratigraphic unit (units B and C in sections b and c).

(k) ***Extension of allostratigraphic units.***—An allostratigraphic unit is extended from its type area by tracing the boundary discontinuities or by tracing or matching the deposits between the discontinuities.

Ranks of Allostratigraphic Units

*Article 59.—**Hierarchy***

The hierarchy of allostratigraphic units, in order of decreasing rank, is allogroup, alloformation, and allomember.

Remarks: (a) ***Alloformation.***—The alloformation is the fundamental unit in allostratigraphic classification. An alloformation may be completely or only partly divided into allomembers, if some useful purpose is served, or it may have no allomembers.

(b) ***Allomember.***—An allomember is the formal allostratigraphic unit next in rank below an alloformation.

(c) ***Allogroup.***—An allogroup is the allostratigraphic unit next in rank above an alloformation. An allogroup is established only if a unit of that rank is essential to elucidation of geologic history. An allogroup may consist entirely of named alloformations or, alternatively, may contain one or more named alloformations which jointly do not comprise the entire allogroup.

(d) ***Changes in rank.***—The principles and procedures for elevation and reduction in rank of formal allostratigraphic units are the same as those in Articles 19b, 19g, and 28.

Allostratigraphic Nomenclature

*Article 60.—**Nomenclature***

The principles and procedures for naming allostratigraphic units are the same as those for naming of lithostratigraphic units (see Articles 7, 30).

Remark: (a) ***Revision.***—Allostratigraphic units may be revised or otherwise modified in accordance with the recommendations in Articles 17 to 20.

Formal Units Distinguished by Age

Geologic-Time Units

Nature and Types

*Article 61.—**Types***

Geologic-time units are conceptual, rather than material, in nature. Two types are recognized: those based on material standards or referents (specific rock sequences or bodies), and those independent of material referents (Fig. 1).

Units Based on Material Referents

*Article 62.—**Type Based on Referents***

Two types of formal geologic-time units based on material referents are recognized: they are isochronous and diachronous units.

*Article 63.—**Isochronous Categories***

Isochronous time units and the material bodies from which they are derived are twofold: geochronologic units (Article 80), which are based on corresponding material chronostratigraphic units (Article 66), and polarity-geochronologic units (Article 88), based on corresponding material polarity-chronostratigraphic units (Article 83).

Remark: (a) ***Extent.***—Isochronous units are applicable worldwide; they may be referred to even in areas lacking a material record of the named span of time. The duration of the time may be represented by a unit-stratotype referent. The beginning and end of the time are represented by point-boundary-stratotypes either in a single stratigraphic sequence or in separate stratotype sections (Articles 8b, 10b)

*Article 64.—**Diachronous Categories***

Diachronic units (Article 91) are time units corresponding to diachronous material allostratigraphic units (Article 58), pedostratigraphic units (Article 55), and most lithostratigraphic (Article 22) and biostratigraphic (Article 48) units.

Remarks: (a) ***Diachroneity.***—Some lithostratigraphic and biostratigraphic units are clearly diachronous, whereas others have boundaries which are not demonstrably diachronous within the resolving power of available dating methods. The latter commonly are treated as isochronous and are used for purposes of chronocorrelation (see biochronozone, Article 75). However, the assumption of isochroneity must be tested continually.

(b) ***Extent.***—Diachronic units are coextensive with the diachronous material stratigraphic units on which they are based and are not used beyond the extent of their material referents.

Units Independent of Material Referents

*Article 65.—**Numerical Divisions of Time***

Isocronous geologic-time units based on numerical divisions of time in years are geochronometric units (Article 96) and have no material referents.

Chronostratigraphic Units

Nature and Boundaries

*Article 66.—**Definition***

A chronostratigraphic unit is a body of rock established to serve as the material reference for all rocks formed during the same span of time. Each of its boundaries is synchronous. The body also serves as the basis for defining the specific interval of time, or geochronologic unit (Article 80), represented by the referent.

Remarks: (a) ***Purposes.***—Chronostratigraphic classification provides a means of establishing the temporally sequential order of rock bodies. Principal purposes are to provide a framework for (1) temporal correlation of the rocks in one area with those in another, (2) placing the rocks of the Earth's crust in a systematic sequence and indicating their relative position and age with respect to earth history as a whole, and (3) constructing an internationally recognized Standard Global Chronostratigraphic Scale.

(b) ***Nature.***—A chronostratigraphic unit is a material unit and consists of a body of strata formed during a specific time span. Such a unit represents all rocks, and only those rocks, formed during that time span.

(c) ***Content.***—A chronostratigraphic unit may be based upon the time span of a biostratigraphic unit, a lithic unit, a magnetopolarity unit, or any other feature of the rock record that has a time range. Or it may be any arbitrary but specified sequence of rocks, provided it has properties allowing chronocorrelation with rock sequences elsewhere.

Article 67.—***Boundaries***

Boundaries of chronostratigraphic units should be defined in a designated stratotype on the basis of observable paleontological or physical features of the rocks.

Remark: (a) ***Emphasis on lower boundaries of chronostratigraphic units.***—Designation of point boundaries for both base and top of chronostratigraphic units is not recommended, because subsequent information on relations between successive units may identify overlaps or gaps. One means of minimizing or eliminating problems of duplication or gaps in chronostratigraphic successions is to define formally as a point-boundary stratotype only the base of the unit. Thus, a chronostratigraphic unit with its base defined at one locality, will have its top defined by the base of an overlying unit at the same, but more commonly another locality (Article 8b).

Article 68.—***Correlation***

Demonstration of time equivalence is required for geographic extension of a chronostratigraphic unit from its type section or area. Boundaries of chronostratigraphic units can be extended only within the limits of resolution of available means of chronocorrelation, which currently include paleontology, numerical dating, remanent magnetism, thermoluminescence, relative-age criteria (examples are superposition and cross-cutting relations), and such indirect and inferential physical criteria as climatic changes, degree of weathering, and relations to unconformities. Ideally, the boundaries of chronostratigraphic units are independent of lithology, fossil content, or other material bases of stratigraphic division, but, in practice, the correlation or geographic extension of these boundaries relies at least in part on such features. Boundaries of chronostratigraphic units commonly are intersected by boundaries of most other kinds of material units.

Ranks of Chronostratigraphic Units

Article 69.—***Hierarchy***

The hierarchy of chronostratigraphic units, in order of decreasing rank, is eonothem, erathem, system, series, and stage. Of these, system is the primary unit of worldwide major rank; its primacy derives from the history of development of stratigraphic classification. All systems and units of higher rank are divided completely into units of the next lower rank. Chronozones are non-hierarchical and commonly lower rank chronostratigraphic units. Stages and chronozones in sum do not necessarily equal the units of next higher rank and need not be contiguous. The rank and magnitude of chronostratigraphic units are related to the time interval represented by the units, rather than to the thickness or areal extent of the rocks on which the units are based.

Article 70.—***Eonothem***

The unit highest in rank is eonothem. The Phanerozoic Eonothem encompasses the Palezoic, Mesozoic, and Cenozoic Erathems. Although older rocks have been assigned heretofore to the Precambrian Eonothem, they also have been assigned recently to other (Archean and Proterozoic) eonothems by the IUGS Precambrian Subcommission. The span of time corresponding to an eonothem is an *eon*.

Article 71.—**Erathem**

An erathem is the formal chronostratigraphic unit of rank next lower to eonothem and consists of several adjacent systems. The span of time corresponding to an erathem is an *era*.

Remark: (a) ***Names.***—Names given to traditional Phanerozoic erathems were based upon major stages in the development of life on Earth: Paleozoic (old), Mesozoic (intermediate), and Cenozoic (recent) life. Although somewhat comparable terms have been applied to Precambrian units, the names and ranks of Precambrian divisions are not yet universally agreed upon and are under consideration by the IUGS Subcommission on Precambrian Stratigraphy.

Article 72.—***System***

The unit of rank next lower to erathem is the system. Rocks encompassed by a system represent a time-span and an episode of Earth history sufficiently great to serve as a worldwide chronostratigraphic reference unit. The temporal equivalent of a system is a *period*.

Remark: (a) ***Subsystem and supersystem.***—Some systems initially established in Europe later were divided or grouped elsewhere into units ranked as systems. *Subsystems* (Mississippian Subsystem of the Carboniferous System) and *supersystems* (Karoo Supersystems) are more appropriate.

*Article 73.—**Series***

Series is a conventional chronostratigraphic unit that ranks below a system and always is a division of a system. A series commonly constitutes a major unit of chronostratigraphic correlation within a province, between provinces, or between continents. Although many European series are being adopted increasingly for dividing systems on other continents, provincial series of regional scope continue to be useful. The temporal equivalent of a series is an *epoch.*

*Article 74.—**Stage***

A stage is a chronostratigraphic unit of smaller scope and rank than a series. It is most commonly of greatest use in inter-continental classification and correlation, although it has the potential for worldwide recognition. The geochronologic equivalent of stage is *age*.

*Remark: (a) **Substage.***—Stages may be, but need not be, divided completely into substages.

*Article 75.—**Chronozone***

A chronozone is a non-hierarchical, but commonly small, formal chronostratigraphic unit, and its boundaries may be independent of those of ranked units. Although a chronozone is an isochronous unit, it may be based on a biostratigraphic unit (example: Cardioceras cordatum Biochronozone), a lithostratigraphic unit (Woodbend lithochronozone), or a magnetopolarity unit (Gilbert Reversed Polarity Chronozone). Modifies (litho-,bio-,polarity) used in formal names of the units need not be repeated in general discussions where the meaning is evident from the context, e.g., *Exus albus* Chronozone.

*Remarks: (a) **Boundaries*** **of chronozones.**—The base and top of a *chronozone* correspond in the unit's stratotype to the observed, defining, physical and paleontological features, but they are extended to other areas by any means available for recognition of synchroneity. The temporal equivalent of a chronozone is a chron.

(b) ***Scope.***—The scope of the non-hierarchical chronozone may range markedly, depending upon the purpose for which it is defined either formally or informally. The informal "biochronozone of the ammonites," for example, represents a duration of time which is enormous and exceeds that of a system. In contrast, a biochronozone defined by a species of limited range, such as the *Exus albus* Chronozone, may represent a duration equal to or briefer than that of a stage.

(c) ***Practical utility.***—Chronozones, especially thin and informal biochronozones and lithochronozones bounded by key beds or other "markers," are the units used most commonly in industry investigations of selected parts of the stratigraphy of economically favorable basins. Such units are useful to define geographic distributions of lithofacies or biofacies, which provide a basis for genetic interpretations and the selection of targets to drill.

Chronostratigraphic Nomenclature

*Article 76.—**Requirements***

Requirements for establishing a formal chronostratigraphic unit include: (i) statement of intention to designate such a unit; (ii) selection of name; (iii) statement of kind and rank of unit; (iv) statement of general concept of unit including historical background, synonymy, previous treatment, and reasons for proposed establishment; (v) description of characterizing physical and/or biological features; (vi) designation and description of boundary type sections, stratotypes, or other kinds of units on which it is based; (vii) correlation and age relations; and (viii) publication in a recognized scientific medium as specified in Article 4.

*Article 77.—**Nomenclature***

A formal chronostratigraphic unit is given a compound name, and the initial letter of all words, except for trivial taxonomic terms, is capitalized. Except for chronozones (Article 75), names proposed for new chronostratigraphic units should not duplicate those for other stratigraphic units. For example, naming a new chronostratigraphic unit simply by adding, "-an" or "-ian" to the name of a lithostratigraphic unit is improper.

Remarks: (a) ***Systems and units of higher rank.***—Names that are generally accepted for systems and units of higher rank have diverse origins, and they also have different kinds of endings (Paleozoic, Cambrian, Cretaceous, Jurassic, Quaternary).

(b) ***Series and units of lower rank.***—Series and units of lower rank are commonly known either by geographic names (Virgilian Series, Ochoan Series) or by names of their encompassing units modified by the capitalized adjectives Upper, Middle and Lower (Lower Ordovician). Names of chronozones are derived from the unit on which they are based (Article 75). For series and stage, a geographic name is preferable because it may be related to a type area. For geographic names, the adjectival endings -an or -ian are recommended (Cincinnatian Series), but it is permissible to use the geographic name without any special ending, if more euphomous. Many series and stage names already in use have been based on lithic units (groups, formations, and members) and bear the names of these units (Wolfcampian Series, (Claiborman Stage). Nevertheless, a stage preferably should have a geographic name not previously used in stratigraphic nomenclature. Use of internationally accepted (mainly European) stage names is preferable to the proliferation of others.

*Article 78.—**Stratotypes***

An ideal stratotype for a chronostratigraphic unit is a completely exposed unbroken and continuous sequence of fossiliferous stratified rocks extending from a well-defined lower boundary to the base of the next higher unit. Unfortunately, few available sequences are sufficiently complete to define stages and units of higher rank, which therefore are best defined by boundary-stratotypes (Article 8b).

Boundary-stratotypes for major chronostratigraphic units ideally should be based on complete sequences of either fossiliferous monofacial marine strata or rocks with other criteria for chronocorrelation to permit widespread tracing of synchronous horizons. Extension of synchronous surface should be based on as many indicator of age as possible.

*Article 79.—**Revision of Units***

Revision of a chronostratigraphic unit without changing its name is allowable but requires as much justification as the establishment of a new unit (Articles 17, 19, and 76). Revision or redefinition of a unit of system or higher rank requires international agreement. If the definition of a chronostratigraphic unit is inadequate, it may be clarified by establishment of boundary stratotypes in a principal reference section.

Geochronologic Units

Nature and Boundaries

*Article 80.—**Definition and Basis***

Geochronologic units are divisions of time traditionally distinguished on the basis of the rock record as expressed by chronostratigraphic unit. A geochronologic unit is not a stratigraphic unit (i.e., it is not a material unit), but it corresponds to the time span of an established chronostratigraphic unit (Articles 65 and 66), and its beginning and ending corresponds to the base and top of the referent.

Ranks and Nomenclature of Geochronologic Units

*Article 81.—**Hierarchy***

The hierarchy of geochronologic units in order of decreasing rank is *eon, era, period, epoch*, and *age*. Chron is a non-hierarchical, but commonly brief, geochronologic unit. Ages in sum do not necessarily equal epochs and need not form a continuum. An eon is the time represented by the rocks constituting an eonothem; era by an erathem; period by a system; epoch by a series; age by a stage; and chron by a chronozone.

*Article 82.—**Nomenclature***

Names for periods and units of lower rank are identical with those of the corresponding chronostratigraphic units; the names of some eras and eons are independently formed. Rules of capitalization for chronostratigraphic units (Article 77) apply to geochronologic units. The adjectives Early, Middle and Late are used for the geochronologic epoch equivalent to the corresponding chronostratigraphic Lower Middle, and Upper series, where these are formally established.

Polarity-Chronostratigraphic Units

Nature and Boundaries

*Article 83.—**Definition***

A polarity-chronostratigraphic unit is a body of rock that contains the primary magnetic-polarity record imposed when the rock was deposited, or crystallized, during a specific interval of geologic time.

Remarks: (a) ***Nature.***—Polarity-chronostratigraphic units depend fundamentally for definition on actual section or sequences, or measurements on individual rock units, and without these standards they are meaningless. They are based on material units, the polarity zones of magnetopolarity classification. Each polarity-chronostratigraphic unit is the record of the time during which the rock formed and the Earth's magnetic field had a designated polarity. Care should be taken to define polarity-chronologic units in terms of polarity-chronostratigraphic units and not vice versa.

(b) ***Principal purposes.***—Two principal purposes are served by polarity-chronostratigraphic classification: (1) correlation of rocks at one place with those of the same age and polarity at other places; and (2) delineation of the polarity history of the Earth' magnetic field.

(c) ***Recognition.***—A polarity-chronostratigraphic unit may be extended geographically from its type locality only with the support of physical and/or paleontologic criteria used to confirm its age.

*Article 84.—**Boundaries***

The boundaries of a polarity Chronozone are placed at polarity-reversal horizons or polarity transition-zones (see Article 45).

Ranks and Nomenclature of Polarity-Chronostratigraphic Units

*Article 85.—**Fundamental Units***

The polarity chronozone consists of rocks of a specified primary polarity and is the fundamental unit of worldwide polarity-chronostratigraphic classification.

Remarks: (a) ***Meaning of term.***—A polarity chronozone is the world-wide body of rock strata that is collectively defined as a polarity-chronostratigraphic unit.

(b) ***Scope.***—Individual polarity zones are the basic building blocks of polarity chronozones. Recognition and definition of polarity chronozones may thus involve step-by-step assembly of carefully dated or correlated individual polarity zones, especially in work with rocks older than the oldest ocean-floor magnetic anomalies. This procedure is the method by which the Brunhes, Matuyama, Gauss, and Gilbert Chronozones were recognized (Cox, Doell, and Dalrymple, 1963) and defined originally (Cox, Doell, and Dalrymple, 1964).

(c) ***Ranks.***—Divisions of polarity chronozones are designated polarity subchronozones. Assemblages of polarity chronozones may be termed polarity superchronozones.

*Article 86.—**Establishing Formal Units***

Requirements for establishing a polarity-chronostratigraphic unit include those specified in Articles 3 and 4, and also (1) definition of boundaries of the unit, with specific references to designated sections and data; (2) distinguishing polarity characteristics, lithologic descriptions, and included fossils; and (3) correlation and age relations.

*Article 87.—**Name***

A formal polarity-chronostratigraphic unit is given a compound name beginning with that for a named geographic feature; the second component indicates the normal, reversed, or mixed polarity of the unit, and the third component is *chronozone*. The initial letter of each term is capitalized. If the same geographic name is used for both a magnetopolarity zone and a polarity-chronostratigraphic unit, the latter should be distinguished by an -an or -ian ending. Example: Tetonian Reversed-Polarity Chronozone.

Remarks: (a) ***Preservation of established name.***—A particularly well-established name should not be displaced, either on the basis of priority, as described in Article 7c, or because it was not taken from a geographic feature. Continued use of Brunhes, Matuyama, Gauss, and Gilbert, for example, is endorsed so long as they remain valid units.

(b) ***Expression of doubt.***—Doubt in the assignment of polarity zones to polarity-chronostratigraphic units should be made explicit if criteria of time equivalence are inconclusive.

Polarity-Chronologic Units

Nature and Boundaries

*Article 88.—**Definition***

Polarity-chronologic units are divisions of geologic time distinguished on the basis of the record of magnetopolarity as embodied in polarity-chronostratigraphic units. No special kind of magnetic time is implied; the designations used are meant to convey the parts of geologic time during which the Earth's magnetic field had a characteristic polarity or sequence of polarities. These units correspond to the time spans represented by polarity chronozones, e.g., Gauss Normal Polarity Chronozone. They are not material units.

Ranks and Nomenclature of Polarity-Chronologic Units

*Article 89.—**Fundamental Unit***

The polarity chron is the fundamental unit of geologic time designating the time span of a polarity chronozone.

*Remark: (a) **Hierarchy.**—*Polarity-chronologic units of decreasing hierarchical ranks are polarity superchron, polarity chron, and polarity subchron.

*Article 90.—**Nomenclature***

Names for polarity chronologic units are identical with those of corresponding polarity-chronostratigraphic units, except that the term chron (or superchron, etc) is substituted for chronozone (or superchronozone, etc.)

Diachronic Units

Nature and Boundaries

*Article 91.-**Definition***

A diachronic unit comprises the unequal spans of time represented either by a specific lithostratigraphic, allostratigraphic, biostratigraphic, or pedostratigraphic unit, or by an assemblage of such units.

*Remarks: (a) **Purpose.**—*Diachronic classification provides (1) a means of comparing the spans of time represented by stratigraphic units with diachronous boundaries at different localities, (2) a basis for broadly establishing in time the beginning and ending of deposition of diachronous stratigraphic units at different sites, (3) a basis for inferring the rate of change in areal extent of depositional processes, (4) a means of determining and comparing rates and durations of deposition at different localities, and (5) a means of comparing temporal and spatial relations of diachronous stratigraphic units (Watson and Wright, 1980).

(b) ***Scope.***—The scope of a diachronic unit is related to (1) the relative magnitude of the transgressive division of time represented by the stratigraphic unit or units on which it is based and (2) the areal extent of those units. A diachronic unit is not extended beyond the geographic limits of the stratigraphic unit or units on which it is based.

(c) ***Basis.***—The basis for a diachronic unit is the diachronous referent.

(d) ***Duration.***—A diachronic unit may be of equal duration at different places despite differences in the times at which it began and ended at those places.

*Article 92.—**Boundaries***

The boundaries of a diachronic unit are the times recorded by the beginning and end of deposition of the material referent at the point under consideration (Figs. 10, 11).

*Remark: (a) **Temporal relations.***—One or both of the boundaries of a diachronic unit are demonstrably time-transgressive. The varying time significance of the boundaries is defined by a series of boundary reference sections (Article 8b, 8c). The duration and age of a diachronic unit differ from place to place (Figs 10, 11).

Ranks and Nomenclature of Diachronic Units

*Article 93.—**Ranks***

A diachron is the fundamental and non-hierarchical diachronic unit. If a hierarchy of diachronic units is needed, the terms episode, phase, span, and cline, in order of decreasing rank, are recommended. The rank of a hierarchical unit is determined by the scope of the unit (Article 91 b), and not by the time span represented by the unit at a particular place.

*Remarks: (a) **Diachron.***—Diachrons may differ greatly in magnitude because they are the spans of time represented by individual or grouped lithostratigraphic, allostratigraphic, biostratigraphic, and (or) pedostratigraphic units.

(b) ***Hierarchical ordering permissible.***—A hierarchy of diachronic units may be defined if the resolution of spatial and temporal relations of diachronous stratigraphic units is sufficiently precise to make the hierarchy useful (Watson and Wright, 1980). Although all hierarchical units of rank lower than episode are part of a unit next higher in rank, not all parts of an episode, phase, or span need be represented by a unit of lower rank.

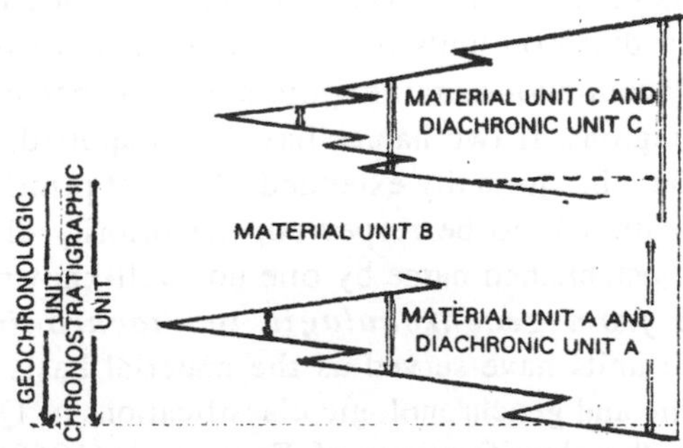

Fig. 10: Comparison of geochronologic, chronostratigraphic, and diachronic units.

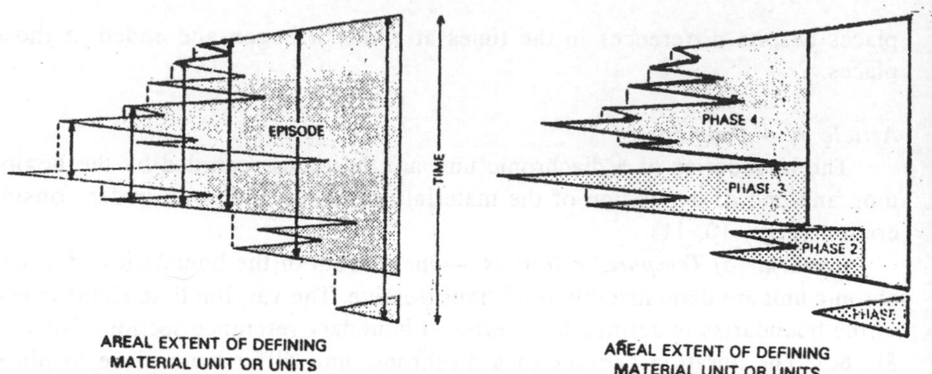

Fig. 11: Schematic relation of phases to an episode.
Parts of a phase similarly may be divided into spans, and spans into clines. Formal definition of spans and clines is unnecesssary in most diachronic unit hierarchies.

(c) ***Episode.***—An episode is the unit of highest rank and greatest scope in hierarchical classification. If the "Wisconsinan Age" were to be redefined as a diachronic unit, it would have the rank of episode.

Article 94.—Name

The name for a diachronic unit should be compound, consisting of a geographic name followed by the term diachron or a hierarchical rank term. Both parts of the compound name are capitalized to indicate formal status. If the diachronic unit is defined by a single stratigraphic unit, the geographic name of the unit may be applied to the diachronic unit. Otherwise, the geographic name of the diachronic unit should not duplicate that of another formal stratigraphic unit. Genetic terms (e.g., alluvial, marine) or climatic terms (e.g., glacial, interglacial) are not included in the names of diachronic units.

Remarks: (a) ***Formal designation of units.***—Diachronic units should be formally defined and named only if such definition is useful.

(b) ***Inter-regional extension of geographic names.***—The geographic name of a diachronic unit may be extended from one region to another if the stratigraphic units on which the diachronic unit is based extend across the regions. If different diachron units in contiguous region eventually prove to be based on laterally continuous stratigraphic units, one name should be applied to the unit in both regions. If two names have been applied, one name should be abandoned and the other formally extended. Rules of priority (Article 7d) apply. Priority in publication is to be respected, but priority alone does not justify displacing a well-established name by one not well-known or commonly used.

(c) ***Change from geochronologic to diachronic classification.***—Lithostratigraphic units have served as the material basis for widely accepted chronostratigraphic and geochronologic classifications of Quaternary nonmarine deposits, such as the classifications of Frye et al. (1968), Willman and Frye

(1970), and Dreimanis and Karrow (1972). In practice, time-parallel horizons have been extended from the stratotypes on the basis of markedly time-transgressive lithostratigraphic and pedostratigraphic unit boundaries. The time ("geochronologic") units defined on the basis of the stratotype sections but extended on the basis of diachronous stratigraphic boundaries, are diachronic units. Geographic names established for such "geochronologic" units may be used in diachronic classification if (1) the chronostratigraphic and geochronologic classifications are formally abandoned and diachronic classifications are proposed to replace the former "geochronologic" classifications, and (2) the units are redefined as formal diachronic units. Preservation of well established names in these specific circumstances retains the intent and purpose of the names and the units, retains the practical significance of the units, enhances communication, and avoids proliferation of nomenclature.

*Article 95.—**Establishing Formal Units***

Requirements for establishing a formal diachronic unit, in addition to those in Article 3, include (1) specification of the nature, stratigraphic relations, and geographic or areal relations of the stratigraphic unit or units that serve as a basis for definition of the unit, and (2) specific designation and description of multiple reference sections that illustrate the temporal and spatial relations of the defining stratigraphic unit or units and the boundaries of the unit or units.

*Remark: (a) **Revision or abandonment.***—Revision or abandonment of the stratigraphic unit or units that serve as the material basis for definition of a diachronic unit may require revision or abandonment of the diachronic unit. Procedure for revision must follow the requirements for establishing a new diachronic unit.

Geochronometric Units

Nature and Boundaries

*Article 96.-**Definition***

Geochronometric units are units established through the direct division of geologic time, expressed in years. Like geochronologic units (Article 80), geochronometric units are abstractions, i.e., they are not material units. Unlike geochronologic units, geochronometric units are not based on the time span of designated chronostratigraphic units (stratotypes), but are simply time divisions of convenient magnitude for the purpose for which they are established, such as the development of a time scale for the Precambrian. Their boundaries are arbitrarily chosen or agreed-upon ages in years.

Ranks and Nomenclature of Geochronometric Units

*Article 97.—**Nomenclature***

Geochronologic rank terms (eon, era, period, epoch, age, and chron) may be used for geochronometric units when such terms are formalized. For example, Archean Eon and Proterozoic Eon, as recognized by the IUGS Subcommission on Precambrian Stratigraphy, are formal geochronometric units in the sense of Article 96, distinguished on the basis of an arbitrarily chosen boundary at 2.5 Ga. Geochronometric units are not defined by, but may have, corresponding chronostratigraphic units (eonothem, erathem, system, series, stage, and chronozone).

Part III:

References[10]

American Commission on Stratigraphic Nomenclature, 1947, Note 1-Organization and objectives of the Stratigraphic Commission: American Association of Petroleum Geologists Bulletin, v. 31, no. 3, p. 513–518.

—, 1961, Code of Stratigraphic Nomenclature: American Association of Petroleum Geologists Bulletin, v. 45, no. 5, p. 645–665.

—, 1970, Code of Stratigraphic Nomenclature (2d ed.): American Association of Petroleum Geologists, Tulsa, Okla., 45 p.

—, 1976, Note 44-Application for addition to code concerning magnetostratigraphic units: American Association of Petroleum Geologists Bulletin, v. 60, no. 2, p. 273–277.

Caster, K.E., 1934, The stratigraphy and paleontology of northwestern Pennyslvania, Part 1, Stratigraphy: Bulletins of American Paleontology, v. 21, 185 p.

Chang, K. H., 1975, Unconformity-bounded stratigraphic units: Geological Society of America Bulletin, v. 86, no. 11, p. 1544–1552.

Committee on Stratigraphic Nomenclature, 1933, Classification and nomenclature of rock unit: Geological Society of America Bulletin, v. 44, no. 2, p. 423–459, and American Association of Petroleum Geologists, Bulletin, v. 17, no. 7, p. 843–868.

Cox, A.V., R.R. Doell, and G.B. Dalrymple, 1963, Geomagnetic polarity epochs and Pleistocene geochronometry: Nature, v. 198, p. 1049–1051.

—, 1964. Reversals of the Earth's magnetic field: Science, v. 144, no. 3626, p. 1537–1543.

Cross, C.W., 1898, Geology of the Telluride area: U.S. Geological Survey 18th Annual Report, pt. 3, p. 759.

Cunning, A.D., J. G.C. M. Fuller, and J. W. Porter, 1959, Separation of strata: Paleozoic limestones of the Williston basin: American Journal of Science, v. 257, no. 10, p. 722–733.

Dreimanis, Aleksis, and P.E., Karrow, 1972, Glacial history of the Great Lakes-St. Lawrence region, the classification of the Wisconsin(an) Stage, and its correlatives: International Geologic Congress, 24th Session, Montreal, 1972, Section 12, Quaternary Geology, p. 5–15.

Dunbar, C.O., and John Rodgers, 1957, Principles of stratigraphy: Wiley, New York, 356 p.

Forgotson, J.M., Jr., 1957, Nature, usage and definition of marker-defined vertically segregated rock units: American Association of Petroleum Geologists Bulletin, v. 41 no. 9, p. 2108–2113.

[10]Readers are reminded of the extensive and noteworthy bibliography of contributors to stratigraphic principles, classification, and terminology cited by the International Stratigraphic Guide (ISSC, 1976, p. 111–187).

Frye, J.C., H.B. Willman, Meyer Rubin, and R.F. Black, 1968, Definition of Wisconsinan Stage: U.S. Geological Survey Bulletin 1274–E, 22 p.

George, T.N., and others, 1969, Recommendations on stratigraphical usage: Geological Society of London, Proceedings no. 1656, p. 139–166.

Harland, W.B., 1977, Essay review [of] International Stratigraphic Guide, 1976: Geology Magazine, v. 114, no. 3, p. 229–235.

—, 1978, Geochronologic scales, in G.V. Cohee et al, eds., Contributions to the Geologic Time Scale: American Association of Petroleum Geologists, Studies in Geology, no. 6, p. 9–32.

Harrison, J.E., and Z.E. Peterman, 1980, North American Commission on Stratigraphic Nomenclature, Note 52—A preliminary proposal for a chronometric time scale for the Precambrian of the United States and Mexico: Geological Society of America Bulletin, v. 91, no. 6, p. 377–380.

Henbest, L.G., 1952, Significance of evolutionary explosions for diastrophic division of Earth history: Journal of Paleonotology, v. 26, p. 299–318.

Henderson, J.B., W.G.E. Caldwell, and J.E. harrison 1980, North American Commission on Stratigraphic Nomenclature, Report 8—Amendment of code concerning terminology for igneous and high-grade metamorphic rocks: Geological Society of American Bulletin, v. 91, no. 6, p. 374–376.

Holland, C.H., and others, 1978, A guide to stratigraphical procedure: Geological Society of London, Special Report 10, p. 1–18.

Huxley, T.H., 1862, The anniversary address: Geological Society of London, Quarterly Journal, v. 18, p. xl–liv.

International Commission on Zoological Nomenclature, 1964: International Code of Zoological Nomenclature adopted by the XV International Congress of Zoology: International Trust for Zoological Nomenclature, London, 176 p.

International Subcommission on Stratigraphic Classification (ISSC). 1976, International Stratigraphic Guide (H.D. Hedberg, ed.): John Wiley and Sons, New York, 200 p.

International Subcommission on Stratigraphic Classification, 1979, Magnetostratigraphy polarity units—a supplementary chapter of the ISSC International Stratigraphic Guide: Geology, v. 7, p. 578–583.

Izett, G.A., and R.E. Wilcox, 1981, Map showing the distribution of the Huckleberry Ridge, Mesa Falls, and Lava Creek volcanic ash beds (Pearlette family ash beds) of Pliocene and Pleistocene age in the western United States and southern Canada: U.S. Geological Survey Miscellaneous Geological Investigations Map I–1325.

Kauffman, E.G., 1969, Cretaceous marine cycles of the Western Interior: Mountain Geologist: Rocky Mountain Association of Geologists, v. 6, no. 4, p. 227–245.

Matthews, R.K., 1974, Dynamic stratigraphy—an introduction to sedimentation and stratigraphy: Prentice Hall, New Jersey, 370 p.

McDougall, Ian, 1977, The present status of the geomagnetic polarity time scale: Research School of Earth Sciences, Australian National University, Publication no. 1288, 34 p.

McElhinny, M.W., 1978, The magnetic polarity time scale; prospects and possibilities in magnetostratigraphy, *in* G.V. Cohee et.al, eds., Contributions to the Geologic Time Scale, American Association of Petroleum Geologists, Studies in Geology, no. 6, p. 57–65.

McIver, N.L., 1972, Cenozoic and Mesozoic stratigraphy of the Nova Scotia shelf: Canadian Journal of Earth Science, v. 9, p. 54–70.

Mclaren, J.J. 1977, The Silurian-Devonian Boundary Committee. A final report, *in*, A. Martinsson, ed., The Silurian-Devonian boundary: IUGS series A, no. 5, p. 1–34.

Morison, R.B., 1967, Principles of Quaternary soil stratigraphy, in R.B. Morrison and H.E. Wright, Jr., eds., Quaternary soils: Reno, Nevada, Center for Water Resources Research, Desert Research Institute, Univ. Nevada, p. 1–69.

North American Commission on Stratigraphic Nomenclature, 1981, Draft North American Stratigraphic Code: Canadian Society of Petroleum Geologists, Calgary, 63 p.

Palmer, A.R., 1965, Biomere-a new kind of biostratigraphic unit: Journal of Paleontology, v. 39, no. 1, p. 149–153.

Parsons, R.B., 1981, Proposed soil-stratigraphic guide, *in*, International Union for Quaternary Research and International Society of Soil Science: INQUA Commission 6 and ISSS Commission 5 working Group, Pedology, Report, p. 6–12.

Pawluk, S. 1978, The pedogenic profile in the stratigraphic section, *in* W. C. Mahaney, ed., Quaternary soils: Norwich, England, GeoAbstracts, Ltd., p. 61–75.

Ruhe, R.V., 1965, Quaternary paleopedology, *in* H.E. Wright, Jr., and D.G. Frey, eds., The Quaternary of the United States: Princeton, N.J., Princeton University Press, p. 755–764.

Schultz, E.H., 1982, The chronosome and supersome-terms proposed for low-rank chronostratigraphic units: Canadian Petroleum Geology, v. 30, no. 1, p. 29–33.

Shaw, A.B., 1964, Time in stratigraphy: McGraw-Hill, New York, 365 p.

Sims, P.K., 1979, Precambrian subdivided: Geotimes, v. 24, no. 12, p. 15.

Sloss,L.L., 1963, Sequences in the cratonic interior of North America: Geological Society of America Bulletin, v. 74, no. 2, p. 94–114.

Tracey, J.I., Jr. and others, 1971, Initial reports of the Deep Sea Drilling Project, v. 8: U.S. Government Printing Office, Washington, 1037 p.

Valentine, K.W.G., and J.B. Dalrymple, 1976, Quaternary buried paleosols: A critical review: Quaternary Research, v. 6, p. 209–222.

Vella, P., 1964, Biostratigraphic units: New Zealand Journal of Geology and Geophysics, v. 7, no. 3, p. 615–625.

Watson, R.A., and H.E. Wright, Jr., 1980, The end of the Pleistocene: A general critique of chronostratigraphic classification: Boreas, v. 9 p. 153–163.

Weiss, M.P., 1979a, Comments and suggestions invited for revision of American Stratigraphic Code: Geologic Society of American, News and Information, v. 1, no. 7, p. 97–99.

—, 1979b, Stratigraphic Commission Note 50-Proposal to change name of Commission: American Association of Petroleum Geologists Bulletin, v. 63, no. 10 p. 1986.

Weller, J.M., 1960, Stratigraphic principles and practice: Harper and Brothers, New York, 725 p.

Willman, H.B., and J.C. Frye, 1970, Pleistocene stratigraphy of Illinois: Illinois State Geological Survey Bulletin 94, 204 p.

Subject Index

Author Index

T - #0463 - 101024 - C0 - 234/156/23 - PB - 9789058092274 - Gloss Lamination